Interactive Student Edition

Reveal
ALGEBRA 1®

Volume 2

Mc
Graw
Hill

mheducation.com/prek-12

Cover: (t to b, l to r) pearleye/E+/Getty Images, Merfin/iStock/Getty Images,
fototrav/E+/Getty Images, rickyd/Shutterstock

Send all inquiries to:
McGraw-Hill Education
8787 Orion Place
Columbus, OH 43240

ISBN: 978-0-07-662599-4 (*Interactive Student Edition*, Volume 1)
MHID: 0-07-662599-0 (*Interactive Student Edition*, Volume 1)
ISBN: 978-0-07-899743-3 (*Interactive Student Edition*, Volume 2)
MHID: 0-07-899743-7 (*Interactive Student Edition*, Volume 2)

Printed in the United States of America.

17 18 19 LWI 27 26 25 24

Contents in Brief

Reveal AGA® Makes Math Meaningful...

Learning on the Go!

The flexible approach of *Reveal Algebra 1, Reveal Geometry,* and *Reveal Algebra 2 (Reveal AGA)* can work for you using digital only or digital and your *Interactive Student Edition* together.

Interactive Student Edition

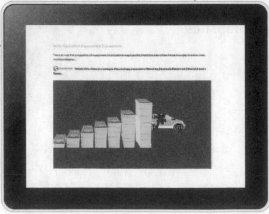

Student Digital Center

...to Reveal YOUR Full Potential!

Reveal AGA® Brings Math to Life in Every Lesson

Reveal AGA is a blended print and digital program that supports access on the go. You'll find the *Interactive Student Edition* aligns to the Student Digital Center, so you can record your digital observations in class and reference your notes later, or access just the digital center, or a combination of both! The Student Digital Center provides access to the interactive lessons, interactive content, animations, videos, and technology-enhanced practice questions.

Write down your username and password here

Username: _____

Password: _____

Go Online!
my.mheducation.com

Web Sketchpad® Powered by The Geometer's Sketchpad®- Dynamic, exploratory, visual activities embedded at point of use within the lesson.

Animations and Videos – Learn by seeing mathematics in action.

Interactive Tools – Get involved in the content by dragging and dropping, selecting, highlighting, and completing tables.

Personal Tutors – See and hear a teacher explain how to solve problems.

eTools – Math tools are available to help you solve problems and develop concepts.

Module 1
Expressions

Module 2

Equations in One Variable

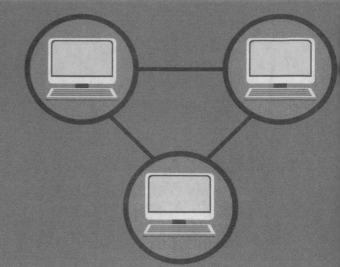

Module 3
Relations and Functions

Module 4
Linear and Nonlinear Functions

Module 5

Creating Linear Equations

Module 6
Linear Inequalities

Module 7

Systems of Linear Equations and Inequalities

Module 8

Exponents and Roots

Module 9
Exponential Functions

Module 10
Polynomials

Module 11

Quadratic Functions

Module 12
Statistics

Systems of Linear Equations and Inequalities

ℯ Essential Question
How are systems of equations useful in the real world?

What Will You Learn?

Place a check mark (✓) in each row that corresponds with how much you already know about each topic **before** starting this module.

KEY

👎 — I don't know.   👈 — I've heard of it.   👍 — I know it!

	Before			After		
	👎	👈	👍	👎	👈	👍
solve systems of equations by graphing						
solve systems of equations by substitution						
solve systems of equations by elimination with addition						
solve systems of equations by elimination with subtraction						
solve systems of equations by elimination with multiplication						
solve systems of inequalities by graphing						

📖 **Foldables** Make this Foldable to help you organize your notes about expressions. Begin with one sheet of paper.

1. **Fold** lengthwise to the holes.

2. **Cut** 6 tabs.

3. **Label** the tabs using the lesson titles.

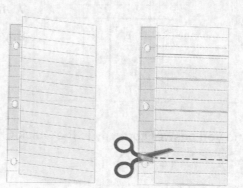

7-1 Graphing Systems of Equations

7-2 Substitution

7-3 Elimination Using Addition and Subtraction

7-4 Elimination Using Multiplication

7-5 Systems of Inequalities

Vocab

What Vocabulary Will You Learn?

Check the box next to each vocabulary term that you may already know.

☐ consistent
☐ dependent
☐ elimination
☐ inconsistent

☐ independent
☐ substitution
☐ system of equations
☐ system of inequalities

Are You Ready?

Complete the Quick Review to see if you are ready to start this module.
Then complete the Quick Check.

Quick Review

Example 1

Name the ordered pair for Q on the coordinate plane.

Follow a vertical line from the point to the x-axis. This gives the x-coordinate, 3.

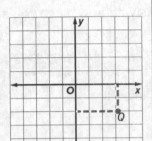

Follow a horizontal line from the point to the y-axis. This gives the y-coordinate, −2.

The ordered pair is (3, −2).

Example 2

Solve $12x + 3y = 36$ for y.

$12x + 3y = 36$	Original equation
$12x + 3y - 12x = 36 - 12x$	Subtract 12x from each side.
$3y = 36 - 12x$	Simplify.
$\dfrac{3y}{3} = \dfrac{36 - 12x}{3}$	Divide each side by 3.
$\dfrac{3y}{3} = \dfrac{36}{3} - \dfrac{12x}{3}$	Express as a difference.
$y = 12 - 4x$	Simplify.

Quick Check

Name the ordered pair for each point.

1. A

2. B

3. C

4. D

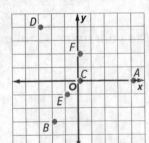

Solve each equation for the variable specified.

5. $2x + 4y = 12$, for x

6. $x = 3y - 9$, for y

7. $m - 2n = 6$, for m

8. $y = mx + b$, for x

How Did You Do?

Which exercises did you answer correctly in the Quick Check? Shade those exercise numbers below.

① ② ③ ④ ⑤ ⑥ ⑦ ⑧

Graphing Systems of Equations

Explore Intersections of Graphs

Online Activity Use graphing technology to complete the Explore.

> **INQUIRY** How can you solve a linear equation by graphing?

Learn Graphs of Systems of Equations

A set of two or more equations with the same variables is called a **system of equations**. An ordered pair that is a solution of both equations is a solution of the system. A system of two linear equations can have one solution, an infinite number of solutions, or no solution.

- A system of equations is **consistent** if it has at least one ordered pair that satisfies both equations.

- If a consistent system of equations has exactly one solution, it is said to be **independent**. The graphs intersect at one point.

- If a consistent system of equations has an infinite number of solutions, it is **dependent**. The graphs are the same line. This means that there are unlimited solutions that satisfy both equations.

- A system of equations is **inconsistent** if it has no ordered pair that satisfies both equations. The graphs are parallel.

Example 1 Consistent Systems

Use the graph to determine the number of solutions the system has. Then state whether the system of equations is *consistent* or *inconsistent* and if it is *independent* or *dependent*.

$y = -3x + 1$
$y = x - 3$

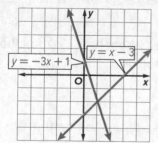

Since the graphs of these two lines intersect at one point, there is exactly one solution. Therefore, the system is _____.

Go Online You can complete an Extra Example online.

Today's Goals
- Determine the number of solutions of a system of linear equations.
- Solve systems of equations by graphing.
- Solve linear equations by graphing systems of equations.
- Use graphing calculators to solve systems of equations.

Today's Vocabulary
system of equations
consistent
independent
dependent
inconsistent

Go Online You may want to complete the Concept Check to check your understanding.

Copyright © McGraw-Hill Education

Example 2 Inconsistent Systems

Use the graph to determine the number of solutions the system has. Then state whether the system of equations is *consistent* or *inconsistent* and if it is *independent* or *dependent*.

$$y = \frac{1}{2}x + 2$$
$$y = \frac{1}{2}x - 1$$

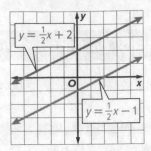

Since the graphs of these two lines are parallel, there is no solution of the system. Therefore, the system is _____.

Check

Determine whether each graph shows a system that is *consistent* or *inconsistent* and if it is *independent* or *dependent*.

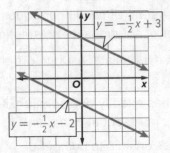 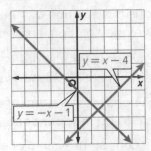

Example 3 Number of Solutions, Equations in Slope-Intercept Form

Determine the number of solutions the system has. Then state whether the system of equations is *consistent* or *inconsistent* and if it is *independent* or *dependent*.

$y = 6x + 10$ Because the slopes are the _____ and the y-intercepts are _____, the lines are _____.

$y = 6x + 4$ The system has _____ solution. Therefore, the system is _____.

Check

Determine the number of solutions the system has. _____

$$y = \frac{4}{5}x - 2$$

$$4y - 5x = 9$$

⚫ **Go Online** You can complete an Extra Example online.

Example 4 Number of Solutions, Equations in Standard Form

Determine the number of solutions the system has. Then state whether the system of equations is *consistent* or *inconsistent* and if it is *independent* or *dependent*.

$4y - 6x = 16$
$9x - 6y = -24$

Write both equations in slope-intercept form.

$4y - 6x = 16$	Original equation
$4y - 6x + 6x = \underline{\hspace{1cm}} + 16$	Isolate the y-term.
$4y = \underline{\hspace{2cm}}$	Simplify.
$\dfrac{4y}{4} = \dfrac{6x}{4} + \dfrac{16}{4}$	Divide by coefficient of y.
$y = \dfrac{3}{2}x + \underline{\hspace{1cm}}$	Simplify.

$9x - 6y = -24$
$9x - 6y - 9x = \underline{\hspace{1cm}} -24$
$-6y = \underline{\hspace{2cm}}$
$\dfrac{-6y}{-6} = \dfrac{-9x}{-6} + \dfrac{-24}{-6}$
$y = \underline{\hspace{1cm}} + 4$

Because the slopes are _____ and the y-intercepts are _____, this is _____ line.

Because the graphs of these two lines are the same, there are infinitely many solutions. Therefore, the system is _____.

Check

Determine the number of solutions the system has. _____

$4x - 8y = 16$

$6x - 12y = 5$

Think About It!

How many solutions will a system have if the slopes are different?

Learn Solving Systems of Equations by Graphing

You can solve a system of equations by graphing each equation carefully on the same coordinate plane. Every point that lies on the line of one equation represents a solution of that equation. Similarly, every point on the line of the second equation in a system represents a solution of that equation. Therefore, the solution of a system of equations is the point at which the graphs intersect.

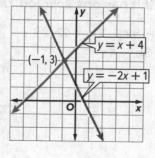

For example, the solution of this system is (−1, 3). That is the point at which the graphs intersect. Since the point of intersection lies on both lines, the ordered pair satisfies each equation in the system.

Example 5 Solve a System by Graphing

Graph the system and determine the number of solutions that it has. If it has one solution, determine its coordinates.

$$y = -2x + 14$$
$$y = \frac{3}{5}x + 1$$

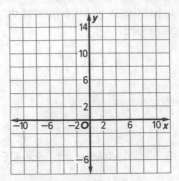

The graphs of the lines appear to intersect at the point (_____). If you substitute 5 for x and 4 for y into the equations, both are true. Therefore, _____ is the solution of the system.

Check

Graph the system of equations.

$$3x + 5y = 10$$
$$x - 5y = -10$$

What is the solution of the system?

(_____)

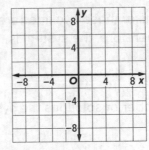

Example 6 Graph and Solve a System of Equations

Graph the system and determine the number of solutions that it has. If it has one solution, determine its coordinates.

$$-3x + 2y = 12$$
$$6x - 4y = 8$$

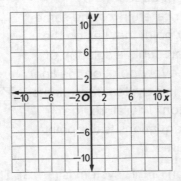

The lines have the same slope but different y-intercepts, so the lines are parallel. Since they do not intersect, this system has _____.

🔵 **Go Online** You can complete an Extra Example online.

🍦 **Think About It!**

Why is it necessary to substitute the values of x and y into both equations to check your solution?

🔵 **Go Online** You can watch a video to see how to use a graphing calculator with this example.

Problem-Solving Tip

Tools When graphing by hand, using graph paper and a straightedge can help you make your graphs more accurate.

🌐 Apply Example 7 Write a System of Equations

POPULATION China and India are the two most populous countries in the world. The populations of these countries have increased steadily in recent years. In 2010, China and India had populations of about 1.34 billion and 1.19 billion, respectively. By 2016, the populations had grown to about 1.38 billion in China and 1.29 billion in India. Predict the approximate year when the populations of the two countries will be the same.

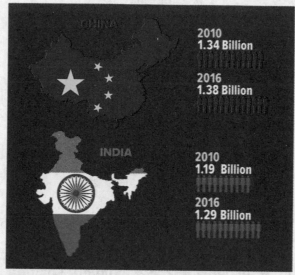

Source: IMF, CEIC

1. What is the task?

Describe the task in your own words. Then list any questions that you may have. How can you find answers to your questions?

_____

_____

_____

_____

2. How will you approach the task? What have you learned that you can use to help you complete the task?

_____

_____

_____

3. What is your solution?

Use your strategy to solve the problem.
Find the average rate of change for the populations of China and India.

China: _____

India: _____

Write a system of equations to represent the situation. Let x = the number of years after 2016. Let y = population.

_____

_____

(continued on the next page)

Study Tip

Assumptions
Populations do not typically increase at a steady rate. However, assuming that the rate of increase is constant allows you to estimate future data.

Study Tip

Labeling Axes It is important to clearly define and label axes in a real-world situation. The intersection does not mean that the population will be the same in year 9. Since the *x*-axis represents the number of years after 2016, you must add the *x*-coordinate of the intersection to 2016 to find the year.

💭 **Think About It!**

Can a value of *x* that is not a whole number be a viable solution? Justify your argument.

Graph the system of equations.

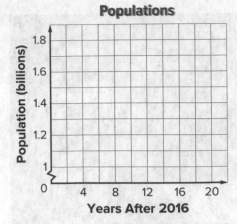

When will China and India have the same population?

_____

4. **How can you know that your solution is reasonable?**

✏️ **Write About It!** Write an argument that can be used to defend your solution.

_____

_____

Check

OLYMPICS The number of men and women participating in the Winter Olympic Games has been steadily increasing in recent years. In the 19th Winter Olympics, 1389 men and 787 women participated. 1660 men and 1121 women participated in the 22nd Winter Olympics.

Part A

Write and graph a system to describe the number of men and women participating if *x* represents the number of Winter Olympics after the 22nd Winter Olympics.

_____

_____

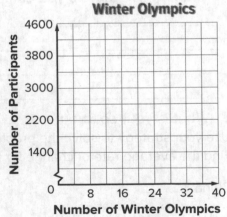

Part B

Use the graph to predict the Winter Olympics when the number of men and women participating will be the same.

_____

🖱 **Go Online** You can complete an Extra Example online.

Learn Using Systems to Solve Linear Equations

Key Concept · Using Systems to Solve Linear Equations	
Step 1	Write a system by setting each expression equal to y.
Step 2	Graph the system.
Step 3	Find the intersection.

Talk About It!

How do you know that the point of intersection satisfies both equations?

Example 8 Use a System to Solve a Linear Equation

Use a system of equations to solve $-6x + 8 = -4$.

Step 1 Write a system.

Write a system of equations. Set each side of $-6x + 8 = -4$ equal to y.

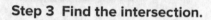

Step 2 Graph the system.

Enter the equations and graph.

Step 3 Find the intersection.

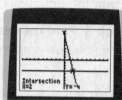

The solution is the x-coordinate of the intersection,

_____.

Go Online to see how to use a graphing calculator with this example.

Step 4 Check your solution.

$-6x + 8 = -4$	Original equation
$-6(\underline{}) + 8 \stackrel{?}{=} -4$	Substitution
$\underline{} + 8 \stackrel{?}{=} -4$	Multiply.
$\underline{} = -4$	Add.

Check

Use a system of equations and your graphing calculator to solve $-3.2x - 5.8 = 2.8x + 7$. Round to the nearest hundredth, if necessary.

$x = $ _____

Go Online You can complete an Extra Example online.

Learn Solving Systems of Equations by Using Graphing Technology

You can use a graphing calculator to graph and solve a system of equations by following these steps.

Step 1 Isolate y in each equation.

Step 2 Graph the system.

Step 3 Find the intersection.

Go Online You can watch a video to see how to graph systems of equations on a graphing calculator.

Example 9 Solve a System of Equations

Solve the system of equations.
$-1.38x - y = 5.13$
$0.62x + 2y = 1.60$

Step 1 Isolate y.

Solve each equation for y.

$$-1.38x - y = 5.13 \qquad \text{First equation}$$
$$-y = \underline{\quad} + \underline{\quad}x \qquad \text{Add } 1.38x \text{ to each side.}$$
$$y = \underline{\quad\quad} - \underline{\quad}x \qquad \text{Multiply each side by } -1.$$

$$0.62x + 2y = 1.60 \qquad \text{Second equation}$$
$$2y = \underline{\quad} - \underline{\quad\quad} \qquad \text{Subtract } 0.62x \text{ from each side.}$$
$$y = \underline{\quad} - \underline{\quad} \qquad \text{Divide each side by 2.}$$

Step 2 Graph the system.
Enter the equations and graph.

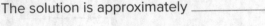 **Go Online** to see how to use a graphing calculator with this example.

Step 3 Find the intersection.
The solution is approximately _____.

Check

What is the solution to the system of equations?

$2.29x - 4.41y = 6.52$
$4.16x + 1.11y = 4.72$

(_____, _____)

🌐 Example 10 Write and Solve a System of Equations

BUSINESS Denzel is starting a food truck business to sell gourmet grilled cheese sandwiches. He has spent $34,000 on the truck, equipment, permits, and other start-up costs. Each sandwich costs about $1.32 to make, and he sells them for $7.

How many sandwiches does Denzel need to sell to start earning a profit?

Let x = the number of _____.
Let ___ = total cost or revenue.

Total Cost: $y = 1.32x$ _____
Total Revenue: $y =$ _____

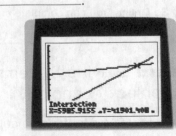

Step 1 Graph the system.
Enter the equations and graph.

Go Online to see how to use a graphing calculator with this example.

Step 2 Find the intersection.
The solution is approximately (_____, _____).
This means that after Denzel has sold _____ sandwiches, he will begin to earn a _____.

🌐 **Go Online** You can complete an Extra Example online.

Practice

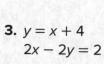 **Go Online** You can complete your homework online.

Examples 1 and 2

Use the graph to determine the number of solutions the system has. Then state whether the system of equations is *consistent* or *inconsistent* and if it is *independent* or *dependent*.

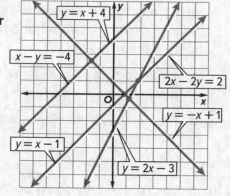

1. $y = x - 1$
$y = -x + 1$

2. $x - y = -4$
$y = x + 4$

3. $y = x + 4$
$2x - 2y = 2$

4. $y = 2x - 3$
$2x - 2y = 2$

Examples 3 and 4

Determine the number of solutions the system has. Then state whether the system of equations is *consistent* or *inconsistent* and if it is *independent* or *dependent*.

5. $y = \frac{1}{2}x$
$y = x + 2$

6. $4x - 6y = 12$
$-2x + 3y = -6$

7. $8x - 4y = 16$
$-5x - 5y = 5$

8. $2x + 3y = 10$
$4x + 6y = 12$

9. $y = -\frac{3}{2}x + 5$
$y = -\frac{2}{3}x + 5$

10. $y = x - 3$
$y = -4x + 3$

Examples 5 and 6

Graph each system and determine the number of solutions it has. If it has one solution, determine its coordinates.

11. $y = -3$
$y = x - 3$

12. $y = 4x + 2$
$y = -2x - 4$

13. $y = x - 6$
$y = x + 2$

14. $x + y = 4$
$3x + 3y = 12$

15. $x - y = -2$
$-x + y = 2$

16. $2x + 3y = 12$
$2x - y = 4$

Example 7

17. AVIATION An air traffic controller manages the flow of aircraft in and out of airport airspace by guiding pilots during takeoff and landing. An air traffic controller is monitoring two planes that are in flight near a local airport. The first plane is at an altitude of 1000 meters and is ascending at a rate of 400 meters per minute. The second plane is at an altitude of 5900 meters and is descending at a rate of 300 meters per minute.

   a. Write and graph a system of equations that represents the altitude of each plane, where x is the amount of time, in minutes, and y is the altitude, in meters.

   b. Predict when the planes will be at the same altitude.

18. STRUCTURE Gustavo sets up tables for a caterer on weekends. Each round table can seat 8 people. Each rectangular table can seat 10 people. One weekend, his boss asked him to set up tables for 124 people. He uses 2 more round tables than rectangular tables. Define variables and write a system of equations to find the number of round tables and rectangular tables Gustavo used. Then solve the system graphically.

Example 8

Write and graph a system of equations to solve each linear equation.

19. $3x + 6 = 6$

20. $2x - 17 = x - 10$

21. $-12x + 90 = 30$

22. $13x - 28 = 24$

23. $2x + 5 = 2x + 5$

24. $x + 1 = x + 3$

Example 9

Solve each system of equations. State the decimal solution to the nearest hundredth.

25. $2.5x + 3.75y = 10.5$
$1.25x - 8.5y = -5.25$

26. $2.2x + 1.8y = -3.6$
$-4.8x + 12.4y = 10.6$

27. $1.12x - 2.24y = 4.96$
$-3.56x - 2.48y = -7.32$

Example 10

28. USE TOOLS An office building has two elevators. One elevator starts out on the 4th floor, 35 feet above the ground, and is descending at a rate of 2.2 feet per second. The other elevator starts out at ground level and is rising at a rate of 1.7 feet per second. Write and solve a system of equations to determine when both elevators will be at the same height. Interpret the solution.

29. USE TOOLS A bookstore makes a profit of $2.50 on each book they sell, and $0.75 on each magazine they sell. One week, the store sold x books and y magazines, for a weekly profit of $450. The total number of publications sold that week was 260. Write and solve a system of equations to determine the number of books and magazines that the bookstore sold that week. Interpret the solution.

Mixed Exercises

Graph each system and determine the number of solutions it has. If it has one solution, determine its coordinates. Then state whether the system is *consistent* or *inconsistent* and if it is *independent* or *dependent*.

30. $2x - y = -3$
$\quad\ 2x + y = -1$

31. $2x + y = 4$
$\quad\ y = -2x - 2$

32. $2y = 5 + x$
$\quad\ 3x - 6y = -15$

33. $2x - y = 5$
$\quad\ x + y = -2$

34. $2y = 1.2x - 10$
$\quad\ 4y = 2.4x$

35. $x = 6 - \frac{3}{8}y$
$\quad\ 4 = \frac{2}{3}x + \frac{1}{4}y$

36. $x - y = 3$
$\quad\ x - 2y = 3$

37. $x + 2y = 4$
$\quad\ y = -\frac{1}{2}x + 2$

38. $y = 2x + 3$
$\quad\ 3y = 6x - 6$

39. $y - x = -1$
$\quad\ x + y = 3$

40. BUSINESS The number of items sold at Store 1 can be represented by $y = 200x + 300$, where x represents the number of days and y represents the number of items sold. The number of items sold at Store 2 can be represented by $y = 200x + 100$, where x represents the number of days and y represents the number of items sold. Look at the graph of the system of equations and determine whether it has *no* solution, *one* solution, or *infinitely many* solutions.

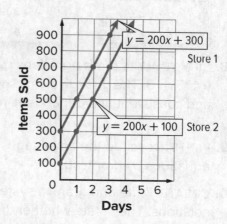

41. USE A MODEL Olivia and her brother William had a bicycle race. Olivia rode at a speed of 20 feet per second, while William rode at a speed of 15 feet per second. Olivia gave William a 150-foot head start, and the race ended in a tie.

a. Define variables and write a system of equations to represent this situation.

b. Graph the system of equations.

c. How far away was the finish line from where Olivia started?

42. CONSTRUCT ARGUMENTS Abhijit uses a calculator to graph the system $y = 1.21x - 3$ and $5y - 5 = 6x$. He concludes that the system has no solution. Do you agree or disagree? Explain your reasoning.

43. USE A MODEL Some days, Cayley walks to school. Other days, she rides her bicycle. When she walks to school, she needs to leave home 15 minutes earlier than when she rides her bicycle. Cayley walked 3 days last week and rode her bike on 2 days. Cayley spent a total of 1 hour 10 minutes going to school last week.

    a. Define variables and write a system of equations to represent this situation.

    b. Graph the system of equations.

    c. How long does it take Cayley to walk to school?

44. REGULARITY Maureen says that if a system of linear equations has three equations, then there could be exactly two solutions to the system. Is Maureen correct? Use a graph to justify your reasoning.

🧠 Higher-Order Thinking Skills

45. PERSEVERE Use graphing to find the solution of the system of equations $2x + 3y = 5$, $3x + 4y = 6$, and $4x + 5y = 7$.

46. ANALYZE Determine whether a system of two linear equations with (0, 0) and (2, 2) as solutions *sometimes*, *always*, or *never* has other solutions. Justify your argument.

47. WHICH ONE DOESN'T BELONG? Which one of the following systems of equations doesn't belong with the other three? Justify your conclusion.

$4x - y = 5$ $-2x + y = -1$	$-x + 4y = 8$ $3x - 6y = 6$	$4x + 2y = 14$ $12x + 6y = 18$	$3x - 2y = 1$ $2x + 3y = 18$

48. CREATE Write three equations such that they form three systems of equations with $y = 5x - 3$. The three systems should be inconsistent, consistent and independent, and consistent and dependent, respectively.

49. WRITE Describe the advantages and disadvantages of solving systems of equations by graphing.

50. CREATE Write and graph a system of equations that has the following number of solutions. Then state whether the system of equations is *consistent* or *inconsistent* and if it is *independent* or *dependent*.

    a. one solution   **b.** infinite solutions   **c.** no solution

51. FIND THE ERROR Store A is offering a 10% discount on the purchase of all electronics in their store. Store B is offering $10 off all the electronics in their store. Francisca and Alan are deciding which offer will save them more money. Is either of them correct? Explain your reasoning.

Francisca	Alan
You can't determine which store has the better offer unless you know the price of the items you want to buy.	Store A has the better offer because 10% of the sale price is a greater discount than $10.

Substitution

Explore Using Substitution

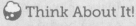 **Online Activity** Use a system of equations to complete the Explore.

> ⊘ ×
>
> ⓠ **INQUIRY** How can you rewrite a system of equations as a single equation with only one variable?

Learn Solving Systems of Equations by Substitution

Exact solutions result when algebraic methods are used to solve systems of equations. One algebraic method is called **substitution**.

Key Concept • Substitution Method	
Step 1	When necessary, solve at least one equation for one variable.
Step 2	Substitute the resulting expression from Step 1 into the other equation to replace the variable. Then solve the equation.
Step 3	Substitute the value from Step 2 into either equation, and solve for the other variable. Write the solution as an ordered pair.

Example 1 Solve a System by Substitution

Use substitution to solve the system of equations.

$3x - y = -7; y = 4x + 11$

Step 1 The second equation is already solved for y.

Step 2 Substitute $4x + 11$ for y in the first equation.

$3x - \underline{\hspace{1.5cm}} = -7$ Substitute $4x + 11$ for y.

$3x - \underline{\hspace{1.5cm}} = -7$ Distributive property

$\underline{\hspace{1.5cm}} = -7$ Combine like terms.

$-x = \underline{\hspace{1cm}}$ Add $\underline{\hspace{1cm}}$ to each side.

$x = \underline{\hspace{1cm}}$ Multiply each side by -1.

Step 3 Substitute -4 for x in either equation to find y.

$y = 4x + 11$ Second equation

$y = 4(\underline{\hspace{1cm}}) + 11$ Substitution

$y = \underline{\hspace{1cm}}$ Simplify.

The solution is ($\underline{\hspace{1cm}}$, $\underline{\hspace{1cm}}$).

 Go Online You can complete an Extra Example online.

Today's Goal
- Solve systems of equations by using the substitution method.

Today's Vocabulary
substitution

⊙ **Think About It!**

In algebra, what does it mean to *substitute*?

⊙ **Think About It!**

How would the substitution process differ if the second equation were $4x - y = -11$?

 Go Online

You can watch a video to see how to use algebra tiles with this example.

Check

Refer to the system of equations.

$3x - 2y = -17$

$y = 2x + 2$

Part A Which expression could be substituted for y in the first equation to find the value of x? _____

A. $2x - 2$ B. $-\frac{3}{2}x + \frac{17}{2}$ C. $2x + 2$ D. $3x - 2y$

Part B What is the solution of the system? _____

A. $(-13, -24)$ B. $(-13, 24)$ C. $(13, -28)$ D. $(13, 28)$

Example 2 Solve and Then Substitute

Use substitution to solve the system of equations.

$5x - 3y = -25$

$x + 4y = 18$

Step 1 Solve the second equation for x since the coefficient is 1.

$x + 4y = 18$	Second equation
_____ $= 18 - 4y$	Subtract _____ from each side.

Step 2 Substitute $18 - 4y$ for x in the first equation.

$5x - 3y = -25$	First equation
5 _____ $-3y = -25$	Substitute $18 - 4y$ for x.
90 _____ $- 3y = -25$	Distributive Property
90 _____ $= -25$	Combine like terms.
$-23y =$ _____	Subtract 90 from each side.
$y =$ _____	Divide each side by _____.

Step 3 Substitute 5 for y in either equation to find x.

$x + 4y = 18$	Second equation
$x + 4($ _____ $) = 18$	Substitute 5 for y.
$x +$ _____ $= 18$	Simplify.
$x =$ _____	Subtract 20 from each side.

The solution is (_____ , _____).

Check

Use substitution to solve the system of equations.

$5x + 3y = 5$

$x + 2y = -13$

_____

 Go Online You can complete an Extra Example online.

Talk About It!

Would the solution of the system be the same if the first step used was to solve the second equation for y? Explain.

Study Tip

Slope-Intercept Form
If both equations are in the form $y = mx + b$, the expressions can simply be set equal to each other and then solved for x. For example, if $y = 2x - 5$ and $y = -4x + 1$, then $2x - 5 = -4x + 1$. The solution for x can then be used to find the value of y.

Example 3 Use Substitution When There are No or Many Solutions

Use substitution to solve the system of equations.

$4x + 2y = -8$

$y = -2x - 4$

Substitute _____ for y in the first equation.

$4x + 2y = -8$	First equation
$4x + 2 \underline{\hspace{1cm}} = -8$	Substitute $-2x - 4$ for y.
$4x \underline{\hspace{1cm}} = -8$	Distributive Property
$\underline{\hspace{1cm}} = -8$	Simplify.

The equation _____ is an identity. Thus, there are an _____ of solutions.

When graphed, the equations are the _____ .

Check

Select the correct statement about the system of equations. _____

$-x + 2y = 2$

$y = \frac{1}{2}x + 1$

A. This system has no solution.

B. This system has one solution at $\left(\frac{2}{3}, \frac{4}{3}\right)$.

C. This system has one solution at $\left(\frac{4}{3}, \frac{2}{3}\right)$.

D. This system has infinitely many solutions.

🌎 Example 4 Write and Solve a System of Equations

TREE PRESERVATION A town ordinance defines an adult tree as having a diameter greater than 10 inches and a sapling as having a diameter less than 10 inches. The ordinance requires that on a new building project, two new trees are planted for each adult tree felled and six new trees are planted for each sapling felled. Last year, there were 167 trees felled, and the community planted 742 replacement trees. How many of each type of tree were felled?

(continued on the next page)

Study Tip

Dependent Systems
There are infinitely many solutions of the system because the equations in slope-intercept form are equivalent, and they have the same graph.

💭 Think About It!

What would a solution like $-8 = 0$ mean? What would it look like on a graph?

Study Tip

Assumptions Using systems of equations to solve real-world problems generally requires assuming that the problem actually has a solution. It is a good idea to graph the lines first, to see if a solution exists, before going through the steps to solve for each variable.

 Go Online

An alternate method is available for this example.

Use a Source

Research replacing trees cut for construction in an area near you. How could you find the number of trees to plant for a project similar to the one in this community?

Let a = the number of adult trees felled, and let t = the number of sapling trees felled.

$a + t = 167$	This equation represents the total number of adult trees a and sapling trees t felled with a sum of 167.
$2a + 6t = 742$	This equation represents the combinations of adult trees a and sapling trees t replaced with a total of 742.

The solution of the system of equations represents the option that meets both of the constraints.

The first equation can easily be solved for a or t.

Solve for a.

Step 1 Solve the first equation for a.

$a + t = 167$	First equation
$a + t \underline{\quad} = 167 \underline{\quad}$	Subtract t from each side.
$a = \underline{\qquad}$	Simplify.

Step 2 Substitute $167 - t$ for a in the second equation.

$2a + 6t = 742$	Second equation
$2 \underline{\qquad} + 6t = 742$	Substitute $167 - t$ for a.
$\underline{\qquad} + 6t = 742$	Distributive Property
$\underline{\quad} + 334 = 742$	Combine like terms.
$4t = \underline{\quad}$	Subtract 334 from each side.
$t = \underline{\quad}$	Divide each side by 4.

Step 3 Substitute 102 for t in either equation to find the value of a.

$a + t = 167$	First equation
$a + \underline{\quad} = 167$	Substitute 102 for t.
$a = \underline{\quad}$	Subtract 102 from each side.

The solution is (\underline{\quad} , \underline{\quad}).

There were \underline{\qquad\qquad} and \underline{\qquad\qquad} felled. Because only whole numbers of trees can be felled, this is a viable solution.

Check

GEOMETRY Kymani has two equal-sized large pitchers and two equal-sized small pitchers. All of the pitchers together hold 40 cups of water. The capacity of one large pitcher minus the capacity of one small pitcher is 12 cups. How many cups can each type of pitcher hold?

Small pitcher = \underline{\quad} cups

Large pitcher = \underline{\quad} cups

 Go Online You can complete an Extra Example online.

Practice

Go Online You can complete your homework online.

Examples 1–3

Use substitution to solve each system of equations.

1. $y = 5x + 1$
$4x + y = 10$

2. $y = 4x + 5$
$2x + y = 17$

3. $y = 3x - 34$
$y = 2x - 5$

4. $y = 3x - 2$
$y = 2x - 5$

5. $2x + y = 3$
$4x + 4y = 8$

6. $3x + 4y = -3$
$x + 2y = -1$

7. $y = -3x + 4$
$-6x - 2y = -8$

8. $-1 = 2x - y$
$8x - 4y = -4$

9. $x = y - 1$
$-x + y = -1$

10. $y = -4x + 11$
$3x + y = 9$

11. $y = -3x + 1$
$2x + y = 1$

12. $3x + y = -5$
$6x + 2y = 10$

13. $5x - y = 5$
$-x + 3y = 13$

14. $2x + y = 4$
$-2x + y = -4$

15. $-5x + 4y = 20$
$10x - 8y = -40$

Example 4

16. **MONEY** Harvey has some $1 bills and some $5 bills. In all, he has 6 bills worth $22. Let x be the number of $1 bills, and let y be the number of $5 bills. Write a system of equations to represent the information, and use substitution to determine how many bills of each denomination Harvey has.

17. **REASONING** Shelby and Calvin are conducting an experiment in chemistry class. They need 5 milliliters of a solution that is 65% acid and 35% distilled water. There is no undiluted acid in the chemistry lab, but they do have two beakers of diluted acid. Beaker A contains 70% acid and 30% distilled water. Beaker B contains 20% acid and 80% distilled water.

   a. Write a system of equations that Shelby and Calvin could use to determine how many milliliters they need to pour from each beaker to make their solution.

   b. Solve your system of equations. How many milliliters from each beaker do Shelby and Calvin need?

Mixed Exercises

Use substitution to solve each system of equations.

18. $y = 3.2x + 1.9$
$2.3x + 2y = 17.72$

19. $y = \frac{1}{4}x - \frac{1}{2}$
$8x + 12y = -\frac{1}{2}$

20. $y = -10x - 6.8$
$-50x - 10.5y = 60.4$

21. **USE A SOURCE** Research population trends in South America. Write and solve a system of equations to predict when the population of two countries will be equal.

22. **REGULARITY** Angle A and angle B are complementary, and their measures have a difference of 20°. What are the measures of the angles? Generalize your method.

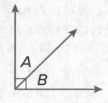

23. STRUCTURE A two-digit number is reduced by 45 when the digits are interchanged. The digit in the tens place of the original number is 1 more than 3 times the digit in the units place. Define variables and write a system of equations to find the original number.

24. USE A MODEL A zoo keeps track of the number of visitors to each exhibit. The table shows the number of visitors for two exhibits on one day.

	Big Cats	Petting Zoo
Adults	x	21
Children	1024	y

a. Three times the total number of adults was 17 less than the number of children who visited the petting zoo. Write an equation to model this relationship.

b. The total number of children who visited the zoo that day was 681 less than 10 times the number of adults who visited the big cats. Write an equation to model this relationship.

c. Solve the system of equations to find the total number of visitors to those two exhibits on that day.

d. Tickets to the zoo cost $25 for adults and $10 for children. The total ticket sales one day were $47,750. The number of children who visited the zoo was 270 more than 6 times the number of adults who visited. Write and solve a system of equations to find the total number of visitors to the zoo that day.

🧠 **Higher-Order Thinking Skills**

25. FIND THE ERROR In the system $a + b = 7$ and $1.29a + 0.49b = 6.63$, a represents pounds of apples and b represents pounds of bananas a person bought. Guillermo and Cara are finding and interpreting the solution. Is either correct? Explain your reasoning.

Guillermo

$1.29a + 0.49b = 6.63$

$1.29a + 0.49(a + 7) = 6.63$

$1.29a + 0.49a + 3.43 = 6.63$

$1.78a = 3.2$

$a = 1.8$

$a + b = 7$, so $b = 5.2$. The solution $(1.8, 5.2)$ means that 1.8 pounds of apples and 5.2 pounds of bananas were bought.

Cara

$1.29a + 0.49b = 6.63$

$1.29(7 - b) + 0.49b = 6.63$

$9.03 - 1.29b + 0.49b = 6.63$

$-0.8b = -2.4$

$b = 3$

The solution $b = 3$ means that 3 pounds of apples and 3 pounds of bananas were bought.

---

26. PERSEVERE A local charity has 60 volunteers. The ratio of boys to girls is 7:5. Find the number of volunteers who are boys and the number who are girls.

27. ANALYZE Compare and contrast the solution of a system found by graphing and the solution of the same system found by substitution.

28. CREATE Create a system of equations that has one solution. Illustrate how the system could represent a real-world situation and describe the significance of the solution in the context of the situation.

29. WRITE Explain how to determine what to substitute when using the substitution method of solving systems of equations.

Elimination Using Addition and Subtraction

Learn Solving Systems of Equations by Elimination with Addition

The **elimination** method involves eliminating a variable by combining the individual equations within a system of equations. One way to combine equations is by using addition.

Key Concept • Elimination Method Using Addition	
Step 1	Write the system so like terms with opposite coefficients are aligned.
Step 2	Add the equations, eliminating one variable. Then solve the equation.
Step 3	Substitute the value from Step 2 into one of the equations and solve for the other variable. Write the solution as an ordered pair.

Example 1 Elimination Using Addition

Use elimination to solve the system of equations.

$3x + 5y = 11$
$5x - 5y = 5$

Step 1 Align terms with opposite coefficients.

Since $5y$ and $-5y$ have opposite coefficients, _____ the equations to eliminate the variable _____.

Step 2 Add the equations.

$3x +$ _____ $y = 11$

(+) $5x - 5y = 5$

_____ = _____   The variable y is eliminated.

$\dfrac{8x}{8} = \dfrac{16}{8}$ Divide each side by 8.

$x =$ _____   Simplify.

Step 3 Solve for the other variable.

$3x + 5y = 11$ First equation
3 _____ $+ 5y = 11$ Replace x with 2.
_____ $+ 5y = 11$ Multiply.
6 _____ $+ 5y = 11$ _____   Subtract 6 from each side.
$5y = 5$ Simplify.
$\dfrac{5y}{5} = \dfrac{5}{5}$ Divide each side by 5.
$y =$ _____   Simplify.

The solution is (_____, _____).

 **Go Online** You can complete an Extra Example online.

Today's Goals
- Solve systems of equations by eliminating a variable using addition.
- Solve systems of equations by eliminating a variable using subtraction.

Today's Vocabulary
elimination

Study Tip

Answer Check You can check your answer by substituting the solution into the equation you did not use in Step 3. If the equality is valid, your solution is correct.

 **Talk About It!**

When graphed, where would these lines intersect? Explain your reasoning.

Check

Use elimination to solve the system of equations. _____

$5x + 13y = 20$

$-5x - 3y = 30$

Example 2 Write and Solve a System of Equations Using Addition

Seven times a number x minus four times another number y is thirteen. Negative seven times a number x plus seven times another number y is fourteen. Find the numbers.

Seven times a number x minus four times another number y is thirteen.	Negative seven times a number x plus seven times another number y is fourteen.
$7x - 4y = 13$	$-7x + 7y = 14$

Steps 1 and 2 Write the equations vertically and add.

$$7x - \underline{\quad}y = 13$$
$$(+)\quad -7x + 7y = \underline{\quad}$$
$$3y = 27 \qquad \text{The variable } x \text{ is eliminated.}$$
$$\frac{3y}{3} = \frac{27}{3} \qquad \text{Divide each side by 3.}$$
$$y = 9 \qquad \text{Simplify.}$$

Step 3 Substitute 9 for y in either equation to find the value of x.

$-7x + 7y = 14$	Second equation
$-7x + 7\underline{\quad} = 14$	Replace y with 9.
$-7x + \underline{\quad} = 14$	Multiply.
$-7x + 63\underline{\quad} = 14\underline{\quad}$	Subtract 63 from each side.
$-7x = -49$	Simplify.
$\frac{-7x}{-7} = \frac{-49}{-7}$	Divide each side by -7.
$x = \underline{\quad}$	Simplify.

The solution is (_____, _____). So, _____ and _____.

Check

Two times a number x minus six times another number y is negative six. Negative two times a number x plus five times another number y is eighteen.

Write the system of equations.

_____

_____

Solve the system of equations. _____

🔎 **Go Online** You can complete an Extra Example online.

Learn Solving Systems of Equations by Elimination with Subtraction

When the coefficients of a variable are the same in two equations, you can eliminate the variable by subtracting one equation from the other.

Key Concept • Elimination Method Using Subtraction	
Step 1	Write the system so like terms with the same coefficients are aligned.
Step 2	Subtract one equation from the other, eliminating one variable. Then solve the equation.
Step 3	Substitute the value from Step 2 into one of the equations and solve for the other variable. Write the solution as an ordered pair.

Example 3 Elimination Using Subtraction

Use elimination to solve the system of equations.

$3x + 6y = 30$
$5x + 6y = 6$

Step 1 Align terms with the same coefficients.

Since $6y$ and $6y$ have the same coefficients, you can _____ the equations to eliminate the variable _____.

Step 2 Subtract the equations.

$$3x + 6y = 30$$
$$(-)5x + 6y = 6$$
$$\overline{\quad\quad -2x = 24}$$
The variable y is eliminated.

$$\frac{-2x}{-2} = \frac{24}{-2}$$
Divide each side by -2.

$$x = \underline{\quad}$$
Simplify.

Step 3 Substitute −12 for x in either equation to find the value of y.

$3x + 6y = 30$ — First equation

$3 \underline{\quad} + 6y = 30$ — Replace x with -12.

$\underline{\quad} + 6y = 30$ — Multiply.

$-36 \underline{\quad} + 6y = 30 \underline{\quad}$ — Add 36 to each side.

$6y = \underline{\quad}$ — Simplify.

$\frac{6y}{6} = \frac{66}{6}$ — Divide each side by 6.

$y = \underline{\quad}$ — Simplify.

The solution is (_____ , _____).

Check

Use elimination to solve the system of equations. _____

$-2x + 3y = 48$
$7x + 3y = 21$

R Go Online You can complete an Extra Example online.

Copyright © McGraw-Hill Education

Study Tip

Adding and Subtracting Equations When the variable you want to eliminate has the same coefficient in the two equations, subtract. When the variable you want to eliminate has opposite coefficients, add.

Watch Out!

Subtracting an Equation When subtracting one equation from another in order to eliminate a variable, do not forget to distribute the negative sign to each term of the expressions on both sides of the equal sign.

Example 4 Write and Solve a System of Equations Using Subtraction

COMPUTERS Mei and Kara build computers from parts and sell them to make a profit. Mei can build a computer in 0.9 hour, and Kara can build one in 1.2 hours. During a typical week, Mei and Kara spend a total of 15 hours building computers. One week, Mei builds twice as many computers, and the two spend a total of 24 hours on their project. How many computers do Mei and Kara each make during a typical week?

Think About It!

What assumption was made about the rates at which Mei and Kara build computers? Why is that assumption made?

Words	Mei's time spent	plus Kara's time spent	is 15 hours.
Variables	Let m = the number of computers that Mei built and k = the number of computers that Kara built.		
Equation	$0.9m$	$+ 1.2k$	$= 15$

Words	Double Mei's time spent	plus Kara's time spent	is 24 hours.
Variables	Let m = the number of computers that Mei built and k = the number of computers that Kara built.		
Equation	$2(0.9m)$	$+ 1.2k$	$= 24$

Steps 1 and 2 Write the equations vertically and subtract.

$$0.9m + 1.2k = 15$$
$$(-) \; 2(0.9m) + 1.2k = 24$$
$$\overline{-0.9m = -9} \qquad \text{The variable } k \text{ is eliminated.}$$
$$\frac{-0.9m}{-0.9} = \frac{-9}{-0.9} \qquad \text{Divide each side by } -0.9.$$
$$m = \underline{} \qquad \text{Simplify.}$$

Step 3 Substitute 10 for m in either equation to find the value of k.

$0.9m + 1.2k = 15$	First equation
$0.9 \,(\underline{}) + 1.2k = 15$	Replace m with 10.
$\underline{} + 1.2k = 15$	Multiply.
$9 \underline{} + 1.2k = 15 \underline{}$	Subtract 9 from each side.
$1.2k = 6$	Simplify.
$\dfrac{1.2k}{1.2} = \dfrac{6}{1.2}$	Divide each side by 1.2.
$k = \underline{}$	Simplify.

During a typical week, Mei builds _____ computers and Kara builds _____. Since they cannot sell part of a computer, it makes sense that they would build a whole number of computers in a week. Therefore, 10 computers and 5 computers are viable solutions.

Go Online You can complete an Extra Example online.

Practice

🔎 **Go Online** You can complete your homework online.

Examples 1, 3

Use elimination to solve each system of equations.

1. $-v + w = 7$
$v + w = 1$

2. $y + z = 4$
$y - z = 8$

3. $-4x + 5y = 17$
$4x + 6y = -6$

4. $5m - 2p = 24$
$3m + 2p = 24$

5. $a + 4b = -4$
$a + 10b = -16$

6. $6r - 6t = 6$
$3r - 6t = 15$

7. $6c - 9d = 111$
$5c - 9d = 103$

8. $11f + 14g = 13$
$11f + 10g = 25$

9. $9x + 6y = 78$
$3x - 6y = -30$

10. $3j + 4k = 23.5$
$8j - 4k = 4$

11. $-3x - 8y = -24$
$3x - 5y = 4.5$

12. $6x - 2y = 1$
$10x - 2y = 5$

13. $x - y = 1$
$x + y = 3$

14. $-x + y = 1$
$x + y = 11$

15. $x + 4y = 11$
$x - 6y = 11$

16. $-x + 3y = 6$
$x + 3y = 18$

17. $3x + 4y = 19$
$3x + 6y = 33$

18. $x + 4y = -8$
$x - 4y = -8$

19. $3x + 4y = 2$
$4x - 4y = 12$

20. $3x - y = -1$
$-3x - y = 5$

21. $2x - 3y = 9$
$-5x - 3y = 30$

22. $x - y = 4$
$2x + y = -4$

23. $3x - y = 26$
$-2x - y = -24$

24. $5x - y = -6$
$-x + y = 2$

25. $6x - 2y = 32$
$4x - 2y = 18$

26. $3x + 2y = -19$
$-3x - 5y = 25$

27. $7x + 4y = 2$
$7x + 2y = 8$

Example 2

28. Twice a number added to another number is 15. The sum of the two numbers is 11. Find the numbers.

29. Twice a number added to another number is −8. The difference of the two numbers is 2. Find the numbers.

30. The difference of two numbers is 2. The sum of the same two numbers is 6. Find the numbers.

Example 4

31. GOVERNMENT The Texas State Legislature is comprised of state senators and state representatives. There is a greater number of representatives than senators. The sum of the number of representatives and the number of senators is 181. The difference of the number of representatives and number of senators is 119.
    a. Write a system of equations to find the number of state representatives, r, and senators, s.
    b. How many senators and how many representatives make up the Texas State Legislature?

32. SPORTS As of 2019, the New York Yankees had won the World Series more than any other team in baseball. The difference of the number of World Series championships won by the Yankees and 2 times the number of World Series championships won by the second-most-winning team, the St. Louis Cardinals, is 5. The sum of the two teams' World Series championships is 38.
    a. Write a system of equations to find the number of World Series championships won by the Yankees, y, and the number of World Series championships won by the Cardinals, x.
    b. How many times has each team won the World Series?

Mixed Exercises

Use elimination to solve each system of equations.

33. $4(x + 2y) = 8$
    $4x + 4y = 12$

34. $3x - 5y = 11$
    $5(x + y) = 5$

35. $4x + 3y = 6$
    $3(x + y) = 7$

36. $0.3x - 2y = -28$
    $0.8x + 2y = 28$

37. $\frac{1}{2}q - 4r = -2$
    $\frac{1}{6}q - 4r = 10$

38. $\frac{1}{2}x + \frac{1}{3}y = -1$
    $-\frac{1}{2}x + \frac{2}{3}y = 10$

39. **REASONING** At the end of a recent WNBA regular season, the difference of the number of wins and losses by the Phoenix Mercury was 12. The difference of the number of wins and two times the number of losses was 1. How many regular season games did the Phoenix Mercury play during that season?

40. **USE A MODEL** Marisol works for a florist that sells two types of bouquets, as shown at the right. On Monday, Marisol used 96 tulips to make the bouquets. On Tuesday, she used 192 tulips to make the same number of Spring Mix bouquets as Monday, but 3 times as many Garden Delight bouquets.

Seasonal Bouquets	
Spring Mix	12 tulips
Garden Delight	16 tulips

a. Write a system of equations that you can use to find how many bouquets of each type Marisol made. Describe what each variable represents.

b. Find the total number of tulips Marisol used to make Garden Delight bouquets on Monday and Tuesday. Explain your answer.

41. **USE A MODEL** Jeremy and Kendrick each bought snacks for their friends at a skating rink. The table shows the number of bags of popcorn and the number of plates of nachos each person bought, as well as the total cost of the snacks.

Name	Bags of Popcorn	Plates of Nachos	Total Cost
Jeremy	4	2	$18.50
Kendrick	7	2	$26.75

a. Write a system of equations that you can use to find the prices of the popcorn and the nachos. Describe what each variable represents.

b. Solve the system of equations and explain what your solution represents.

42. **USE A MODEL** The table shows the time Erin spent jogging and walking this weekend and the total distance she covered each day. Erin always jogs at the same rate and always walks at the same rate.

Day	Time Jogging	Time Walking	Total Distance
Saturday	15 min	30 min	3.5 mi
Sunday	1 h	30 min	8 mi

a. Write a system of equations that you can use to represent this situation. Describe what each variable represents.

b. Solve the system by elimination. Show your work. Then interpret the solution.

43. STRUCTURE Consider the system of equations $0.4x - 2y = 6$ and $0.8x + 2y = 0$.

    a. What is the first step to solving the system of equations by elimination? Explain your reasoning.

    b. What is the solution, as an ordered pair?

    c. Graph the equations on a coordinate plane. Explain how you can use the graph to check your solution.

    d. How would the solution change if the second equation was $x - 5y = 15$? Explain.

    e. How would the solution change if the second equation was $0.4x - 2y = 0$? Explain.

Higher-Order Thinking Skills

44. ANALYZE Mikasi says that if you solve a system of equations using elimination by addition and the result is $0 = 0$, then the solution of the system is $(0, 0)$. Provide a counterexample to show that his statement is false. Justify your argument.

45. ANALYZE Reece says that if you solve a system of equations using elimination by addition and the result is $0 = 2$, then the solution of the system is $(0, 2)$. Provide a counterexample to show that her statement is false. Justify your argument.

46. CREATE Create a system of equations that can be solved by using addition to eliminate one variable. Formulate a general rule for creating such systems.

47. CREATE The solution of a system of equations is $(-3, 2)$. One equation in the system is $x + 4y = 5$. Find a second equation for the system such that the system can be solved using elimination by addition. Explain how you derived this equation.

48. PERSEVERE The sum of the digits of a two-digit number is 8. The result of subtracting the units digit from the tens digit is −4. Define variables and write the system of equations that you would use to find the number. Then solve the system and find the number.

49. WRITE Describe when it would be most beneficial to use elimination to solve a system of equations.

_Copyright © McGraw-Hill Education

412 **Module 7** • Systems of Linear Equations and Inequalities

Elimination Using Multiplication

Explore Graphing and Elimination Using Multiplication

Online Activity Use an interactive tool to complete the Explore.

> **INQUIRY** How can you produce a new system of equations with the same solution as the given system?

Learn Solving Systems of Equations by Elimination with Multiplication

Key Concept • Elimination Method Using Multiplication

Step 1 Multiply at least one equation by a constant to get two equations that contain opposite terms.

Step 2 Add the equations, eliminating one variable. Then solve the equation.

Step 3 Substitute the value from Step 2 into one of the equations and solve for the other variable. Write the solution as an ordered pair.

Example 1 Elimination Using Multiplication

Use elimination to solve the system of equations.

$10x + 5y = 30$
$5x - 3y = -7$

Step 1 Multiply an equation by a constant.

The coefficients of x will be opposites if the second equation is multiplied by -2.

$5x - 3y = -7$ Second equation.

$\underline{\quad}(5x - 3y) = \underline{\quad}(-7)$ Multiply each side by -2.

$\underline{\qquad\qquad}$ Simplify.

Step 2 Add the equations.

$10x + 5y = 30$
$(+) -10x + 6y = 14$
$\overline{\qquad\qquad}$
$11y = 44$ The variable x is eliminated.

$\underline{\quad} = $ Divide each side by 11.

$y = \underline{\quad}$ Simplify.

(continued on the next page)

Today's Goal
• Solve systems of equations by eliminating a variable using multiplication and addition.

Think About It!
How does the process of solving a system of equations by elimination using multiplication differ from elimination using just addition?

Study Tip
Common Factors If the coefficients of a variable are not the same, or are opposites, and they share a greatest common factor greater than 1, then that variable is the easiest to eliminate using multiplication. For example, the system in this example has two variables, x and y. The coefficients of the y-variable expressions are 5 and -3, which share no common factor greater than 1. However, the coefficients of the x-variable expressions are 10 and 5, which share a common factor of 5. Thus, the x-variable requires fewer steps to eliminate.

Would you get the solution (1, 4) if you eliminated the *y*-variable instead of the *x*-variable? If no, explain your reasoning. If yes, explain why the *y*-variable was selected for elimination instead of the *x*-variable.

Step 3 Substitute 4 for *y* in either equation to find the value of *x*.

$$10x + 5y = 30$$ First equation

$$10x + 5\underline{\quad} = 30$$ Replace *y* with 4.

$$10x + \underline{\quad} = 30$$ Multiply.

$$10x + 20 - \underline{\quad} = 30 - \underline{\quad}$$ Subtract 20 from each side.

$$\frac{10x}{10} = \frac{10}{10}$$ Divide each side by 10.

$$x = \underline{\quad}$$ Simplify.

The solution is (_____).

Check

Use elimination to solve the system of equations.

$$13x + 14y = 59$$

$$4x + 7y = 37$$

_____

Example 2 Multiply Both Equations to Eliminate a Variable

Use elimination to solve the system of equations.

$$3x + 4y = -22$$
$$-2x + 3y = -8$$

Step 1 Multiply both equations by a constant.

$3x + 4y = -22$	Original equation	$-2x + 3y = -8$
$2(3x + 4y) = 2(-22)$	Multiply by a constant.	$3(-2x + 3y) = 3(-8)$
$2(3x) + 2(4y) = 2(-22)$	Distributive Property	$3(-2x) + 3(3y) = 3(-8)$
$6x + 8y = -44$	Simplify	$-6x + 9y = -24$

Step 2 Add the equations.

$$6x + 8y = -44$$
$$(+)\,{-6x + 9y = -24}$$
_____

$$17y = -68$$ The variable x is eliminated.

$$\frac{17y}{17} = \frac{-68}{17}$$ Divide each side by 17.

$$y = -4$$ Simplify.

🔘 **Go Online** You can complete an Extra Example online.

Step 3 Use substitution to find the value of *x*.

_____	First equation
_____	Replace *y* with −4.
_____	Multiply.
_____	Add 16 to each side.
_____	Divide each side by 3.
_____	Simplify.

The solution is (_____).

Check

Use elimination to solve the system of equations.
$11x - 6y = 25$
$3x + 9y = 60$

_____

🌐 Example 3 Write and Solve a System Using Multiplication

COMICS **Jorge's comic book collection consists of single issues that cost $4 each and paperback collections that cost $12 each. He has 100 books in all. His collection cost him $616. Write and solve a system of equations to determine how many single issues and paperbacks Jorge has in his collection.**

Complete the table to write the system of equations. Let *c* = the number of single issue comics and *p* = the number of paperback collections.

The number of single issue comics	plus	the number of paperback collections	is	100 books
_____	+	_____	=	_____
$4 per single issue comic	plus	$12 per paperback collection	is	$616
_____	+	_____	=	_____

(continued on the next page)

Math History Minute

German mathematician **Carl Friedrich Gauss (1777-1855)** contributed significantly to many fields, including number theory, algebra, and statistics. The elimination method is related to the Gaussian elimination method, an algorithm for solving systems of linear equations that was known to Chinese mathematicians as early as 179 B.C.

Step 1 Multiply an equation by a constant.

$$c + p = 100 \qquad \text{First equation}$$

$$\underline{\quad}(c + p) = \underline{\quad}(100) \qquad \text{Multiply each side by } -4.$$

$$\underline{\quad}c + \underline{\quad}p = \underline{\quad}(100) \qquad \text{Distributive Property}$$

$$\underline{\quad}c + \underline{\quad}p = \underline{\quad\quad} \qquad \text{Simplify.}$$

Step 2 Add the equations.

$$-4c - 4p = -400$$
$$\underline{(+)4c + 12p = 616}$$
$$\frac{8p}{8} = \frac{216}{8} \qquad \text{The variable } c \text{ is eliminated.}$$
$$\text{Divide each side by 8.}$$
$$p = \underline{\quad} \qquad \text{Simplify.}$$

Step 3 Use substitution to find the value of c.

Substitute 27 for p in either equation to find the value of c.

$$c + p = 100 \qquad \text{First equation.}$$

$$c + \underline{\quad} = 100 \qquad \text{Replace } p \text{ with 27.}$$

$$c + 27 - \underline{\quad} = 100 - \underline{\quad} \qquad \text{Subtract 27 from each side.}$$

$$c = \underline{\quad} \qquad \text{Simplify.}$$

Jorge has ___ single issue comics and ___ paperback collections.

Check

SOFTWARE A software company releases two products: a home version of their photo editor, which costs $20, and a professional version, which costs $45. The company sells 1000 copies of the photo editing software, earning a total revenue of $38,075. Write and solve a system of equations to determine how many home versions and professional versions of the software the company sold.

Let h = the number of home versions sold and p = the number of professional versions sold.

Part A Write the system of equations.

_____

_____

Part B Solve the system of equations.

_____

The software company sold _____ home versions and _____ professional versions of their photo editor.

Go Online You can complete an Extra Example online.

Practice

Examples 1 and 2

Use elimination to solve each system of equations.

1. $x + y = 2$
$-3x + 4y = 15$

2. $x - y = -8$
$7x + 5y = 16$

3. $x + 5y = 17$
$-4x + 3y = 24$

4. $6x + y = -39$
$3x + 2y = -15$

5. $2x + 5y = 11$
$4x + 3y = 1$

6. $3x - 3y = -6$
$-5x + 6y = 12$

7. $3x + 4y = 29$
$6x + 5y = 43$

8. $8x + 3y = 4$
$-7x + 5y = -34$

9. $8x + 3y = -7$
$7x + 2y = -3$

10. $4x + 7y = -80$
$3x + 5y = -58$

11. $12x - 3y = -3$
$6x + y = 1$

12. $-4x + 2y = 0$
$10x + 3y = 8$

Example 3

13. SPORTS The Fan Cost Index (FCI) tracks the average costs for attending sporting events, including tickets, drinks, food, parking, programs, and souvenirs. According to the FCI, a family of four would spend a total of $592.30 to attend two Major League Baseball (MLB) games and one National Basketball Association (NBA) game. The family would spend $691.31 to attend one MLB and two NBA games.

a. Write a system of equations to find the family's costs for each kind of game according to the FCI.

b. Solve the system of equations to find the cost for a family of four to attend each kind of game according to the FCI.

14. ART Mr. Santos, the curator of the children's museum, recently made two purchases of firing clay and polymer clay for a visiting artist to sculpt. Use the table to find the cost of each product per kilogram.

Firing Clay (kg)	Polymer Clay (kg)	Total Cost
5	24	$64.05
25	8	$51.45

a. Write a system of equations to find the cost of each product per kilogram.

b. Solve the system of equations to find the cost of each product per kilogram.

Mixed Exercises

15. Two times a number plus three times another number equals 13. The sum of the two numbers is 7. What are the numbers?

16. Four times a number minus twice another number is −16. The sum of the two numbers is −1. Find the numbers.

17. **FUNDRAISING** Trisha and Byron are washing and vacuuming cars to raise money for a class trip. Trisha raised $38 by washing 5 cars and vacuuming 4 cars. Byron raised $28 by washing 4 cars and vacuuming 2 cars. Find the amount they charged to wash a car and vacuum a car.

18. **STRUCTURE** Consider the system of equations $-2x + 3y = -5$ and $3x - 4y = 6$.
    a. Describe two different ways to solve the system by elimination.
    b. Explain why multiplying the first equation by 6 and the second equation by 5 and then adding is not useful for solving the system.
    c. Solve the system by elimination.

19. **REASONING** The owner of a juice stand wants to make a new juice drink. He would like to mix Tropical Breeze, t, and Kona Cooler, k, to make 10 quarts of a new drink that is 40% pineapple juice.

Juice Drinks	
Tropical Breeze	20% pineapple juice
Kona Cooler	50% pineapple juice

    a. Make a table to help write a system of equations that the owner of the juice stand can solve to determine the amount of each drink he should use to make the new drink.
    b. Solve the system and explain what your solution represents.
    c. Explain how you know your answer is correct.

20. **USE A MODEL** Marlene works as a cashier at a grocery store. At the end of the day, she has a total of 125 five-dollar bills and ten-dollar bills. The total value of these bills is $990.

    a. Write a system of equations that you can use to find the number of five-dollar bills and the number of ten-dollars bills. Describe what each variable represents.

    b. Solve the system and explain what your solution represents.

21. **FIND THE ERROR** Jason and Daniela are solving a system of equations. Is either of them correct? Explain your reasoning.

22. **ANALYZE** Determine whether the following statement is *true* or *false*: *A system of linear equations will only have infinitely many solutions if the equations have the same coefficients*. Justify your argument.

23. **CREATE** Write a system of equations that can be solved by multiplying one equation by −3 and then adding the two equations together.

24. **PERSEVERE** The solution of the system $4x + 5y = 2$ and $6x - 2y = b$ is $(3, a)$. Find the values of a and b. Discuss the steps you used.

25. **WRITE** Why is substitution sometimes more helpful than elimination, and vice versa?

Jason

$2r + 7t = 11$
$r - 9t = -7$
$2r + 7t = 11$
$(-) 2r - 18t = -14$
$25t = 25$
$t = 1$
$2r + 7t = 11$
$2r + 7(1) = 11$
$2r + 7 = 11$
$2r = 4$
$\frac{2r}{2} = \frac{4}{2}$
$r = 2$
The solution is (2, 1).

Daniela

$2r + 7t = 11$
$(-) r - 9t = -7$
$r = 18$
$2r + 7t = 11$
$2(18) + 7t = 11$
$36 + 7t = 11$
$7t = -25$
$\frac{7t}{7} = \frac{25}{7}$
$t = -3.6$
The solution is (18, -3.6).

Systems of Inequalities

Explore Solutions of Systems of Inequalities

 Online Activity Use an interactive tool to complete the Explore.

> ✕
>
> ② **INQUIRY** How are the solutions of a system of inequalities represented on a graph?

Learn Solving Systems of Inequalities by Graphing

A set of two or more inequalities with the same variables is a **system of inequalities**. The solution of a system of inequalities with two variables is the set of ordered pairs that satisfy all of the inequalities in the system. The solution is represented by the overlap, or intersection, of the graphs of the inequalities.

Example 1 Solve by Graphing

Solve the system of inequalities by graphing.

$x - 2y > -6$

$y < 3x$

Step 1 Graph one inequality of the system.
The boundary of $x - 2y > -6$ is _____ and _____ in the solution. The half-plane is shaded below the boundary, in yellow, to indicate solutions of $x - 2y > -6$.

Step 2 Graph the second inequality of the system.
The boundary of $y < 3x$ is _____ and _____ in the solution. The half-plane is shaded to the right of the boundary, in blue, to indicate solutions of $y < 3x$.

Solution
The solution of the system is the set of ordered pairs in the _____ of the graphs of $x - 2y > -6$ and $y < 3x$. The region is shaded green.

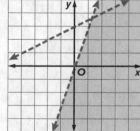

 Go Online You can complete an Extra Example online.

Today's Goal
- Solve systems of linear inequalities by graphing.

Today's Vocabulary
system of inequalities

 Go Online
You can watch a video to see how to solve a system of linear inequalities.

😎 **Think About It!**
The boundaries for the system $y > 3$, $y \le -2x + 1$ intersect at $(-1, 3)$. Is $(-1, 3)$ included in the solution? Explain.

 Go Online
You can watch a video to see how to use a graphing calculator with this example.

Check

Graph the system of inequalities.

$\frac{1}{3}x + 2 < y$

$x \geq -3$

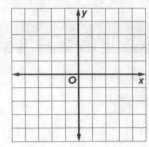

Example 2 Solve by Graphing, No Solution

Solve the system of inequalities by graphing.

$-3x + 4y > 0$

$3x - 4y \geq 8$

Step 1 Graph one inequality of the system.

The boundary of $-3x + 4y > 0$ is _____ and _____ in the solution. The half-plane is shaded above the boundary, in yellow, to indicate solutions of $-3x + 4y > 0$.

Step 2 Graph the second inequality of the system.

The boundary of $3x - 4y \geq 8$ is _____ and _____ in the solution. The half-plane is shaded below the boundary, in blue, to indicate solutions of $3x - 4y \geq 8$.

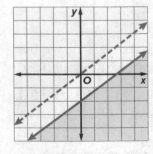

Solution

The graphs of $-3x + 4y = 0$ and $3x - 4y = 8$ are parallel lines. The regions _____ at any point, so the system has

_____.

 Go Online You can complete an Extra Example online.

💬 **Talk About It!**

Is it possible for a system of inequalities that has boundaries with different slopes to have no solution? Justify your argument.

Study Tip

Shaded Regions When graphing more than one region, it is helpful to use a different color of pencil or a different pattern for each region. This will make it easier to see where the regions intersect and find possible solutions.

Check

Graph the system of inequalities.

$x + 3 < y$

$3x - 3y \geq 12$

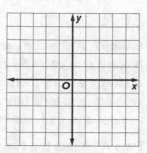

Example 3 Apply Systems of Inequalities

SEWING **A family and consumer sciences class is making pillows and blankets to donate to a local shelter. The class has 40 yards of fabric to use. Pillows require 1.25 yards of fabric and take 1 hour to make. Blankets use 4 yards of fabric and take 2.5 hours to make. The class has 28 hours of class time left for the semester. Determine the number of pillows and blankets the class can make for the shelter.**

Part A Define the variables, and write a system of inequalities to represent the situation.

Let p represent the number of pillows the class can make.
Let b represent the number of _____ the class can make.

Fabric and time are two constraints on the numbers of pillows and blankets the class can make.

Because _____ use 1.25 yards of fabric and _____ use 4 yards of fabric, the inequality that represents the fabric constraint is _____.

Making a pillow takes 1 hour and making a blanket takes _____ hours. Because the class has only 28 hours left, the inequality that represents the time constraint is _____.

Part B Graph the system.

Part C Find a viable solution.

Only whole-number solutions make sense in this situation. One possible solution is (_____); _____ blankets and _____ pillows can be made.

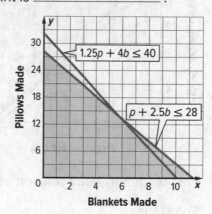

 Go Online You can complete an Extra Example online.

Think About It!

Can the class make 2 blankets and 24 pillows? Explain.

Think About It!

Could the graph be represented with "Pillows Made" as the x-axis and "Blankets Made" as the y-axis? Explain.

Check

BAKERY Aisha can work up to 20 hours per week. Working at a bakery, she earns $7 per hour most of the time and $8.50 per hour during the early morning shift. Aisha needs to earn at least $150 this week to pay for a trip with her friends. Determine the number of regular and early morning hours that Aisha could work.

Part A Select the correct system and graph. Let r = regular hours and m = early morning hours. ___

A. $r < 20$

$7r + 8.5m \geq 150$

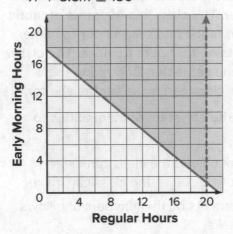

B. $r + m \leq 20$

$r + m \leq 150$

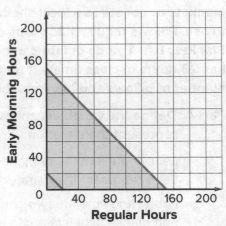

C. $r + m \leq 20$

$7r + 8.5m \geq 150$

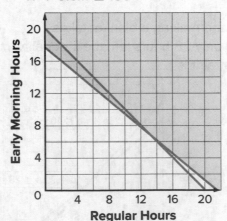

D. $7r + 8.5m > 20$

$7r + 8.5m \geq 150$

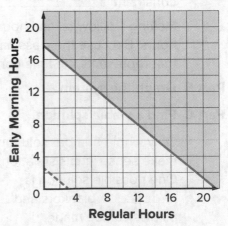

Part B Identify each solution as *viable* or *nonviable*.

Solutions

(2, 17) _____   (10, 10) _____

(4, 7) _____   (18, 6) _____

(5, 15) _____   (21, 6) _____

(8, 13) _____

Practice

◤ **Go Online** You can complete your homework online.

Examples 1 and 2

Solve each system of inequalities by graphing.

1. $y < 6$
   $y > x + 3$

2. $y \geq 0$
   $y \leq x - 5$

3. $y \leq x + 10$
   $y > 6x + 2$

4. $y \geq x + 10$
   $y \leq x - 3$

5. $y < 5x - 5$
   $y > 5x + 9$

6. $y \geq 3x - 5$
   $3x - y > -4$

7. $x > -1$
   $y \leq -3$

8. $y > 2$
   $x < -2$

9. $y > x + 3$
   $y \leq -1$

10. $x < 2$
   $y - x \leq 2$

11. $x + y \leq -1$
   $x + y \geq 3$

12. $y - x > 4$
   $x + y > 2$

Example 3

13. FITNESS Diego started an exercise program in which each week he walks from 9 to 12 miles and works out at the gym from 4.5 to 6 hours.

   a. Write a system of inequalities to represent this situation. Define your variables.

   b. Graph the system.

   c. List three viable solutions.

14. SOUVENIRS Emiliana wants to buy turquoise stones on her trip to New Mexico to give to at least 4 of her friends. The gift shop sells stones for either $4 or $6 per stone. Emiliana has no more than $30 to spend.

   a. Write a system of inequalities to represent this situation. Define your variables.

   b. Graph the system.

   c. List three viable solutions.

Mixed Exercises

Write a system of inequalities for each graph.

15.
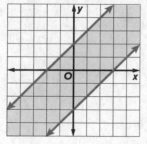

16.
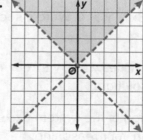

17.
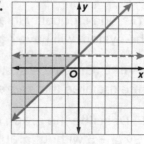

18. **PRECISION** Write a system of inequalities to represent the graph shown at the right.

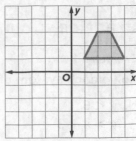

19. **CONSTRUCT ARGUMENTS** Is (2.5, 1) a solution of the system of inequalities $2x + 3y > 8$ and $4x - 5y \geq 2$? Justify your argument. Then explain how you can tell if the point is a solution without graphing the inequality.

20. **PETS** Priya's Pet Store never has more than a combined total of 20 cats and dogs and never more than 8 cats. This is represented by the inequalities $x + y \leq 20$ and $x \leq 8$. Represent the number of cats and dogs that can be at the store on a graph. Solve the system of inequalities by graphing.

21. **FUNDRAISING** The baseball team plans to sell tins of popcorn and peanuts. The players have $900 to spend on products and can order up to 200 tins. They want to order at least as many tins of popcorn as tins of peanuts. A tin of popcorn costs $3, and a tin of peanuts costs $4. Define the variables and write a system of inequalities to represent this situation. Then list any constraints for the variables.

22. **BUSINESS** For maximum efficiency, a factory must have at least 100 workers, but no more than 200 workers on a shift. The factory also must manufacture at least 30 units per worker.

   a. Let x be the number of workers and let y be the number of units. Write a system of inequalities expressing the conditions in the problem.

   b. Graph the systems of inequalities.

   c. Find three possible solutions.

23. **DESIGN** LaShawn designs Web sites for local businesses. He charges $25 an hour to build a Web site and charges $15 an hour to update Web sites once he builds them. He wants to earn at least $100 every week, but he does not want to work more than 6 hours each week. What is a possible number of hours LaShawn can spend each week building Web sites x and updating Web sites y that will allow him to attain his goals? Write your answer as an ordered pair.

🍪 **Higher-Order Thinking Skills**

24. **PERSEVERE** Create a system of inequalities equivalent to $|x| \leq 4$.

25. **ANALYZE** State whether the following statement is *sometimes*, *always*, or *never* true. Justify your argument.

   Systems of inequalities with parallel boundaries have no solutions.

26. **CREATE** One inequality in a system is $3x - y > 4$. Write a second inequality so that the system will have no solution.

27. **PERSEVERE** Graph the system of inequalities $y \geq 1$, $y \leq x + 4$, and $y \leq -x + 4$. Estimate the area that represents the solution.

28. **WRITE** Describe the graph of the solution of the system $6x - 3y \leq -5$ and $6x - 3y \geq -5$ without graphing. Explain your reasoning.

Essential Question
How are systems of equations useful in the real world?

Module Summary

Lesson 7-1
Graphing Systems of Equations

- When you solve a system of equations with $y = f(x)$ and $y = g(x)$, the solution is an ordered pair that satisfies both equations. Thus, the x-coordinate of the intersection of $y = f(x)$ and $y = g(x)$ is the value of x where $f(x) = g(x)$.

- A system of equations is consistent if it has at least one ordered pair that satisfies both equations.

- A system of equations is independent if it has exactly one solution.

- A system of equations is dependent if it has an infinite number of solutions.

- A system of equations is inconsistent if it has no ordered pair that satisfies both equations.

Lessons 7-2 through 7-4
Solving Systems of Equations Algebraically

- To use the substitution method, solve at least one equation for one variable. Substitute the resulting expression into the other equation to replace the variable. Then solve the equation. Substitute this value into either equation, and solve for the other variable. Write the solution as an ordered pair.

- To use the elimination method, write the system so like terms with opposite coefficients are aligned. Add or subtract the equations,

eliminating one variable. Then solve the equation. Substitute this value into one of the equations and solve for the other variable. Write the solution as an ordered pair. You may need to multiply at least one equation by a constant to get two equations that contain opposite terms.

Lesson 7-5
Systems of Inequalities

- A set of two or more inequalities with the same variables is called a system of inequalities.

- The solution of a system of inequalities with two variables is the set of ordered pairs that satisfy all of the inequalities in the system. The solution is represented by the overlap, or intersection, of the graphs of the inequalities.

Study Organizer

 Foldables

Use your Foldable to review this module. Working with a partner can be helpful. Ask for clarification of concepts as needed.

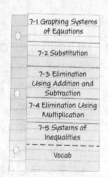

Test Practice

1. MULTI-SELECT Use the graph. Which systems of equations are consistent and independent? (Lesson 7-1)

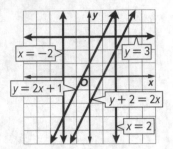

(A) $y = 2x + 1$
   $y + 2 = 2x$

(B) $y = 2x + 1$
   $y = 3$

(C) $y = 3$
   $x = -2$

(D) $x = 2$
   $y + 2 = 2x$

(E) $x = -2$
   $x = 2$

2. OPEN RESPONSE Consider the system of equations. (Lesson 7-1)

$8x + 2y = 8$
$y = -4x + 4$

How many solutions are there for the system? Is the system dependent or independent?

3. MULTIPLE CHOICE Which system of equations can be entered into a graphing calculator to solve $3.5x + 18 = -5.8x + 30$? (Lesson 7-1)

(A) $y = 3.5x$
   $y = -5.8x$

(B) $y = 3.5x + 18$
   $y = -5.8x + 30$

(C) $0 = 3.5x + 18$
   $0 = -5.8x + 30$

(D) $y = 9.3x - 12$

4. MULTIPLE CHOICE Use a system of equations and a graphing calculator to solve $6.9x + 4.3 = 4.7x + 8$. Round your answer to the nearest hundredth, if necessary. (Lesson 7-1)

(A) 1.06

(B) 1.68

(C) 2.14

(D) 5.59

5. OPEN RESPONSE Taylan is selling plastic and wooden frames. He sold 7 total frames. The number of plastic frames Taylan sold was 5 less than twice the number of wooden frames. How many of each type of frame did Taylan sell? (Lesson 7-2)

6. MULTIPLE CHOICE Consider the system of equations.

$3x - 2y = 0$
$x + y = 10$

What is the solution of the system? (Lesson 7-2)

(A) The solution to the system is $(20, -10)$.

(B) The solution to the system is $(3, 7)$.

(C) The solution to the system is $(4, 6)$.

(D) The solution to the system is $(6, 4)$.

7. OPEN RESPONSE Determine whether the system has *no solution, one solution,* or *infinitely many solutions*. If the system has one solution, name it. (Lesson 7-2)

$x + y = 5$
$3x + 2y = 8$

8. OPEN RESPONSE The sum of the measures of two complementary angles is 90 degrees. Angles P and Q are complementary, and the measure of angle P is 6 degrees more than twice the measure of angle Q.

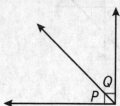

Write a system of equations and use substitution to find the measure of angles P and Q. (Lesson 7-2)

9. OPEN RESPONSE Solve the system of equations. (Lesson 7-3)

$3x + y = 34$

$0.5x - y = 1$

10. MULTIPLE CHOICE A rectangle is x inches wide and $3y$ inches long. The sum of the length and width is 36 inches and the difference between the length and twice the width is 12 inches. Find the length and width. (Lesson 7-3)

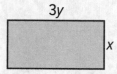

Ⓐ width: 8 inches; length: 28 inches

Ⓑ width: 8 inches; length: 9.3 inches

Ⓒ width: 12 inches; length: 24 inches

Ⓓ width: 15 inches; length: 21 inches

11. MULTI-SELECT Select all of the ways the system can be solved. (Lesson 7-4)

$9x - 2y = 4$
$3x + 8y = -12$

Ⓐ Multiply the first equation by 4, then add the equations.

Ⓑ Multiply the second equation by 3, then subtract the equations.

Ⓒ Multiply the first equation by 3, then add the equations.

Ⓓ Multiply the second equation by 3, then add the equations.

Ⓔ Multiply the first equation by 3, then subtract the equations.

12. OPEN RESPONSE Solve the system of equations. (Lesson 7-4)

$2x + 5y = 5$
$3x + 4y = -3$

13. OPEN RESPONSE Solve the system of equations. (Lesson 7-4)

$2r - t = 7$
$r - t = 1$

[ ]

14. OPEN RESPONSE It takes 3 hours to paddle a kayak 12 miles downstream and 4 hours for the return trip upstream. Find the rate of the kayak in still water.

Let k = the rate of the kayak in still water and c = the rate of the current. (Lesson 7-4)

	r	t	d	$rt = d$
Downstream	$k + c$	3	12	$3(k + c) = 12$
Upstream	$k - c$	4	12	$4(k - c) = 12$

[ ]

15. MULTIPLE CHOICE The graph shows the solution to the given system of inequalities. (Lesson 7-5)

$-x + 2y \leq 1$
$-3x + 2y \geq 2$

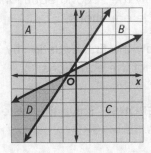

In what region is the solution set?

Ⓐ A

Ⓑ B

Ⓒ C

Ⓓ D

16. MULTIPLE CHOICE Which graph represents the solution of the system of inequalities? (Lesson 7-5)

$x - y \geq 2$
$2x + y > -3$

Ⓐ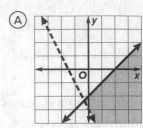

Ⓑ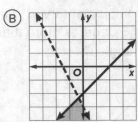

Ⓒ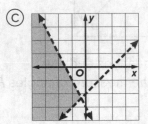

Ⓓ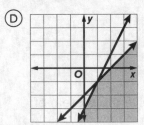

17. MULTIPLE CHOICE Diana wants to build a rectangular pen for her goats. The length of the pen should be at least 50 feet, and the perimeter of the pen should be no more than 190 feet. What is a viable solution for the dimensions of the pen? (Lesson 7-5)

Ⓐ 29 feet by 40 feet

Ⓑ 29 feet by 76 feet

Ⓒ 29 feet by 65 feet

Ⓓ 29 feet by 29 feet

e Essential Question

How do you perform operations and represent real-world situations with exponents?

What will you learn?

Place a check mark (✓) in each row that corresponds with how much you already know about each topic **before** starting this module.

KEY

👎 — I don't know.   👍 — I've heard of it.   👍 — I know it!

	Before			After		
	👎	👍	👍	👎	👍	👍
use the Product of Powers Property						
use the Power of a Power Property						
use the Power of a Product Property						
use the Quotient of Powers Property						
use the Power of a Quotient Property						
simplify expressions with zero exponents						
simplify expressions with negative exponents						
use rational exponents to solve problems						
simplify radical expressions						
solve exponential equations						

📖 **Foldables** Make this Foldable to help you organize your notes about exponents and roots. Begin with seven sheets of notebook paper.

1. **Arrange** the paper into a stack.

2. **Staple** along the left side. Starting with the second sheet of paper, **cut** along the right side to form tabs.

3. **Label** the cover sheet "Exponents and Roots" and label each tab with a lesson number.

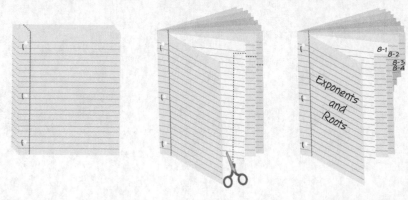

What Vocabulary Will You Learn?

Check the box next to each vocabulary term that you may already know.

☐ cube root
☐ exponential equation
☐ index
☐ monomial
☐ negative exponent

☐ *n*th root
☐ perfect cube
☐ perfect square
☐ principal square root
☐ radical expression

☐ radicand
☐ rational exponent
☐ square root

Are You Ready?

Complete the Quick Review to see if you are ready to start this module.
Then complete the Quick Check.

Quick Review

Example 1

Write $5 \cdot 5 \cdot 5 \cdot 5 + y \cdot y$ using exponents.

There are 4 factors of 5.

There are 2 factors of y.

$5 \cdot 5 \cdot 5 \cdot 5 = 5^4$

$y \cdot y = y^2$

$5 \cdot 5 \cdot 5 \cdot 5 + y \cdot y = 5^4 + y^2$

Example 2

Evaluate $\left(\frac{5}{7}\right)^2$.

$\left(\frac{5}{7}\right)^2 = \frac{5}{7} \cdot \frac{5}{7}$ Expand the expression.

$= \frac{5 \cdot 5}{7 \cdot 7}$ Multiply the numerators and multiply the denominators.

$= \frac{25}{49}$ Simplify.

Quick Check

Write each expression using exponents.

1. $4 \cdot 4 \cdot 4 \cdot 4 \cdot 4$

2. $x \cdot x \cdot x$

3. $m \cdot m \cdot m \cdot p \cdot p \cdot p \cdot p \cdot p$

4. $\left(\frac{1}{5}\right) \cdot \left(\frac{1}{5}\right) \cdot \left(\frac{1}{5}\right) \cdot \left(\frac{1}{5}\right) \cdot \left(\frac{1}{5}\right) \cdot \left(\frac{1}{5}\right)$

Evaluate each expression.

5. 2^5

6. $(-5)^2$

7. $\left(\frac{1}{2}\right)^4$

8. $(-4)^3$

How Did You Do?

Which exercises did you answer correctly in the Quick Check? Shade those exercise numbers below.

Multiplication Properties of Exponents

Explore Products of Powers

 Online Activity Use an interactive tool to complete the Explore.

> @ **INQUIRY** How can you determine the product of two powers a^m and a^p?   ×

Learn Product of Powers

A **monomial** is a number, a variable, or a product of a number and one or more variables. It has only one term. The term a^4 is a monomial. An expression that involves division by a variable is not a monomial. The term $\frac{ab}{c}$ is not a monomial.

Key Concept • Product of Powers	
Words	To multiply two powers that have the same base, add their exponents.
Symbols	For any real number a and any integers m and p, $a^m \cdot a^p = a^{m+p}$.
Examples	$b^2 \cdot b^4 = b^{2+4}$ or b^6; $d^3 \cdot d^7 = d^{3+7}$ or d^{10}

Example 1 Product of Powers

Simplify each expression.

a. $(3n^4)(4n^7)$

$(3n^4)(4n^7) = (3 \cdot 4)(\underline{\hspace{1cm}})$ Group coefficients and variables.

$= (3 \cdot 4)(\underline{\hspace{1cm}})$ Product of Powers

$= \underline{\hspace{1cm}}$ Simplify.

b. $(7xy^2)(2x^4y^3)$

$(7xy^2)(2x^4y^3) = (7 \cdot 2)(\underline{\hspace{1cm}})(\underline{\hspace{1cm}})$ Group coefficients and variables.

$= (7 \cdot 2)(\underline{\hspace{1cm}})(\underline{\hspace{1cm}})$ Product of Powers

$= \underline{\hspace{1cm}}$ Simplify.

Check

Simplify $(7n^7)(-7n)$.   Simplify $(11x^6y^6)(xy^9)$.

Today's Goals
- Find products of monomials.
- Find the power of a power.
- Find the power of a product.

Today's Vocabulary
monomial

Think About It!
How does the process for simplifying $b^2 \cdot b^4$ differ from simplifying $b^2 + b^4$?

Study Tip
Terminology Recall that for the expression b^2, b is called the *base* and the 2 is called the *exponent* or *power*.

Think About It!
What properties allow you to group the coefficients and the variables in the first step of each solution?

Problem-Solving Tip

Identify Subgoals
Before you can find the distance that US 708 travels, you need to write the rate and time in scientific notation. To convert a number from standard form to scientific notation, rewrite it as a number a such that $1 \le a < 10$ multiplied by a power of 10. Thus, 100 in scientific notation is 1×10^2. Conversely, 1×10^2 in standard form is 100.

Study Tip

Assumptions Assuming that US 708 travels at exactly 2,700,000 miles per hour for the entire year allows us to approximate the distance it will travel. While 2,700,000 miles per hour is only an approximation of the star's speed, and the speed may not be constant, using this constant rate allows for a reasonable estimate.

🌐 **Example 2** Product of Powers and Scientific Notation

STARS **The fastest recorded star in the Milky Way galaxy is US 708, which travels 2,700,000 miles per hour. How far does US 708 travel in 1 year? Write your answer in scientific notation and round to the nearest tenth. (Hint: 1 year = 8760 hours)**

Step 1 Convert the speed of US 708 and the number of hours in a year to scientific notation.

Speed of US 708

$2{,}700{,}000 = 2.7 \times \underline{\hspace{2cm}}$ $1 \le 2.7 < 10$

$= 2.7 \times \underline{\hspace{2cm}}$ $1{,}000{,}000 = 10^6$

Hours in a Year

$8760 = 8.76 \times \underline{\hspace{2cm}}$ $1 \le 8.76 < 10$

$= 8.76 \times \underline{\hspace{1cm}}$ $1000 = 10^3$

Step 2 Use the formula $d = rt$ to find approximately how far US 708 travels in a year.

$d = rt$ distance = rate · time

$d = (\underline{\hspace{2cm}}) \times (8.76 \times 10^3)$ $r = 2.7 \times 10^6$ mph, $t = 8.76 \times 10^3$ hrs

$= (2.7 \times 8.76) \times (\underline{\hspace{2cm}})$ Comm. and Assoc. Properties

$= \underline{\hspace{2cm}} \times (10^6 \times 10^3)$ Multiply.

$= 23.652 \times \underline{\hspace{2cm}}$ Product of Powers

$= 23.652 \times \underline{\hspace{2cm}}$ Simplify.

$= 2.3652 \times 10^1 \times 10^9$ $23.652 = 2.3652 \times 10^1$

$= 2.3652 \times 10^{1+9}$ Product of Powers

$= \underline{\hspace{2cm}}$ Simplify.

$\approx 2.4 \times 10^{10}$ Round to the nearest tenth.

Check

COMPUTERS Typically, a 4K computer monitor describes a screen with a resolution of 3840 × 2160. This means that the display's width is 3840 pixels and the height is 2160 pixels. How many pixels are there on a typical 4K computer monitor? _____

A. 2.2×10^3 pixels

B. 3.8×10^3 pixels

C. 8.3×10^6 pixels

D. 8.3×10^7 pixels

🌐 **Go Online** You can complete an Extra Example online.

Learn Power of a Power

Key Concept • Power of a Power	
Words	To find a power of a power, multiply the exponents.
Symbols	For any real number a and any integers m and p, $(a^m)^p = a^{mp}$.
Examples	$(b^2)^4 = b^{2 \cdot 4}$ or b^8; $(d^3)^7 = d^{3 \cdot 7}$ or d^{21}

Example 3 Power of a Power

Simplify each expression.

a. $(n^5)^3$

$(n^5)^3 = $ _____   Power of a Power

$= $ _____   Simplify.

b. $(x^4)^2$

$(x^4)^2 = $ _____   Power of a Power

$= $ _____   Simplify.

Check

Simplify $(n^4)^{11}$.

_____

Simplify $(x^7)^3$.

_____

Think About It!

Is $(x^4)^2$ equivalent to $(x^2)^4$? Explain your reasoning.

Learn Power of a Product

Key Concept • Power of a Product	
Words	To find a power of a product, find the power of each factor and multiply.
Symbols	For any real numbers a and b and any integer m, $(ab)^m = a^m b^m$.
Examples	$(-5x^2y)^3 = (-5)^3x^6y^3$ or $-125x^6y^3$

Think About It!

Is $2(xy)^3$ equivalent to $(2xy)^3$? Explain your reasoning.

Example 4 Power of a Product

Simplify each expression.

a. $(3x^5y^2)^5$

$(3x^5y^2)^5 = $ _____   Power of a Product

$= $ _____   Simplify.

(continued on the next page)

 Go Online You can complete an Extra Example online.

Copyright © McGraw-Hill Education

b. $(-5ab^4)^2$

$(-5ab^4)^2 = $ _____   Power of a Product

$= $ _____   Simplify.

Check

Simplify $(4x^3y^2)^3$.

_____

Simplify $(-3a^3b)^6$.

_____

⊕ Example 5 Power of a Product and Area

If the side of each smaller square is x inches, and the side of the whole canvas is s inches, then what is the area of the painting in terms of x?

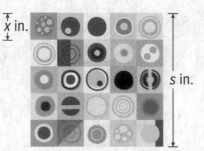

x in.

s in.

The side length of the canvas, s, can also be described as $5x$.

$\begin{aligned} A &= s^2 & \text{Area of a square} \\ &= \underline{\qquad} & s = 5x \\ &= 5^2 x^{1 \cdot 2} & \text{Power of a Product} \\ &= \underline{\qquad} & \text{Simplify.} \end{aligned}$

Check

GAME DESIGN Jeanine is designing an early version of a game and wants to use a cube to stand in for the character, which she will design later. She bases the dimensions of the cube around the height of the character, which she defines as $\frac{1}{8}x$ pixels, where x is the height of the total game screen. What is the volume of the cube in terms of x? _____

A. $\frac{1}{64}x^2$

B. $\frac{1}{512}x^3$

C. $\frac{1}{8}x^3$

D. $512x^3$

 Go Online You can complete an Extra Example online.

Practice

Go Online You can complete your homework online.

Examples 1, 3, and 4

Simplify each expression.

1. $(q^2)(2q^4)$

2. $(-2u^2)(6u^6)$

3. $(9w^2x^8)(w^6x^4)$

4. $(y^6z^9)(6y^4z^2)$

5. $(b^8c^6d^5)(7b^6c^2d)$

6. $(14fg^2h^2)(3f^4g^2h^2)$

7. $(j^5k^7)^4$

8. $(n^3p)^4$

9. $[(2^2)^2]^2$

10. $[(3^2)^2]^4$

11. $[(4r^2t)^3]^2$

12. $[(-2xy^2)^3]^2$

13. $(y^2z)(yz^2)$

14. $(\ell^2k^2)(\ell^3k)$

15. $(-5m^3)(3m^8)$

16. $(-2c^4d)(-4cd)$

17. $(3pr^2)^2$

18. $(2b^3c^4)^2$

Example 2

19. **COMMUNITY SERVICE** During the school year, each student planted 1.2×10^2 flowers as part of a community service project. If there are 1.5×10^3 students in the school, how many flowers did they plant in total?

20. **CHERRIES** A farmer has 350 cherry trees on his farm. If each cherry tree yields 6200 cherries per year, how many cherries can the farmer harvest per year? Write your answer in scientific notation.

21. **SUN** The temperature of the Sun's surface is about 5.5×10^3 degrees Celsius. The temperature of the Sun's core is about 2.7×10^3 times hotter than the surface. What is the approximate temperature of the Sun's core?

22. **SALES** An automobile company sold 2.3 million new cars in a year. If the average price per car was $21,000, how much money did the company make that year? Write your answer in scientific notation.

23. **APPLE JUICE** An apple juice company produced 8.3 billion individual-size bottles of apple juice last year. If each bottle contains 355 milliliters of juice, how many milliliters of apple juice did the company produce?

Example 5

24. **BLOCKS** Building blocks are in the shape of a cube. The dimensions of the building blocks are based on the length of the packaging box, which is defined as $\frac{1}{12}x^2$ centimeters, where x is the length of the packaging box. What is the volume of a building block in terms of x?

25. FIELDS A field is in the shape of a square. The length of the field is $5x^3y^5$. What is the area of the field in terms of x and y?

26. PICTURE FRAME Michelle is purchasing a new picture frame to hang in her bedroom. The picture frame is square shaped. What is the area of the interior of the picture frame in terms of c and d?

$\leftarrow 5c^3d^2 \rightarrow$

27. SHIPPING A shipping box is in the shape of a cube. What is the volume of the shipping box in terms of m and n?

$4m^8n^4$

Mixed Exercises

Simplify each expression.

28. $(2a^3)^4(a^3)^3$

29. $(c^3)^2(-3c^5)^2$

30. $(2gh^4)^3[(-2g^4h)^3]^2$

31. $(5k^2m)^3[(4km^4)^2]^2$

32. $(p^5r^2)^4(-7p^3r^4)(6pr^3)$

33. $(5x^2y)^2(2xy^3z)^3(4xyz)$

34. $(5a^2b^3c^4)^4(6a^3b^4c^2)$

35. $(10xy^5z^3)(3x^4y^6z^3)$

36. $(0.5x^3)^2$

37. $(0.4h^5)^3$

38. $\left(-\dfrac{3}{4c}\right)^3$

39. $\left(\dfrac{4}{5}a^2\right)^2$

40. $(8y^3)(-3x^2y^2)\left(\dfrac{3}{8}xy^4\right)$

41. $\left(\dfrac{4}{7}m\right)^2 (49m)(17p)\left(\dfrac{1}{34}p^5\right)$

42. $(-3r^3w^4)^3(2rw)^2(-3r^2)^3(4rw^2)^3(2r^2w^3)$ **43.** $(3ab^2c)^2(-2a^2b^4)^2(a^4c^2)^3(a^2b^4c^5)^2(2a^3b^2c^4)^3$

STRUCTURE Determine whether each pair of expressions is equivalent. Write *yes* or *no*.

44. $(a^7)(a^7)(a^7)(a^7)$ and a^{2401}

45. $(-j^9)(-j^9)(-j^9)(-j^9)$ and $-j^{27}$

46. $(5p^3q)(4p^5q^9)$ and $20p^8q^{10}$

47. $(6w^5)^2$ and $36w^{25}$

48. $(x^{10})^3$ and $(x^3)^{10}$

49. $[(2n^2)^3]^2$ and $(4n^4)^3$

50. GRAVITY An object that has been falling for x seconds has dropped at an average speed of $16x$ feet per second. If the object is dropped from a great height, its total distance traveled is the product of the average rate times the time. Write a simplified expression to show the distance the object has traveled after x seconds.

51. ELECTRICITY An electrician uses the formula $W = I^2R$, where W is the power in watts, I is the current in amperes, and R is the resistance in ohms.

   a. Find the power in a household circuit that has $2x^2$ amperes of current and $5x^3$ ohms of resistance.

   b. If the current is reduced by one-half, what happens to the power?

52. MOLECULES A glass of water contains 0.25 liter of water. If 1 milliliter of water contains 3.3×10^{22} water molecules, how many water molecules are there in the glass of water?

53. CIVIL ENGINEERING A developer is planning a sidewalk for a new development. The sidewalk can be installed in rectangular sections that have a fixed width of 3 feet and a length that can vary. Assuming that each section is the same length, express the area of a 4-section sidewalk as a monomial.

54. USE A SOURCE Suppose each student in your school is required to complete 60 hours of community service before they graduate. Find the number of students in the senior class at your school. Based on this number, calculate total minimum number of community service hours completed by the senior class by the time they graduate. Write your answer in scientific notation.

55. REGULARITY Recall that both multiplication and addition are commutative and associative. Multiplication also distributes over addition.

   a. What would it mean for the operation of raising one number to an exponent to be commutative? Decide and explain whether this operation is commutative.

   b. What would it mean for the operation of raising one number to an exponent to be associative? Decide and explain whether this operation is associative.

   c. What would it mean for the process of raising one exponent to another to distribute over addition? Decide and explain whether this is true.

   d. What would it mean for the process of raising one exponent to another to distribute over multiplication? Decide and explain whether this is true.

56. STATE YOUR ASSUMPTION Consider the expressions $(m^p)(m^q)$ and $(m^p)^q$.

    a. What must be true about p and q for $(m^p)(m^q) = (m^p)^q$? Give an example.

    b. Assuming m, p, and q are all positive integers, can $(m^p)(m^q) > (m^p)^q$ ever be true? Explain.

57. PERSEVERE For any nonzero real numbers a and b and any integers m and t, simplify the expression $\left(\dfrac{-a^m}{b^t}\right)^{2t}$. Explain.

58. ANALYZE Consider the equations in the table.

    a. For each equation, write the related expression and record the power of x.

Equation	Related Expression	Power of x	Linear or Nonlinear
$y = x$			
$y = x^2$			
$y = x^3$			

    b. Graph each equation using a graphing calculator.

    c. Classify each graph as *linear* or *nonlinear*.

    d. Explain how to determine whether an equation, or its related expression, is linear or nonlinear without graphing.

59. CREATE Write three different expressions that can be simplified to x^6.

60. WRITE Write a product of powers that is positive for all nonzero values of the variable. Explain your reasoning.

61. FIND THE ERROR Jade and Sal are each writing an equivalent form of g^5gh^6. Is either correct? Explain your reasoning.

Jade	Sal
$g^5gh^6 =$ $(g \cdot g \cdot g \cdot g \cdot g) \cdot (g) \cdot (h \cdot h \cdot h \cdot h \cdot h \cdot h)$	$g^5gh^6 =$ $(g \cdot g \cdot g \cdot g \cdot g) \cdot (gh \cdot gh \cdot gh \cdot gh \cdot gh \cdot gh)$

Division Properties of Exponents

Copyright © McGraw-Hill Education

Explore Quotients of Powers

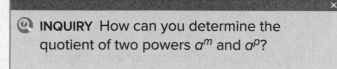 **Online Activity** Use an interactive tool to complete the Explore.

> **INQUIRY** How can you determine the quotient of two powers a^m and a^p?

Learn Quotient of Powers

You can use repeated multiplication and the principles for reducing fractions to simplify the quotients of monomials with the same base, like $\dfrac{2^8}{2^3}$. First, expand the numerator and the denominator. Then, divide the common factors.

$$\frac{2^8}{2^3} = \frac{\overbrace{2 \cdot 2 \cdot 2 \cdot 2 \cdot 2 \cdot 2 \cdot 2 \cdot 2}^{8 \text{ factors}}}{\underbrace{2 \cdot 2 \cdot 2}_{3 \text{ factors}}}$$

$$= 2 \cdot 2 \cdot 2 \cdot 2 \cdot 2$$

$$= 2^5$$

$$\frac{r^5}{r^4} = \frac{\overbrace{r \cdot r \cdot r \cdot r \cdot r}^{5 \text{ factors}}}{\underbrace{r \cdot r \cdot r \cdot r}_{4 \text{ factors}}}$$

$$= r$$

These examples demonstrate the Quotient of Powers Property.

Key Concept • Quotient of Powers	
Words	To divide two powers with the same base, subtract the exponents.
Symbols	For any nonzero number a, and any integers m and p, $\dfrac{a^m}{a^p} = a^{m-p}$.
Examples	$\dfrac{b^{12}}{b^9} = b^{12-9} = b^3$; $\dfrac{w^6}{w^2} = w^{6-2} = w^4$

 Go Online You can complete an Extra Example online.

Today's Goals
- Find quotients of monomials.
- Find powers of quotients.

Think About It!
What steps would you take to simplify $\dfrac{10^{12}}{10^9}$?

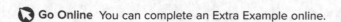

Think About It!

In the expression $\frac{b^5c^7}{b^2c}$, why can you not subtract $5 - 7$ in the numerator and $2 - 1$ in the denominator and simplify?

Think About It!

Why must you assume that the denominator does not equal 0?

Example 1 Quotient of Powers

Simplify $\frac{b^5c^7}{b^2c}$. **Assume that the denominator does not equal zero.**

Step 1 Group powers with the same base.

$$\frac{b^5c^7}{b^2c} = \left(\underline{}\right)\left(\underline{}\right)$$

Step 2 Use the Quotient of Powers Property.

$$\left(\frac{b^5}{b^2}\right)\left(\frac{c^7}{c}\right) = (\underline{})(c^{7-1}) \quad \text{Subtract the exponents in each group.}$$

$$= \underline{} \quad \text{Simplify.}$$

Check

Simplify $\frac{a^2b^9\,cd^4}{b^8cd}$. Assume that the denominator does not equal zero. _____

A. abd^3

B. a^2bcd^3

C. a^2bcd^4

D. a^2bd^3

🌐 Example 2 Apply Division of Monomials

CHEMISTRY At sea level, there are about 10^{25} molecules in a cubic liter of air. In the stratosphere, about 30 kilometers above the Earth's surface, the same cubic liter of air has about 10^{23} molecules. Approximately how many times as many molecules are there in a cubic liter of air at sea level as there are in the stratosphere?

$$\frac{a^m}{a^p} = a^{m-p} = \underline{} = \underline{}$$

There are about _____ times as many molecules in a cubic liter of air at sea level as there are in the stratosphere.

Check

SHOPPING Canada's West Edmonton Mall claims to have the largest parking lot in the world, with about 3^9 parking spaces. California's Glendale Galleria has about 3^8 parking spaces.

The West Edmonton Mall has _____ times as many parking spaces as the Glendale Galleria.

🅑 **Go Online** You can complete an Extra Example online.

Learn Power of a Quotient

You can use the Product of Powers Property to find the powers of quotients for monomials.

$$\left(\frac{2}{5}\right)^3 = \overbrace{\left(\frac{2}{5}\right)\left(\frac{2}{5}\right)\left(\frac{2}{5}\right)}^{3 \text{ factors}} = \underbrace{\frac{2 \cdot 2 \cdot 2}{5 \cdot 5 \cdot 5}}_{3 \text{ factors}} = \frac{2^3}{5^3}$$

$$\left(\frac{p}{q}\right)^2 = \overbrace{\left(\frac{p}{q}\right)\left(\frac{p}{q}\right)}^{2 \text{ factors}} = \overbrace{\frac{p \cdot p}{\underbrace{q \cdot q}_{2 \text{ factors}}}}^{2 \text{ factors}} = \frac{p^2}{q^2}$$

These examples demonstrate the Power of a Quotient Property.

Key Concept • Power of a Quotient	
Words	To find the power of a quotient, find the power of the numerator and the power of the denominator.
Symbols	For any real numbers a and $b \neq 0$, and any integer m, $\left(\frac{a}{b}\right)^m = \frac{a^m}{b^m}$.
Examples	$\left(\frac{1}{4}\right)^5 = \frac{1^5}{4^5}$; $\left(\frac{c}{d}\right)^6 = \frac{c^6}{d^6}$

Example 3 Power of a Quotient

Simplify $\left(\frac{5a^2}{6}\right)^3$.

$$\left(\frac{5a^2}{6}\right)^3 = \frac{(5a^2)^3}{6^3} \qquad \text{Power of a Quotient}$$

$$= \frac{5^3(a^2)^3}{6^3} \qquad \text{Power of a Product}$$

$$= \underline{\hspace{2cm}} \qquad \text{Power of a Power}$$

Check

Simplify $\left(\frac{5j^3l^2}{7k^5}\right)^4$. Assume that the denominator does not equal zero. ___

A. $\dfrac{5j^{12}l^8}{7k^{20}}$

B. $\dfrac{625j^{12}l^8}{2401k^{20}}$

C. $\dfrac{625j^7l^6}{2401k^9}$

D. $\dfrac{625j^{12}l^8}{7k^5}$

Go Online You can complete an Extra Example online.

💭 **Think About It!**

How would you simplify $\left(\frac{1}{2}\right)^2$?

Study Tip

Power Rules with Variables The power rules apply to variables and numbers. For example, $\left(\frac{4m}{2n}\right)^4 = \frac{(4m)^4}{(2n)^4} = \frac{4^4m^4}{2^4n^4} = \frac{256m^4}{16n^4}$ or $\frac{16m^4}{n^4}$.

Think About It!

To how many factors does the exponent 2 need to be applied? Name them.

Example 4 Power of a Quotient with Variables

Simplify $\left(\dfrac{x^4 y}{xyz}\right)^2$. **Assume that the denominator does not equal zero.**

Write the appropriate justification next to each step.

$$\left(\frac{x^4 y}{xyz}\right)^2 = \frac{(x^4 y)^2}{(xyz)^2} \qquad \underline{\hspace{4cm}}$$

$$= \frac{(x^4)^2 y^2}{x^2 y^2 z^2} \qquad \underline{\hspace{4cm}}$$

$$= \frac{x^8 y^2}{x^2 y^2 z^2} \qquad \underline{\hspace{4cm}}$$

$$= \frac{x^6}{z^2} \qquad \underline{\hspace{4cm}}$$

Check

Simplify $\left(\dfrac{4m^2 n^2 p^2}{3mp}\right)^4$. Assume that the denominator does not equal zero. _____

A. $\dfrac{256mn^2 p}{81}$

B. $\dfrac{256m^4 n^6 p^4}{81}$

C. $\dfrac{256m^4 n^8 p^4}{81}$

D. $\dfrac{256m^4 n^8 p^4}{81mp}$

Go Online to practice what you've learned about simplifying expressions involving exponents in the Put It All Together over Lessons 8-1 and 8-2.

Pause and Reflect

Did you struggle with anything in this lesson? If so, how did you deal with it?

Record your observations here.

Go Online You can complete an Extra Example online.

Practice

🅝 **Go Online** You can complete your homework online.

Examples 1, 3, and 4

Simplify each expression. Assume that no denominator equals zero.

1. $\dfrac{m^4p^2}{m^2p}$

2. $\dfrac{p^{12}t^3r}{p^2tr}$

3. $\dfrac{c^4d^4f^3}{c^2d^4f^3}$

4. $\left(\dfrac{3xy^4}{5z^2}\right)^2$

5. $\left(\dfrac{p^2t^7}{10}\right)^3$

6. $\dfrac{a^7b^8c^8}{a^5bc^7}$

7. $\left(\dfrac{3np^3}{7q^2}\right)^2$

8. $\left(\dfrac{2r^3t^6}{5u^9}\right)^4$

9. $\left(\dfrac{3m^5r^3}{4p^8}\right)^4$

10. $\dfrac{p^{12}t^7r^2}{p^2t^7r}$

11. $\dfrac{k^4m^3p^2}{k^2m^2}$

12. $\dfrac{m^7p^2}{m^3p^2}$

13. $\dfrac{32x^3y^2z^5}{-8xyz^2}$

14. $\left(\dfrac{4p^7}{7r^2}\right)^2$

15. $\dfrac{9d^7}{3d^6}$

16. $\dfrac{12n^5}{36n}$

17. $\dfrac{w^4x^3}{w^4x}$

18. $\dfrac{a^3b^5}{ab^2}$

Example 2

19. **SPACE** The Moon is approximately 25^4 kilometers away from Earth on average. The Olympus Mons volcano on Mars stands 25 kilometers high. How many Olympus Mons volcanoes, stacked on top of one another, would fit between the surface of Earth and the Moon?

20. **GEOMETRY** Write the ratio of the area of a circle with radius r to the circumference of the same circle.

21. **COMBINATIONS** The number of four-letter combinations that can be formed with the English alphabet is 26^4. The number of six-letter combinations that can be formed is 26^6. How many times more six-letter combinations can be formed than four-letter combinations?

22. **BLOOD COUNT** A lab technician draws a sample of blood. A cubic millimeter of the blood contains 22^5 red blood cells and 22^3 white blood cells. How many times more red blood cells are there than white blood cells?

Mixed Exercises

Simplify each expression. Assume that no denominator equals zero.

23. $\dfrac{-4w^{12}}{12w^3}$

24. $\dfrac{13r^7}{39r^4}$

25. $\left(\dfrac{2a^4c^3}{5b^2d^2}\right)^2$

26. $\dfrac{m^6}{m^2n^3}$

27. $\left(\dfrac{24a^{11}b^{16}c^6}{18a^6b^6c^6}\right)^3$

28. $\left(\dfrac{q^2r^3}{qr^2}\right)^5$

29. $\dfrac{-16x^7}{-6xy^3}$

30. $\dfrac{21d^{18}f^6}{7d^{11}f^5}$

31. $\left(\dfrac{2x^3y^2z^5}{3xyz}\right)^3$

32. $\dfrac{8c^4d^2f^9}{4cd^2f^3}$

33. $\left(\dfrac{7c^4}{14d^2}\right)^6$

34. $\left(\dfrac{6j^5}{7m^6n^3}\right)^2$

35. **SOUND** Decibels are used to measure sound. The softest sound that can be heard is rated at 0 decibels, or a relative loudness of 1. Ordinary conversation is rated at about 60 decibels, or a relative loudness of 10^6. A stock car race is rated at about 130 decibels, or a relative loudness of 10^{13}. How many times greater is the relative loudness of a stock car race than the relative loudness of ordinary conversation?

36. **COMPUTERS** The byte is the fundamental unit of computer processing. Almost all aspects of a computer's performance and specifications are measured in bytes or multiples of bytes. The byte is based on powers of 2, as shown in the table. How many times greater is a megabyte than a kilobyte?

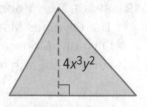

Memory Term	Number of Bytes
kilobyte	2^{10}
megabyte	2^{20}
gigabyte	2^{30}

37. **AREA** The area of the triangle shown is $6x^5y^3$. Find the base of the triangle.

$4x^3y^2$

38. **AREA** The area of the rectangle in the figure is $32xy^3$ square units. Find the width of the rectangle.

$8xy$

39. **USE A MODEL** An investment is expected to increase in value by 4% every year.

   a. Write an expression that represents the value of the investment after t years if the initial value was n dollars.

   b. By what percent does the value of the investment change between the end of year 2 and the end of year 8? Round your answer to the nearest tenth of a percent, and show your work.

40. INVESTMENTS A poor investment is expected to decrease in value by 5% every year.

   a. If the initial value of the investment was $100, what does it mean for it to decrease in value by 5% in the first year?

   b. Erik claims that rather than multiplying by 0.05 and subtracting, we can simply multiply the investment by 0.95. Is he correct? Explain.

   c. Write an expression that represents the value of the investment after t years if the initial value was n dollars.

   d. By what percent does the value of the investment change between the end of year 2 and the end of year 10? Round your answer to the nearest tenth of a percent, and show your work.

41. PAPER FOLDING If you fold a sheet of paper in half, you have a thickness of 2 sheets. Folding again, you have a thickness of 4 sheets. Fold the paper in half one more time. How many times thicker is a sheet that has been folded 3 times than a sheet that has not been folded?

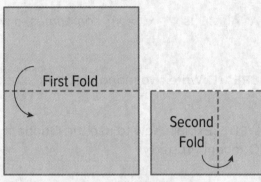

First Fold

Second Fold

42. USE TOOLS Create a table in a computer spreadsheet program to show the powers of 2^1 through 2^{10}. Use the formula functions in the program to show the Quotient of Powers Property is true for five different division problems using the powers you already entered.

🧠 **Higher-Order Thinking Skills**

43. CONSTRUCT ARGUMENTS Is $\left(\frac{x}{y^2}\right)^3$ the same as $\left(\frac{x}{y^3}\right)^2$? Justify your argument.

44. STRUCTURE Find the value of x that makes $\frac{8^{5x}}{8^{4x+1}} = 8^9$ true. Explain.

45. REASONING Which expression does not have the same answer as the others when simplified using exponent rules?

   A. $\left(\frac{3x^4y}{5}\right)^2$ **B.** $\frac{-18x^{14}y^{10}}{-50x^6y^8}$ **C.** $\frac{-3^2x^9y^5}{(-5)^2xy^3}$

46. PRECISION What error did the student make in simplifying the expression $\frac{5^4}{5} = 1^4 = 1$? What is the correct value of the expression?

47. REGULARITY Simplify the expression $\frac{a^{x+y}}{a^y}$.

48. CREATE Write three quotients of powers expressions that are equivalent to 2^5.

49. REGULARITY Consider the equation $\frac{3^h}{3^k} = 3^2$.

    a. Find two numbers h and k that satisfy the equation.

    b. Are there any other pairs of numbers that satisfy the equation? Explain.

50. FIND THE ERROR Kathryn and Salvador used different methods to simplify $\left(\frac{p^9}{p^5}\right)^2$. Is either correct? Explain your reasoning.

Kathryn	Salvador
$\left(\frac{p^9}{p^5}\right)^2 = \frac{p^{18}}{p^{10}} = p^8$	$\left(\frac{p^9}{p^5}\right)^2 = (p^4)^2 = p^8$

51. ANALYZE Is $x^y \cdot x^z = x^{yz}$ sometimes, always, or never true? Justify your argument.

52. CREATE Write two monomials with a quotient of $24a^2b^3$.

53. WRITE Explain how to use the Quotient of Powers Property and the Power of a Quotient Property.

54. PERSEVERE Is the expression $\left(\frac{p^7}{p^3}\right)^5$ positive or negative for all nonzero values of p? Explain.

55. WHICH ONE DOESN'T BELONG? Which quotient does *not* belong with the other three? Justify your conclusion.

$\frac{6^7}{6^2}$	$\frac{(-7)^3}{(-7)^2}$	$\frac{6^5}{6^3}$	$\frac{(-3)^8}{(-2)^4}$

56. FIND THE ERROR Andrew and Mateo are trying to simplify the expression $\frac{9^5}{9^3}$. Is either correct? Explain your reasoning.

Andrew	Mateo
$\frac{9^5}{9^3} = 9^2$	$\frac{9^5}{9^3} = \frac{1}{9^2}$

Negative Exponents

Today's Goals
- Simplify expressions containing zero and negative exponents.
- Simplify expressions containing negative exponents.

Today's Vocabulary
negative exponent

Learn Zero Exponent

Key Concept • Zero Exponent Property	
Words	Any nonzero number raised to the zero power is equal to 1.
Symbols	For any nonzero number a, $a^0 = 1$.
Examples	$30^0 = 1$; $\left(\frac{x}{y}\right)^0 = 1$; $\left(\frac{8}{3}\right)^0 = 1$

Example 1 Zero Exponent

Simplify each expression. Assume that no denominator equals zero.

a. $\left(-\dfrac{8m^2np^8}{9k^2mn^4}\right)^0$

$$\left(-\dfrac{8m^2np^8}{9k^2mn^4}\right)^0 = \underline{\quad} \qquad a^0 = 1$$

b. $\dfrac{a^4b^0}{a^2}$

$$\dfrac{a^4b^0}{a^2} = \dfrac{a^4(1)}{a^2} \qquad b^0 = 1$$

$$= a^2 \qquad \text{Quotient of Powers}$$

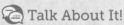

💬 Talk About It!

Why is the answer to part **b** not 1?

Check

Select the simplified form of $\dfrac{56g^2hj^{11}}{8g^0h^0j^0}$. Assume that the denominator does not equal zero. _____

A. $7g^2hj^{11}$

B. $7gj^{10}$

C. $7ghj^{10}$

D. $7g^2hj^{10}$

Explore Simplifying Expressions with Negative Exponents

 Online Activity Use an interactive tool to complete the Explore.

> **@ INQUIRY** How can you simplify expressions with negative exponents?

 **Go Online** An alternate method is available for this example.

 Go Online You can complete an Extra Example online.

Learn Negative Exponents

Any nonzero real number can be raised to a negative power. That power is called a **negative exponent**.

Key Concept • Negative Exponent Property	
Words	For any nonzero number a and any integer n, a^{-n} is the reciprocal of a^n. Also, the reciprocal of a^{-n} is a^n.
Symbols	For any nonzero number a and any integer n, $a^{-n} = \frac{1}{a^n}$.
Examples	$3^{-5} = \frac{1}{3^5} = \frac{1}{243}$; $\frac{1}{m^{-2}} = m^2$

An expression is considered simplified when:

• it contains only positive exponents.

• each base appears exactly once.

• there are no powers of powers.

• all fractions are in simplest form.

Example 2 Negative Exponents

Simplify $\frac{a^3 b^{-4}}{c^{-2}}$. Assume that the denominator does not equal zero.

$$\frac{a^3 b^{-4}}{c^{-2}} = \left(\frac{a^3}{1}\right)\left(\frac{b^{-4}}{1}\right)\left(\frac{1}{c^{-2}}\right) \qquad \text{Write as a product of fractions.}$$

$$= \left(\frac{a^3}{1}\right)\left(\frac{1}{b^4}\right)\left(\frac{c^2}{1}\right) \qquad a^{-n} \text{ is } \frac{1}{a^n}; \frac{1}{a^{-n}} = a^n$$

$$= \underline{\hspace{2cm}} \qquad \text{Multiply.}$$

Check

Select the simplified form of $\frac{42 r^{-2} t^{-6} u^3}{14 r^2 u^{-3}}$. Assume that the denominator does not equal zero. _____

A. $\frac{3}{r^4 t^6 u^6}$

B. $\frac{3u^6}{r^4 t^6}$

C. $\frac{3 t^6 u^6}{r^4}$

D. $3 r^4 t^6 u^6$

🐢 **Think About It!**

Describe how to simplify an expression of the form a^{-n}.

Math History Minute

In the 15th century, French physician **Nicolas Chuquet (c. 1445–1488)** wrote a book called *Triparty en la science des nombres*, or *A Three-Part Book on the Science of Numbers*, which contained early notation for zero and negative exponents.

🔎 **Go Online** You can complete an Extra Example online.

Example 3 Simplify an Expression with Negative Exponents

Simplify $\dfrac{3g^{-3}h^2}{-36g^3hj^{-4}}$. Assume that the denominator does not equal zero.

$$\dfrac{3g^{-3}h^2}{-36g^3hj^{-4}} = \left(-\dfrac{3}{36}\right)\left(\dfrac{g^{-3}}{g^3}\right)\left(\dfrac{h^2}{h}\right)\left(\dfrac{1}{j^{-4}}\right)$$ Group powers with the same base.

$$= \left(-\dfrac{1}{12}\right)(\underline{\hspace{1cm}})(\underline{\hspace{1cm}})(\underline{\hspace{0.5cm}})$$ Quotient of Powers and Negative Exponent Properties

$$= \left(-\dfrac{1}{12}\right)\underline{\hspace{1cm}}j^4$$ Simplify.

$$= \left(-\dfrac{1}{12}\right)\left(\dfrac{1}{g^6}\right)hj^4$$ Negative Exponent Property

$$= \underline{\hspace{1cm}}$$ Multiply.

Think About It!
Why do you not leave the answer as $\left(-\dfrac{1}{12}\right)g^{-6}hj^4$? Justify your argument.

Check

Select the simplified form of $\dfrac{35h^0j^{-5}k^{-2}}{14h^2m^5}$. Assume that the denominator does not equal zero. _____

A. $\dfrac{5}{2hj^5m^5k^2}$ B. $\dfrac{5h^2}{2hj^5m^5k^2}$

C. $\dfrac{5}{2h^2j^5m^5k^2}$ D. $\dfrac{5h^2m^5}{2j^5k^2}$

Order of magnitude is used to compare measures and to estimate and perform rough calculations. The **order of magnitude** of a quantity is the number rounded to the nearest power of 10. For example, the power of 10 closest to 105,000,000 is 10^8, or 100,000,000. So the order of magnitude of 105,000,000 is 10^8.

🌐 Apply Example 4 Apply Properties of Exponents

SPEED The maximum speed of a peregrine falcon is about **90 meters per second. The maximum speed of a garden snail is about 0.01 meter per second. How many orders of magnitude as fast as a peregrine falcon is a garden snail?**

1. **What is the task?**

Describe the task in your own words. Then list any questions that you may have. How can you find answers to your questions?

_____

_____

_____

(continued on the next page)

▶ **Go Online** You can complete an Extra Example online.

2. How will you approach the task? What have you learned that you can use to help you complete the task?

_____

_____

_____

3. What is your solution?

Use your strategy to solve the problem.

The maximum speed of a snail is 0.01 m/s. So, the order of magnitude of the speed of the snail is _____.

The maximum speed of a falcon is close to 100 m/s. So, the order of magnitude of the speed of the falcon is _____.

What is the ratio of the order of magnitude of the snail to the order of magnitude of the falcon?

$$\frac{\text{order of magnitude of snail}}{\text{order of magnitude of falcon}} = \frac{10^{-2}}{10^2}$$

A snail is approximately _____ times as fast as a falcon, or a snail is _____ orders of magnitude as fast as a falcon.

4. How can you know that your solution is reasonable?

Write About It! Write an argument that can be used to defend your solution.

_____

_____

_____

Check

LENGTH The radius of Earth is about 10,000,000 meters. The radius of a virus is about 0.0000001 meter. Approximately how many orders of magnitude as long as the radius of a virus is the radius of Earth?

Part A Using estimation, how many times greater is the radius of Earth than the radius of a virus? _____

    A. 0.0001

    B. 10

    C. 1,000,000

    D. 100,000,000,000,000

Part B The radius of Earth is about _____ orders of magnitude greater than the radius of a virus.

Go Online You can complete an Extra Example online.

Practice

Go Online You can complete your homework online.

Examples 1–3

Simplify each expression. Assume that no denominator equals zero.

1. $\dfrac{r^6 n^{-7}}{r^4 n^2}$

2. $\dfrac{h^3}{h^{-6}}$

3. $\dfrac{f^{-7}}{f^4}$

4. $\left(\dfrac{16p^5 w^2}{2p^3 w^3}\right)^0$

5. $\dfrac{f^{-5} g^4}{h^{-2}}$

6. $\dfrac{15x^6 y^{-9}}{5xy^{-11}}$

7. $\dfrac{-15t^0 u^{-1}}{5u^3}$

8. $\dfrac{(z^2 w^{-1})^3}{(z^3 w^2)^2}$

9. $\dfrac{-10m^{-1} y^0 r}{-14m^{-7} y^{-3} r^{-4}}$

10. $\dfrac{51x^{-1} y^3}{17x^2 y}$

11. $\dfrac{3m^{-3} r^4 p^2}{12t^4}$

12. $\left(\dfrac{3t^6 u^2 v^5}{9tuv^{21}}\right)^0$

13. $\dfrac{x^{-4} y^9}{z^{-2}}$

14. $\left(-\dfrac{5f^9 g^4 h^2}{fg^2 h^3}\right)^0$

15. $\dfrac{p^4 t^{-3}}{r^{-2}}$

16. $-\dfrac{5c^2 d^5}{8cd^5 f^0}$

17. $\dfrac{-2f^3 g^2 h^0}{8f^2 g^2}$

18. $\dfrac{g^0 h^7 j^{-2}}{g^{-5} h^0 j^{-2}}$

Example 4

19. **METRIC MEASUREMENT** Consider a dust mite that measures 10^{-3} millimeters in length and a gecko that measures 10 centimeters long. How many orders of magnitude as long as the mite is the gecko?

20. **CHEMISTRY** The nucleus of a certain atom is 10^{-13} centimeters across. If the nucleus of a different atom is 10^{-11} centimeters across, how many orders of magnitude as great is the second nucleus?

21. **WEIGHT** A paper clip weighs about 10^{-3} kilograms. A draft horse weighs about 10^3 kilograms. How many orders of magnitude as heavy is a draft horse than a paper clip?

22. **GDP** Gross Domestic Product (GDP) is a measure of a country's wealth. Indonesia had an estimated GDP in 2017 of 1,020,515,000,000. Madagascar had an estimated GDP of 10,372,000,000 in 2017. About how many orders of magnitude as great is Indonesia's GDP?

23. **GROWTH** An old oak tree is 1012 inches tall. A younger oak tree near it is 98 inches tall. About how many orders of magnitude as tall is the old oak tree?

Mixed Exercises

Simplify each expression. Assume that no denominator equals zero.

24. $\dfrac{3wy^{-2}}{(w^{-1}y)^3}$

25. $\dfrac{(4k^3m^2)^3}{(5k^2m^{-3})^{-2}}$

26. $\dfrac{-12c^3d^0f^{-2}}{6c^5d^{-3}f^4}$

27. $\dfrac{20qr^{-2}t^{-5}}{4q^0r^4t^{-2}}$

28. $\dfrac{(5pr^{-2})^{-2}}{(3p^{-1}r)^3}$

29. $\dfrac{(2g^3h^{-2})^2}{(g^2h^0)^{-3}}$

30. $\left(\dfrac{2a^{-2}b^4c^2}{-4a^{-2}b^{-5}c^{-7}}\right)^{-1}$

31. $\left(\dfrac{-3x^{-6}y^{-1}z^{-2}}{6x^{-2}yz^{-5}}\right)^{-2}$

32. $\left(\dfrac{4^0c^2d^3f}{2c^{-4}d^{-5}}\right)^{-3}$

33. $\dfrac{(16x^2y^{-1})^0}{(4x^0y^{-4}z)^{-2}}$

34. **RATIOS** Yvonne is comparing the weights of a semi-truck trailer tire and a mobile home. A semi-truck trailer tire weighs about 10^2 pounds, and a mobile home weighs about 10^4 pounds. What is the ratio of the weight of a semi-truck trailer tire compared to the weight of a mobile home? Write your answer as a monomial.

35. **MARKETING** Jana's marketing plan has a goal that each person who hears of their new product tells 10 people about it, each of those 10 people tells another 10, and so on. The level of spread is the number of times this repeats, as shown in the table. How many orders of magnitude greater is the level 4 spread compared to the level 1 spread?

Level	0	1	2	3	4
Spread	1	10	100	1000	10,000
Powers	10^0	10^1	10^2	10^3	10^4

36. **ASTRONOMY** The diagram shows the distance from Earth to the Sun and the distance from Saturn to the Sun. Approximately what portion of the distance from Saturn to the Sun is the distance from Earth to the Sun?

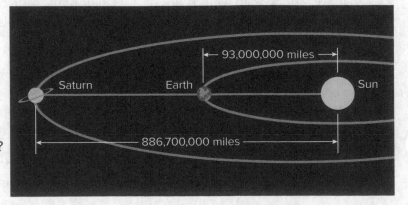

37. **COMPUTERS** In 1995, standard capacity for a personal computer hard drive was 40 megabytes (MB). In 2010, a standard hard drive capacity was 500 gigabytes (GB). Refer to the table.

   a. The newer hard drives have about how many times the capacity of the 1995 drives?

   b. Predict the hard drive capacity in the year 2025 if the capacity continues to grow by the same factor you found in part **a**.

   c. One kilobyte of memory is what fraction of one terabyte?

Memory Capacity Approximate Conversions
8 bits = 1 byte
10^3 bytes = 1 kilobyte
10^3 kilobytes = 1 megabyte
10^3 megabytes = 1 gigabyte
10^3 gigabytes = 1 terabyte
10^3 terabytes = 1 petabyte

38. **MICROSCOPES** Cheveyo is looking at a slide of red blood cells under a microscope. One of the red blood cells has an actual radius of about one millionth of a meter, or 10^{-6} meters. Through the microscope, the radius of that red blood cell appears to be 10^{-2} meters. How many times larger does the microscope make the blood cells appear?

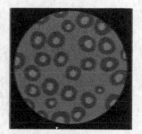

39. **STRUCTURE** Are $4n^5$ and $4n^{-5}$ reciprocals? Explain.

40. **REGULARITY** Explain why $3^0 = 1$.

41. **STRUCTURE** Use the Power of a Power Property to rewrite $\dfrac{1}{p^{-3}}$ with only positive exponents.

42. **REASONING** Angelo believes that $y^6 \geq y^{-6}$ for all nonzero values of y. Do you agree? Explain.

43. **BACTERIA** The bacterial population in a petri dish was measured at 1500 per cubic centimeter Monday morning. Each day the population doubled from the previous day. This can be modeled by $P = 1500(2^d)$, where d is the day and P is the population. If the population was first measured when $d = 0$, find the population of the bacteria the day before, when $d = -1$.

44. **VEHICLE DEPRECIATION** Jay buys a used car for $12,000. If the car loses about 10% of its value every year, the value of the car can be modeled by the equation $V = 12{,}000(0.90^y)$, where y is the number of years and V is the value of the car.

   a. Use the equation to determine the value of Jay's car 2 years before he purchased it.

   b. How much money did Jay save by waiting to purchase the car?

45. Consider the expression $\left(\dfrac{4a^3}{2a^{-2}}\right)^4$.

   a. Simplify $\left(\dfrac{4a^3}{2a^{-2}}\right)^4$ by using the Quotient of Powers Property first. Then use the Power of a Power Property.

   b. Simplify $\left(\dfrac{4a^3}{2a^{-2}}\right)^4$ by using the Power of a Quotient Property first. Then use the Quotient of Powers Property.

   c. Write a statement that generalizes the results of part **a** and part **b**.

46. REASONING Dalip evaluated -5^0 on his calculator.

   a. What answer did the calculator give Dalip when he evaluated -5^0?

   b. Is this a counterexample that contradicts the Zero Exponent Property? Explain.

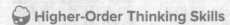 **Higher-Order Thinking Skills**

47. FIND THE ERROR Colleen and Tyler are using different steps to simplify $\left(\dfrac{x^2}{x^9}\right)^3$. Their work is shown in the boxes.

   a. Which student is correct? Explain your reasoning.

   b. Which student's answer is in simplest form?

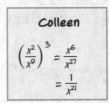
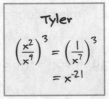

48. PERSEVERE Use the Quotient of Powers Property to explain why $x^{-n} = \dfrac{1}{x^n}$.

49. FIND THE ERROR Lola claims that we never need to use the Division Property of Exponents. She says that exponents in the denominator can be written as negative exponents, and then the Multiplication Property of Exponents can be used instead. Is Lola correct? If so, give an example. If not, explain why.

50. WRITE Explain how to evaluate the expression 0^0.

51. CREATE Measure something large and something very small in the same units. Write and solve a problem comparing the order of magnitude of the two measurements.

52. ANALYZE Consider the expression, $\dfrac{a^b c^{-d}}{w^x y^{-z}}$, where all values are nonnegative. Make a conjecture about why only c and w cannot be zero.

53. ANALYZE Dale wrote the value of several powers of 2 on a piece of paper. Describe the pattern shown. Then explain how to use the pattern to find 2^0.

$$2^4 = 2 \times 2 \times 2 \times 2 = 16$$
$$2^3 = 2 \times 2 \times 2 = 8$$
$$2^2 = 2 \times 2 = 4$$
$$2^1 = 2 = 2$$

Rational Exponents

Explore Expressions with Rational Exponents

Online Activity Use an interactive tool to complete the Explore.

> ×
>
> @ **INQUIRY** How can you simplify expressions with rational exponents?

Learn *n*th Roots

Not all exponents are integers. Exponents that are expressed as a fraction are called **rational exponents**. One of the most commonly used rational exponents is $\frac{1}{2}$. Expressions with an exponent of $\frac{1}{2}$ can also be represented as a square root.

Key Concept • Powers of One Half	
Words	For any nonnegative real number b, $b^{\frac{1}{2}} = \sqrt{b}$.
Example	$64^{\frac{1}{2}} = \sqrt{64}$ or 8

You are likely familiar with the square root of a number a. In the same way, you can find other roots of numbers. For example, if $a^3 = b$, then a is the cube root of b, and if $a^n = b$ for a positive integer n, then a is an **nth root** of b.

For example, $\sqrt[5]{18}$ is read as *the fifth root of 18*. In this example, 5 is the **index** and 18 is the **radicand**, which is the expression inside the radical symbol.

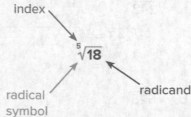

Key Concept • *n*th Roots	
Words	For any real numbers a and b and any positive integer n, if $a^n = b$, then a is an nth root of b.
Example	Because $3^4 = 81$, 3 is a fourth root of 81; $\sqrt[4]{81} = 3$

 Go Online You can complete an Extra Example online.

Today's Goals
- Rewrite expressions involving *n*th roots and rational exponents.
- Rewrite expressions involving powers of *n*th roots and rational exponents.

Today's Vocabulary
rational exponent

*n*th root

index

radicand

Go Online You may want to complete the Concept Check to check your understanding.

Some numbers have more than one *n*th root. Because $3^2 = 9$ and $(-3)^2 = 9$, both 3 and -3 are square roots of 9. Similarly, because $2^4 = 16$ and $(-2)^4 = 16$, both 2 and -2 are fourth roots of 16. When there is more than one real root and *n* is even, the nonnegative root is called the *principal root*. Radical symbols indicate principal roots, so $\sqrt[4]{16} = 2$.

Key Concept • Rational Exponents	
Words	For any real number *b* and any positive integer *n*, $b^{\frac{1}{n}} = \sqrt[n]{b}$, except when $b < 0$ and *n* is even.
Example	$32^{\frac{1}{5}} = \sqrt[5]{32} = \sqrt[5]{2 \cdot 2 \cdot 2 \cdot 2 \cdot 2}$ or 2

Example 1 Radical and Exponential Forms

Write each expression in radical or exponential form.

a. $10^{\frac{1}{2}}$

By the definition of $b^{\frac{1}{2}}$, $10^{\frac{1}{2}} = $ _____.

b. $\sqrt{7k^2}$

Because $7k^2$ is all under the radical symbol, both 7 and k^2 must be raised to the $\frac{1}{2}$ power. So, $\sqrt{7k^2} = $ _____.

Check

Write each expression in radical or exponential form.

a. $\sqrt{16p} = $ _____

b. $34^{\frac{1}{2}} = $ _____

c. $8\sqrt{p} = $ _____

d. $(3p)^{\frac{1}{2}} = $ _____

Example 2 Evaluate *n*th Roots

Evaluate each expression.

a. $\sqrt[5]{243}$

$\sqrt[5]{243} = \sqrt[5]{\underline{\hspace{2cm}}}$ $243 = 3 \cdot 3 \cdot 3 \cdot 3 \cdot 3$

$= \underline{\hspace{1cm}}$ Simplify.

b. $\sqrt[4]{625}$

$\sqrt[4]{625} = \sqrt[4]{\underline{\hspace{2cm}}}$ $625 = 5 \cdot 5 \cdot 5 \cdot 5$

$= \underline{\hspace{1cm}}$ Simplify.

Check

Evaluate each expression.

a. $\sqrt[4]{4096}$

b. $\sqrt[5]{16,807}$

🌀 **Go Online** You can complete an Extra Example online.

Think About It!

How do $(6x)^{\frac{1}{2}}$ and $6x^{\frac{1}{2}}$ differ?

Study Tip

Calculators You can use a calculator to find *n*th roots. On many calculators, enter the value of *n*, press MATH, and choose $\sqrt[x]{\ }$. Then enter the radicand and compute.

Example 3 Evaluate Exponential Expressions with Rational Exponents

Evaluate $4096^{\frac{1}{6}}$.

$$4096^{\frac{1}{6}} = \underline{\qquad}$$

$$\sqrt[6]{4096} = \sqrt[6]{\underline{\qquad}}$$

$$= \underline{\qquad}$$

$b^{\frac{1}{n}} = \sqrt[n]{b}$

$4096 = 4 \cdot 4 \cdot 4 \cdot 4 \cdot 4 \cdot 4$

Simplify.

Talk About It!

Malik says that $\sqrt[3]{64} = \sqrt[3]{8 \cdot 8} = 8$. Is he correct? If not, correct his mistake.

Check

Evaluate each expression.

a. $\left(\frac{81}{256}\right)^{\frac{1}{4}}$ _____

b. $100{,}000^{\frac{1}{5}}$ _____

Learn Powers of nth Roots

Key Concept • Powers of nth Roots	
Words	For any real number b and any integers m and $n > 1$, $b^{\frac{m}{n}} = \left(\sqrt[n]{b}\right)^m$ or $\sqrt[n]{b^m}$, except when $b < 0$ and n is even.
Examples	$216^{\frac{4}{3}} = \left(\sqrt[3]{216}\right)^4 = 6^4$ or 1296
	$32^{\frac{2}{5}} = \sqrt[5]{32^2} = \sqrt[5]{1024} = \sqrt[5]{4 \cdot 4 \cdot 4 \cdot 4 \cdot 4}$ or 4

Sometimes, the exponent can be simplified before it is evaluated. For example,

$$4^{\frac{15}{5}} = 4^3 \qquad\qquad \frac{15}{5} = 3$$

$$= 4 \cdot 4 \cdot 4 \qquad\qquad 4^3 = 4 \cdot 4 \cdot 4$$

$$= 64 \qquad\qquad\qquad \text{Simplify.}$$

Think About It!

Determine the value of $b^{\frac{m}{n}}$ if $m = n$. Explain your reasoning.

Example 4 Evaluate Expressions of Powers of nth Roots

Evaluate each expression.

a. $729^{\frac{5}{6}}$

$$729^{\frac{5}{6}} = \left(\sqrt[6]{729}\right)^5$$

$$= \left(\sqrt[6]{\underline{\qquad}}\right)^5$$

$$= \left(\sqrt[6]{\underline{\quad}}\right)^5$$

$$= \underline{\quad} \text{ or } \underline{\quad}$$

$b^{\frac{m}{n}} = \left(\sqrt[n]{b}\right)^m$

$729 = 3 \cdot 3 \cdot 3 \cdot 3 \cdot 3 \cdot 3$

$3 \cdot 3 \cdot 3 \cdot 3 \cdot 3 \cdot 3 = 3^6$

$\sqrt[6]{3^6} = 3$

(continued on the next page)

Go Online You can watch a video to learn how to evaluate expressions involving rational exponents.

b. $\left(\dfrac{36}{49}\right)^{\frac{3}{2}}$

$$\left(\dfrac{36}{49}\right)^{\frac{3}{2}} = \dfrac{36^{\frac{3}{2}}}{49^{\frac{3}{2}}} \qquad \text{Power of a Quotient}$$

$$= \dfrac{(\sqrt{36})^3}{(\sqrt{49})^3} \qquad b^{\frac{m}{n}} = \left(\sqrt[n]{b}\right)^m$$

$$= \dfrac{6^3}{7^3} \qquad \sqrt{36} = 6 \text{ and } \sqrt{49} = 7$$

$$= \underline{\hspace{1cm}} \qquad \text{Simplify.}$$

Check

Evaluate each expression.

a. $49^{\frac{3}{2}}$ _____

b. $243^{\frac{3}{5}}$ _____

Use a Source

Find the weight of another insect, in milligrams, and determine its resting metabolism. Provide the name, weight, and resting metabolism of the insect, and compare it to the ebony jewelwing damselfly.

🌐 Example 5 Apply Rational Exponents

BIOLOGY **For insects, the resting metabolic rate can be determined by $r = 4.14m^{\frac{2}{3}}$, where r is the resting metabolic rate in cubic milliliters of oxygen per hour and m is the body mass of the insect in milligrams. Determine the resting metabolic rate of a 125-mg ebony jewelwing damselfly.**

$$r = 4.14m^{\frac{2}{3}} \qquad \text{Original equation}$$

$$= 4.14(\underline{\hspace{0.8cm}})^{\frac{2}{3}} \qquad m = 125$$

$$= 4.14 \underline{\hspace{1cm}} \qquad 125^{\frac{2}{3}} = \left(\sqrt[3]{125}\right)^2$$

$$= 4.14(5)^2 \qquad \sqrt[3]{125} = 5$$

$$= 4.14 \cdot \underline{\hspace{1cm}} \qquad \text{Simplify.}$$

$$= \underline{\hspace{1cm}} \qquad \text{Simplify.}$$

The resting metabolic rate of a ebony jewelwing damselfly is _____ cubic milliliters of oxygen per hour.

Check

PROFITS The profit p, in thousands of dollars, of a company is modeled by the equation $p = 3.7c^{\frac{6}{5}}$, where c is the number of customers in thousands. Determine the profits of the company if they have 32,000 customers. _____

A. $142,080

B. $236,800

C. $307,625.29

D. $942,717,779.90

🌐 **Go Online** You can complete an Extra Example online.

Practice

🔁 **Go Online** You can complete your homework online.

Example 1

Write each expression in radical or exponential form.

1. $15^{\frac{1}{2}}$

2. $24^{\frac{1}{2}}$

3. $4k^{\frac{1}{2}}$

4. $(12y)^{\frac{1}{2}}$

5. $\sqrt{26}$

6. $\sqrt{44}$

7. $2\sqrt{ab}$

8. $\sqrt{3xyz}$

Examples 2–4

Evaluate each expression.

9. $\left(\frac{1}{16}\right)^{\frac{1}{4}}$

10. $\sqrt[5]{3125}$

11. $729^{\frac{1}{3}}$

12. $\left(\frac{1}{32}\right)^{\frac{1}{5}}$

13. $\sqrt[6]{4096}$

14. $1024^{\frac{1}{5}}$

15. $\left(\frac{16}{625}\right)^{\frac{1}{4}}$

16. $\sqrt[6]{15,625}$

17. $117,649^{\frac{1}{6}}$

18. $\sqrt[4]{\frac{16}{81}}$

19. $\left(\frac{1}{81}\right)^{\frac{1}{4}}$

20. $\left(\frac{3125}{32}\right)^{\frac{1}{5}}$

21. $729^{\frac{5}{6}}$

22. $256^{\frac{3}{8}}$

23. $125^{\frac{4}{3}}$

24. $49^{\frac{5}{2}}$

25. $\left(\frac{9}{100}\right)^{\frac{3}{2}}$

26. $\left(\frac{8}{125}\right)^{\frac{4}{3}}$

Example 5

27. **VELOCITY** The velocity v in feet per second of a freely falling object that has fallen h feet can be represented by $v = 8h^{\frac{1}{2}}$. Find the velocity of an object if it has fallen a distance of 144 feet.

28. **GEOMETRY** The surface area S of a cube in square inches can be determined by $S = 6V^{\frac{2}{3}}$, where V is the volume of the cube in cubic inches. Find the surface area of a cube that has a volume of 4096 cubic inches.

29. PLANETS The average distance d in astronomical units that a planet is from the Sun can be modeled by $d = t^{\frac{2}{3}}$, where t is the number of Earth years that it takes for the planet to orbit the Sun. Find the average distance a planet is from the Sun if the planet has an orbit of 27 Earth years.

30. BIOLOGY The relationship between the mass m in kilograms of an organism and its metabolism P in Calories per day can be represented by $P = 73.3\sqrt[4]{m^3}$. Find the metabolism of an organism that has a mass of 16 kilograms.

31. PROFIT The profit P of a company, in thousands of dollars, can be modeled by $P = 12.75\sqrt[5]{c^2}$, where c is the number of customers in hundreds. What is the profit of the company if the company has 3200 customers?

32. TIRE MARKS When a driver applies the brakes, the tires lock but the car will continue to slide, leaving skid marks on the road. You can approximate the speed at which a car was traveling on a dry road based on the length of a skid mark left by the car using the formula Speed $= (30 \cdot \text{length} \cdot 0.75)^{\frac{1}{2}}$, where speed is measured in miles per hour and length is measured in feet. At approximately what speed was a car traveling if it left a 50-foot long skid mark? Round to the nearest tenth.

Mixed Exercises

Write each expression in radical form, or write each radical in exponential form.

33. $17^{\frac{1}{3}}$

34. $q^{\frac{1}{4}}$

35. $7b^{\frac{1}{3}}$

36. $m^{\frac{2}{3}}$

37. $\sqrt[3]{29}$

38. $\sqrt[5]{h}$

39. $2\sqrt[3]{a}$

40. $\sqrt[3]{xy^2}$

Simplify.

41. $\sqrt[3]{0.027}$

42. $\sqrt[4]{\dfrac{n^4}{16}}$

43. $a^{\frac{1}{3}} \cdot a^{\frac{2}{3}}$

44. $c^{\frac{1}{2}} \cdot c^{\frac{3}{2}}$

45. $(8^2)^{\frac{2}{3}}$

46. $\left(y^{\frac{3}{4}}\right)^{\frac{1}{2}}$

47. $9^{-\frac{1}{2}}$

48. $16^{-\frac{3}{2}}$

49. $(3^2)^{-\frac{3}{2}}$

50. $\left(81^{\frac{1}{4}}\right)^{-2}$

51. $k^{-\frac{1}{2}}$

52. $\left(d^{\frac{4}{3}}\right)^0$

53. USE A MODEL In economics, the Cobb-Douglas production function is commonly used to relate input to output. The general form of the function is $P = bL^{ax}K^y$. The values in the table are given for the U.S. economy for the years 1900 and 1920. For both parts, assume a is 1.

		1900	1920
L	Labor input	105	194
K	Capital output	107	407
b	Total factor productivity	1.01	1.01
x	Output elasticity of labor	$\frac{3}{4}$	$\frac{3}{4}$
y	Output elasticity of capital	$\frac{1}{4}$	$\frac{1}{4}$

a. Find the total production P for 1900, to the nearest whole number.
b. Find the total production P for 1920, to the nearest whole number.

54. BASKETBALL The formula $S = 4\pi\left(\frac{3V}{4\pi}\right)^{\frac{2}{3}}$ can be used to find the surface area of a sphere, where V represents its volume. A regulation basketball has a volume of about 456 cubic inches. How much leather is needed (surface area) to make a regulation basketball? Round your answer to the nearest tenth.

55. BIOLOGY The function $h(x) = 0.4x^{\frac{2}{3}}$ can be used to find the height h in meters of a female giraffe with mass x kilograms. What is the height of a female giraffe with a mass of 32.8 kilograms? Round your answer to the nearest tenth of a meter.

56. GRAPH The graph of $y = x^{\frac{2}{5}}$ is shown over the interval $0 < x < 10$. Find the values to complete the table. Round to the nearest tenth if necessary.

x	0		5	
y		1		2.3

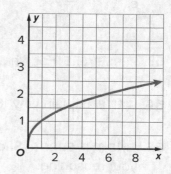

57. CONSTRUCT ARGUMENTS Use the properties of exponents to show which of the expressions are equivalent. Explain your reasoning.

a. $p^{\frac{f}{k}}$ b. $p^{\frac{f}{5}}$ c. $\sqrt[k]{p^f}$ d. $\left(\sqrt[k]{p}\right)^f$

58. USE TOOLS The amount of material needed to make a party hat in the shape of a cone can be found by calculating the hat's lateral area. The formula $LA = \pi r(h^2 + r^2)^{\frac{1}{2}}$ represents the lateral area of a cone for a given height h and radius r. Use your calculator to find the lateral area of a party hat that has a radius of 2 inches and a height of 7 inches. Round the answer to the nearest tenth of a square inch.

59. GEOMETRY The surface area S of a cylinder in square centimeters can be determined by $S = 2\left(\frac{V}{h}\right) + 2\pi\left(\frac{V}{\pi h}\right)^{\frac{1}{2}}h$, where V is the volume of the cylinder in cubic centimeters and h is the height of the cylinder is centimeters. Find the surface area of a cylinder that has a volume of 160π cubic centimeters and a height of 10 centimeters. Write the surface area in terms of π.

60. PENDULUMS A pendulum is a weight hanging from a point of suspension so that it can swing freely. The formula $T = 2\pi\left(\frac{L}{32}\right)^{\frac{1}{2}}$ gives the time T in seconds it takes for a pendulum of length L in feet to complete one full cycle. How long will it take the pendulum shown to complete one full cycle? Write the length of time in terms of π.

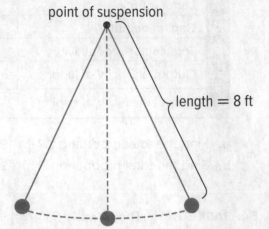
point of suspension

length = 8 ft

🧠 **Higher-Order Thinking Skills**

61. CREATE Write two different expressions with rational exponents equal to $\sqrt{2}$.

62. ANALYZE Determine whether each statement is *sometimes*, *always*, or *never* true. Assume that $x > 0$. Justify your argument.

a. $x^2 = x^{\frac{1}{2}}$

b. $x^{-2} = x^{\frac{1}{2}}$

c. $x^{\frac{1}{3}} = x^{\frac{1}{2}}$

d. $\sqrt{x} = x^{\frac{1}{2}}$

e. $\left(x^{\frac{1}{2}}\right)^2 = x$

f. $x^{\frac{1}{2}} \cdot x^2 = x$

63. PERSEVERE For what values of x is $x = x^{\frac{1}{3}}$?

64. WRITE Explain why $16^{\frac{1}{4}}$ will be less than 3.

65. FIND THE ERROR Sachi and Makayla are evaluating $27^{\frac{2}{3}}$. Is either correct? Explain your reasoning.

Sachi
$27^{\frac{2}{3}} = \sqrt[3]{27^2}$
$= \sqrt[3]{729}$
$= 9$

Makayla
$27^{\frac{2}{3}} = \sqrt[3]{27^2}$
$= 3^2$
$= 9$

66. WHICH ONE DOESN'T BELONG? Consider the four expressions shown. Which one does not belong? Explain your reasoning.

$\left(\sqrt[4]{w}\right)^5$	$\sqrt[4]{w^5}$	$w^{\frac{4}{5}}$	$w^{\frac{5}{4}}$

Simplifying Radical Expressions

Today's Goals
- Simplify square roots.
- Simplify cube roots.

Today's Vocabulary
radical expression
square root
perfect square
principal square root
cube root
perfect cube

Explore Square Roots and Negative Numbers

▶ **Online Activity** Use an interactive tool to complete the Explore.

> ⊘ **INQUIRY** How can you simplify algebraic expressions with square roots?

Learn Simplifying Square Root Expressions

A **radical expression** contains a radical such as a square root. A **square root** of a number is a value that, when multiplied by itself, gives the number. So, because $7 \cdot 7 = 49$, a square root of 49 is 7. A number like 49 is sometimes called a perfect square. A **perfect square** is a rational number with a square root that is a rational number. Recall that the square root of a positive number has two possible solutions. For $\sqrt{a^2}$ where $a > 0$, $\sqrt{a^2} = \pm a$, because $a^2 = (-a)^2$. The positive solution a is called the **principal square root**.

Key Concept • Product Property of Square Roots

Words	For any nonnegative real numbers a and b, the square root of ab is the square root of a times the square root of b.
Symbols	$\sqrt{ab} = \sqrt{a} \cdot \sqrt{b}$ if $a \geq 0$ and $b \geq 0$
Example	$\sqrt{4 \cdot 16} = \sqrt{4} \cdot \sqrt{16} = 2 \cdot 4$ or 8

Key Concept • Quotient Property of Square Roots

Words	For any real numbers a and b, where $a \geq 0$ and $b > 0$, the square root of $\frac{a}{b}$ is equal to the square root of a divided by the square root of b.
Symbols	$\sqrt{\frac{a}{b}} = \frac{\sqrt{a}}{\sqrt{b}}$ if $a \geq 0$ and $b > 0$
Example	$\sqrt{\frac{100}{4}} = \sqrt{25} = 5$ or $\sqrt{\frac{100}{4}} = \frac{\sqrt{100}}{\sqrt{4}} = \frac{10}{2} = 5$

A square root is in simplest form if these three conditions are true.
- The square root contains no perfect square factors other than 1.
- The square root contains no fractions.
- There are no square roots in the denominator of a fraction.

▶ **Go Online** You can complete an Extra Example online.

Study Tip
Prime Factorization If you are trying to determine the prime factorization of a number and the number is even, then you know that the prime factorization contains at least one 2. Divide the number by 2 and repeat until you have an odd number, and then see if the odd number can be factored any further as a product of primes.

Example 1 Simplify Square Roots

Simplify $\sqrt{72}$.

$$\sqrt{72} = \sqrt{2 \cdot 2 \cdot 2 \cdot 3 \cdot 3} \qquad \text{Prime factorization of 72}$$

$$= \sqrt{2} \cdot \sqrt{2} \cdot \sqrt{2} \cdot \sqrt{3} \cdot \sqrt{3} \qquad \text{Product Property of Square Roots}$$

$$= \sqrt{2^2} \cdot \sqrt{2} \cdot \sqrt{3^2} \qquad \text{Product Property of Square Roots}$$

$$= 2 \cdot \sqrt{} \cdot \underline{} = \underline{}\sqrt{2} \qquad \text{Simplify.}$$

💭 **Think About It!**

Why can $\sqrt{2^4}$ be simplified to 4?

Example 2 Multiply Square Roots

Simplify $\sqrt{5} \cdot \sqrt{48}$.

$$\sqrt{5} \cdot \sqrt{48} = \sqrt{5} \cdot (\sqrt{2} \cdot \sqrt{2} \cdot \sqrt{2} \cdot \sqrt{2} \cdot \sqrt{3}) \qquad \begin{array}{l}\text{Prime factorizations}\\ \text{of 5 and 48}\end{array}$$

$$= \sqrt{5} \cdot (\sqrt{2^2} \cdot \sqrt{2^2} \cdot \sqrt{3}) \qquad \begin{array}{l}\text{Product Property of}\\ \text{Square Roots}\end{array}$$

$$= 4\sqrt{} \cdot \sqrt{} = \underline{}\sqrt{15} \qquad \text{Simplify.}$$

Check

Simplify $\sqrt{8} \cdot \sqrt{14}$. _____

Example 3 Divide Square Roots

Simplify $\sqrt{\dfrac{24}{27}}$.

$$\sqrt{\frac{24}{27}} = \sqrt{\frac{8}{9}} \qquad \text{Reduce the radicand.}$$

$$= \frac{\sqrt{8}}{\sqrt{9}} \qquad \text{Quotient Property of Square Roots}$$

$$= \frac{2\sqrt{2}}{3} \qquad \text{Product Property of Square Roots}$$

Example 4 Simplify Square Roots with Variables

Simplify $\sqrt{525x^4y^5z^5}$.

$$\sqrt{525x^4y^5z^5} = \sqrt{3 \cdot 5^2 \cdot 7 \cdot x^4 \cdot y^5 \cdot z^5}$$

$$= \sqrt{3} \cdot \sqrt{5^2} \cdot \sqrt{7} \cdot \sqrt{x^4} \cdot \sqrt{y^5} \cdot \sqrt{z^5}$$

$$= \sqrt{3} \cdot 5 \cdot \sqrt{7} \cdot \sqrt{x^4} \cdot \sqrt{y^4} \cdot \sqrt{y} \cdot \sqrt{z^4} \cdot \sqrt{z}$$

$$= \underline{} \cdot 5 \cdot \underline{} \cdot x^2 \cdot y^2 \cdot \underline{} \cdot z^2 \cdot \underline{}$$

$$= \underline{}$$

💭 **Think About It!**

Why does the final answer include $\sqrt{21}$ when $\sqrt{21}$ does not appear in the previous step?

Check

Simplify $\sqrt{147x^3y^4z^5}$. _____

🔵 **Go Online** You can complete an Extra Example online.

🌐 Example 5 Write and Solve a Radical Equation

FINANCE Sarah uses an online calculator to determine how much interest she would accrue in a savings account after two years. She plans to invest $750, and the calculator determines that she could have $780 after two years. She can find the interest rate r of the savings account by using the formula $r = \sqrt{\frac{a}{p}} - 1$, where a is the amount after two years and p is the initial investment. What is the interest rate?

$$r = \sqrt{\frac{a}{p}} - 1 \qquad \text{Original formula}$$

$$= \sqrt{\frac{780}{750}} - 1 \qquad a = 780; p = 750$$

$$= \sqrt{\frac{26}{25}} - 1 \qquad \text{Simplify the radicand.}$$

$$= \frac{\sqrt{26}}{\sqrt{25}} - 1 \qquad \text{Quotient Property of Square Roots}$$

$$= \frac{\sqrt{26}}{5} - 1 \qquad \text{Simplify.}$$

$$\approx 0.02 \qquad \text{Simplify.}$$

The interest rate is approximately _____%.

💬 **Talk About It!**

Could this equation be solved if $\sqrt{\frac{a}{p}}$ could not be simplified?

Learn Simplifying Cube Root Expressions

The **cube root** of a number is the value that, when multiplied by itself twice, gives the number. A **perfect cube** is a rational number with a cube root that is an integer. So, because $6 \cdot 6 \cdot 6 = 216$, the cube root of 216 is 6 and 216 is a perfect cube.

💬 **Talk About It!**

What must be true for a cube root expression to be in simplest form?

Key Concept • Product Property of Radicals

Words	For any nonnegative real numbers a and b, the nth root of ab is equal to the nth root of a times the nth root of b.
Symbols	$\sqrt[n]{ab} = \sqrt[n]{a} \cdot \sqrt[n]{b}$ if $a \geq 0$ and $b \geq 0$
Examples	$\sqrt[3]{8 \cdot 64} = \sqrt[3]{512}$ or 8; $\sqrt[3]{8 \cdot 64} = \sqrt[3]{8} \cdot \sqrt[3]{64} = 2 \cdot 4$ or 8

Key Concept • Quotient Property of Radicals

Words	For any real numbers a and b, where $a \geq 0$ and $b > 0$, the nth root of $\frac{a}{b}$ is equal to the nth root of a divided by the nth root of b.
Symbols	$\sqrt[n]{\frac{a}{b}} = \frac{\sqrt[n]{a}}{\sqrt[n]{b}}$ if $a \geq 0$ and $b > 0$
Examples	$\sqrt[3]{\frac{216}{8}} = \sqrt[3]{27}$ or 3; $\sqrt[3]{\frac{216}{8}} = \frac{\sqrt[3]{216}}{\sqrt[3]{8}} = \frac{6}{2}$ or 3

Watch Out!

Cube Roots Remember, $\sqrt[3]{64}$ is not the same as $3\sqrt{64}$. The former means *the cube root of 64* or 4, and the latter means *3 times the square root of 64* or 24.

Example 6 Find Cube Roots

Simplify $\sqrt[3]{343}$.

$$\sqrt[3]{343} = \sqrt[3]{7 \cdot 7 \cdot 7} \qquad \text{Prime factorization of 343}$$

$$= \_\_\_ \qquad \text{Simplify.}$$

🔵 **Go Online** You can complete an Extra Example online.

Copyright © McGraw-Hill Education

Think About It!

Explain why $\sqrt[3]{3} \cdot \sqrt[3]{3} \cdot \sqrt[3]{3}$ simplifies to 3.

 Think About It!

Why are you able to simplify the radicand before finding the cube root?

 Think About It!

How would your answer change if you simplified $\sqrt[3]{\dfrac{15x^2y^3z^4}{81x^4}}$ instead of $\sqrt[3]{\dfrac{15x^2y^3z^4}{81x^5}}$? Is this answer in simplest form? Why or why not?

Watch Out!

Simplifying Cube Roots Be sure to simplify the fraction before finding the cube root.

Example 7 Simplify Cube Roots

Simplify $\sqrt[3]{135}$.

$\sqrt[3]{135} = \sqrt[3]{3 \cdot 3 \cdot 3 \cdot 5}$ — Prime factorization of 135

$= \sqrt[3]{3} \cdot \sqrt[3]{3} \cdot \sqrt[3]{3} \cdot \sqrt[3]{5}$ — Product Property of Radicals

$= \underline{\hspace{1cm}}$ — Simplify.

Example 8 Multiply Cube Roots

Simplify $\sqrt[3]{2} \cdot \sqrt[3]{36}$.

$\sqrt[3]{2} \cdot \sqrt[3]{36} = \sqrt[3]{2} \cdot \left(\sqrt[3]{2 \cdot 2 \cdot 3 \cdot 3} \right)$ — Prime factorizations of 2 and 36

$= \underline{\hspace{0.7cm}} \cdot \sqrt[3]{2} \cdot \sqrt[3]{2} \cdot \sqrt[3]{3} \cdot \underline{\hspace{0.7cm}}$ — Product Property of Radicals

$= \underline{\hspace{0.7cm}} \cdot \sqrt[3]{9}$ — Product Property of Radicals

$= \underline{\hspace{1cm}}$ — Simplify.

Example 9 Divide Cube Roots

Simplify $\sqrt[3]{\dfrac{168}{375}}$.

$\sqrt[3]{\dfrac{168}{375}} = \sqrt[3]{\dfrac{2 \cdot 2 \cdot 2 \cdot 3 \cdot 7}{3 \cdot 5 \cdot 5 \cdot 5}}$ — Prime factorizations of 168 and 375

$= \sqrt[3]{\dfrac{2 \cdot 2 \cdot 2 \cdot 7}{5 \cdot 5 \cdot 5}}$ — Simplify the radicand.

$= \dfrac{\sqrt[3]{2 \cdot 2 \cdot 2 \cdot 7}}{\sqrt[3]{5 \cdot 5 \cdot 5}}$ — Quotient Property of Radicals

$= \dfrac{\sqrt[3]{2} \cdot \sqrt[3]{2} \cdot \sqrt[3]{2} \cdot \sqrt[3]{7}}{\sqrt[3]{5} \cdot \sqrt[3]{5} \cdot \sqrt[3]{5}}$ — Product Property of Radicals

$= \dfrac{2\sqrt[3]{7}}{5}$ — Simplify.

Example 10 Simplify Cube Roots with Variables

Simplify $\sqrt[3]{\dfrac{15x^2y^3z^4}{81x^5}}$.

$\sqrt[3]{\dfrac{15x^2y^3z^4}{81x^5}} = \sqrt[3]{\dfrac{3 \cdot 5 \cdot x \cdot x \cdot y \cdot y \cdot y \cdot z \cdot z \cdot z \cdot z}{3 \cdot 3 \cdot 3 \cdot 3 \cdot x \cdot x \cdot x \cdot x \cdot x}}$

$= \sqrt[3]{\dfrac{5 \cdot y \cdot y \cdot y \cdot z \cdot z \cdot z \cdot z}{3 \cdot 3 \cdot 3 \cdot x \cdot x \cdot x}}$

$= \dfrac{\sqrt[3]{5 \cdot y \cdot y \cdot y \cdot z \cdot z \cdot z \cdot z}}{\sqrt[3]{3 \cdot 3 \cdot 3 \cdot x \cdot x \cdot x}}$

$= \dfrac{\sqrt[3]{5} \cdot \sqrt[3]{y} \cdot \sqrt[3]{y} \cdot \sqrt[3]{y} \cdot \sqrt[3]{z} \cdot \sqrt[3]{z} \cdot \sqrt[3]{z} \cdot \sqrt[3]{z}}{\sqrt[3]{3} \cdot \sqrt[3]{3} \cdot \sqrt[3]{3} \cdot \sqrt[3]{x} \cdot \sqrt[3]{x} \cdot \sqrt[3]{x}}$

$= \underline{\hspace{1cm}}$

Go Online You can complete an Extra Example online.

Practice

Go Online You can complete your homework online.

Examples 1–4

Simplify each expression.

1. $\sqrt{52}$

2. $\sqrt{56}$

3. $\sqrt{72}$

4. $\sqrt{162}$

5. $\sqrt{243}$

6. $\sqrt{245}$

7. $\sqrt{5} \cdot \sqrt{10}$

8. $\sqrt{10} \cdot \sqrt{20}$

9. $\sqrt{2} \cdot \sqrt{10}$

10. $\sqrt{5} \cdot \sqrt{60}$

11. $\sqrt{8} \cdot \sqrt{12}$

12. $\sqrt{11} \cdot \sqrt{22}$

13. $\sqrt{\frac{75}{49}}$

14. $\sqrt{\frac{8}{81}}$

15. $\sqrt{\frac{192}{216}}$

16. $\sqrt{\frac{88}{18}}$

17. $\sqrt{\frac{84}{121}}$

18. $\sqrt{\frac{75}{20}}$

19. $\sqrt{16b^4}$

20. $\sqrt{40x^4y^8}$

21. $\sqrt{81a^{12}d^4}$

22. $\sqrt{75qr^3}$

23. $\sqrt{28a^4b^3}$

24. $\sqrt{w^{13}x^5z^8}$

Example 5

25. **NATURE** In 2010, an earthquake below the ocean floor initiated a devastating tsunami in Sumatra. Scientists can approximate the velocity V in feet per second of a tsunami in water of depth d feet with the formula $V = \sqrt{16} \cdot \sqrt{d}$. Determine the velocity of a tsunami in 300 feet of water. Write your answer in simplified radical form.

26. **PHYSICAL SCIENCE** The average velocity V of gas molecules is represented by the formula $V = \sqrt{\frac{3kt}{m}}$. The velocity is measured in meters per second, t is the temperature in Kelvin, and m is the molar mass of the gas in kilograms per mole. The variable k represents a value called the molar gas constant, which is about 8.3. Write a simplified radical expression that represents the average velocity of gas molecules that have a temperature of 300 Kelvin and a mass of 0.045 kilogram per mole. Round to the nearest meter per second.

27. CLOCKS A grandfather clock has a pendulum that swings back and forth. The time t in seconds it takes the pendulum to complete one full cycle is given by the formula $t = 0.2\sqrt{p}$, where p is the length of the pendulum in centimeters. How long is a full cycle if the pendulum is 22 cm long? Round your answer to the nearest tenth of a second.

28. HORIZON The distance, d, in miles that a person can see to the horizon can be modeled by the formula $d = \sqrt{\dfrac{3h}{2}}$ where h is the person's height above sea level in feet. How far, in miles, to the horizon can a person see if they are 60 feet above sea level? Write your answer in simplified radical form.

Examples 6–10

Simplify each expression.

29. $\sqrt[3]{8}$

30. $\sqrt[3]{125}$

31. $\sqrt[3]{216}$

32. $\sqrt[3]{64}$

33. $\sqrt[3]{27}$

34. $\sqrt[3]{1000}$

35. $\sqrt[3]{243}$

36. $\sqrt[3]{4000}$

37. $\sqrt[3]{162}$

38. $\sqrt[3]{875}$

39. $\sqrt[3]{80}$

40. $\sqrt[3]{384}$

41. $\sqrt[3]{3} \cdot \sqrt[3]{32}$

42. $\sqrt[3]{24} \cdot \sqrt[3]{18}$

43. $\sqrt[3]{12} \cdot \sqrt[3]{6}$

44. $\sqrt[3]{16} \cdot \sqrt[3]{20}$

45. $\sqrt[3]{9} \cdot \sqrt[3]{3}$

46. $\sqrt[3]{2} \cdot \sqrt[3]{6}$

47. $\sqrt[3]{\dfrac{162}{375}}$

48. $\sqrt[3]{\dfrac{648}{750}}$

49. $\sqrt[3]{\dfrac{18}{128}}$

50. $\sqrt[3]{\dfrac{162}{3}}$

51. $\sqrt[3]{\dfrac{192}{3}}$

52. $\sqrt[3]{\dfrac{500}{2}}$

53. $\sqrt[3]{54b^8}$

54. $\sqrt[3]{10yz^5} \cdot \sqrt[3]{4y^3}$

55. $\sqrt[3]{\dfrac{192x^7}{1029x^{10}}}$

56. $\sqrt[3]{\dfrac{640a^3b^8}{5ab^4}}$

57. $\sqrt[3]{54x^2y^8} \cdot \sqrt[3]{5x^5y^4}$

58. $\sqrt[3]{25p^2r^5} \cdot \sqrt[3]{30q^3r}$

Mixed Exercises

Simplify each expression.

59. $\sqrt{245m^9n^5}$

60. $2\sqrt{5} \cdot 7\sqrt{10}$

61. $\sqrt{\dfrac{96d^4e^2f^8}{75de^6}}$

62. $-11\sqrt[3]{250}$

63. $\sqrt[3]{320x^{14}y^{17}z^{20}}$

64. $\sqrt[3]{45k^4m^{10}} \cdot \sqrt[3]{32k^7m^3}$

65. $\sqrt[3]{\dfrac{264w^6xy^7}{3w^6x^4y^3}}$

66. $\sqrt[3]{\dfrac{48a^7}{125b^9}}$

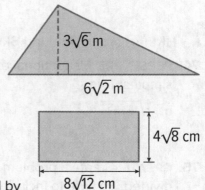

3√6 m

6√2 m

67. AREA The base of a triangle measures $6\sqrt{2}$ meters and the height measures $3\sqrt{6}$ meters. What is the area?

68. AREA The length of a rectangle measures $8\sqrt{12}$ centimeters and the width measures $4\sqrt{8}$ centimeters. What is the area of the rectangle?

4√8 cm

8√12 cm

69. MEAN The geometric mean of two numbers h and k can be found by evaluating $\sqrt{h \cdot k}$. Find the geometric mean of 32 and 14 in simplified radical form.

70. THEATRE CREW Fernanda oversees painting props for a theatre production. She needs to create a cube that will be used as a prop by the actors. She has enough paint to cover 5 square yards. The formula $s = \sqrt{\dfrac{A}{6}}$ gives the longest side length s in yards of a cubic prop Fernanda can make, where A is the surface area to be covered with paint. What is the longest side length of a cube Fernanda could make and not have to purchase any more paint? Round to the nearest tenth of a yard.

71. SCALE MODEL While on vacation, Deon decided to purchase a scale model of the Empire State Building. Before making the purchase, Deon wants to determine the maximum height of a model that will fit inside his suitcase. Deon's suitcase measures 22 inches long by 14 inches wide by 9 inches tall. The diagonal length D is given by $D = \sqrt{L^2 + W^2 + H^2}$, where L is the length in inches, W is the width in inches, and H is the height in inches. Find the length of the tallest scale model that will fit inside Deon's suitcase. Round your answer to the nearest tenth of an inch.

72. BALLOONS The radius of a sphere r is given in terms of its volume V by the formula $r = \sqrt[3]{\dfrac{0.75V}{\pi}}$. By how many inches has the radius of a spherical balloon increased when the amount of air in the balloon is increased from 4.5 cubic feet to 4.7 cubic feet? Round your answer to the nearest hundredth.

73. REGULARITY The Product Property of Square Roots and the Quotient Property of Square Roots can be written in symbols as $\sqrt{ab} = \sqrt{a} \cdot \sqrt{b}$ and $\sqrt{\frac{a}{b}} = \frac{\sqrt{a}}{\sqrt{b}}$, respectively.

a. Explain the Product Property of Square Roots and discuss any limitations of a and b for this property.

b. Explain the Quotient Property of Square Roots and discuss any limitations of a and b for this property.

c. Discuss any similarities of the two properties.

Higher-Order Thinking Skills

74. PERSEVERE Use rational exponents to find an equivalent radical expression in simplest form.

a. $\sqrt[4]{16m^{32}}$ b. $\left(\sqrt{x}\right)\left(\sqrt[3]{x}\right)$ c. $\sqrt[3]{\sqrt{b}}$

75. CREATE Create a problem where two square roots are being either multiplied or divided. Be sure to include at least one variable in your problem. Solve your problem.

76. PERSEVERE Margarita takes a number, subtracts 4, multiplies by 4, takes the square root, and takes the reciprocal to get $\frac{1}{2}$. What number did she start with? Write a formula to describe the process.

77. ANALYZE Find a counterexample to show that the following statement is false. *If you take the square root of a number, the result will always be less than the original number.*

78. ANALYZE Order the expressions from least to greatest. $\sqrt{47}, 9, \sqrt[3]{421}, \sqrt{85}$

79. PERSEVERE If the area of a rectangle is $144\sqrt{5}$ square inches, what are possible dimensions of the rectangle? Explain your reasoning.

80. WRITE Describe the required conditions for a radical expression to be in simplest form.

Operations With Radical Expressions

Learn Adding and Subtracting Radical Expressions

To add or subtract radical expressions, the radicands must be alike in the same way that monomial terms must be alike to add or subtract.

Monomials

$$5a + 3a = (5 + 3)a$$

$$= 8a$$

$$7b - 2b = (7 - 2)b$$

$$= 5b$$

Radical Expressions

$$5\sqrt{3} + 3\sqrt{3} = (5 + 3)\sqrt{3}$$

$$= 8\sqrt{3}$$

$$7\sqrt{5} - 2\sqrt{5} = (7 - 2)\sqrt{5}$$

$$= 5\sqrt{5}$$

Notice that when adding and subtracting radical expressions, the radicand does not change. This is the same as when adding or subtracting monomials.

Example 1 Add and Subtract Expressions with Like Radicands

Simplify $4\sqrt{5} + 3\sqrt{7} - 2\sqrt{5} + 7\sqrt{7}$.

$$4\sqrt{5} + 3\sqrt{7} - 2\sqrt{5} + 7\sqrt{7} = (\underline{\hspace{1cm}})\sqrt{5} + (\underline{\hspace{1cm}})\sqrt{7}$$

$$= \underline{\hspace{2cm}}$$

Check

Simplify $17\sqrt{19} - 14\sqrt{6} + 11\sqrt{6} - 3\sqrt{19}$. _____

A. 0

B. $14\sqrt{19} - 3\sqrt{6}$

C. $11\sqrt{25}$

D. $28\sqrt{19} - 17\sqrt{6}$

 Go Online You can complete an Extra Example online.

Today's Goals
- Add and subtract radical expressions.
- Multiply radical expressions.

💭 Think About It!
Can you simplify $6\sqrt{3} + 7\sqrt{5}$? If so, simplify it, and if not, explain why.

Watch Out!
Radicands Before applying the Distributive Property, be certain that you group expressions with like radicands.

💭 Think About It!
Jaylen tries to simplify $3\sqrt{5} - 3\sqrt{5}$ and determines that the simplest form is $0\sqrt{5}$. Is he correct? Why or why not?

💬 **Talk About It!**

If a radical expression has addition or subtraction with unlike radicands, what must you check before determining whether the terms can be added or subtracted? Explain your reasoning.

Example 2 Add and Subtract Expressions with Unlike Radicands

Simplify $3\sqrt{75} - 6\sqrt{48} - \sqrt{27}$.

$3\sqrt{75} - 6\sqrt{48} - \sqrt{27}$

$\quad = 3(\underline{\quad} \cdot \sqrt{3}) - 6(\underline{\quad} \cdot \sqrt{3}) - (\underline{\quad} \cdot \sqrt{3})$ Product Property of Square Roots

$\quad = 3(\underline{\quad}\sqrt{3}) - 6(\underline{\quad}\sqrt{3}) - \underline{\quad}\sqrt{3}$ Simplify.

$\quad = \underline{\quad}\sqrt{3} - \underline{\quad}\sqrt{3} - \underline{\quad}\sqrt{3}$ Multiply.

$\quad = (15 - 24 - 3)\sqrt{3}$ Distributive Property

$\quad = \underline{\quad\quad}$ Simplify.

Check

Simplify $-11\sqrt{50} + 2\sqrt{32} - 18\sqrt{8}$.

_____

A. $99\sqrt{2}$

B. $-83\sqrt{2}$

C. $-27\sqrt{2}$

D. $-11\sqrt{50} - 14\sqrt{8}$

🌐 **Example 3** Use Radical Expressions

DECORATIONS **Kate is decorating a rectangular pavilion for a party and wants to string lights along the edge of the roof. Two of the sides have a length of $13\sqrt{5}$ feet, and the other two sides have a length of $6\sqrt{45}$ feet. How many feet of lights will Kate need to decorate the entire pavilion?**

Because there are two sides $13\sqrt{5}$ feet long and two sides $6\sqrt{45}$ feet long, the expression $13\sqrt{5} + 13\sqrt{5} + 6\sqrt{45} + 6\sqrt{45}$ represents the total perimeter of the roof.

$13\sqrt{5} + 13\sqrt{5} + 6\sqrt{45} + 6\sqrt{45} = (\underline{\quad\quad})\sqrt{5} + (\underline{\quad\quad})\sqrt{45}$

$\quad\quad\quad\quad\quad\quad\quad\quad\quad\quad = \underline{\quad}\sqrt{5} + \underline{\quad}\sqrt{45}$

$\quad\quad\quad\quad\quad\quad\quad\quad\quad\quad = 26\sqrt{5} + 12(\underline{\quad\quad})$

$\quad\quad\quad\quad\quad\quad\quad\quad\quad\quad = 26\sqrt{5} + 36\sqrt{5}$

$\quad\quad\quad\quad\quad\quad\quad\quad\quad\quad = \underline{\quad\quad}$

Kate will need _____ feet of lights to decorate the pavilion.

🔵 **Go Online** You can complete an Extra Example online.

Check

CONSTRUCTION A business is adding wheelchair ramps to their building. The three ramps will require $10\sqrt{28}$ ft^3, $13\sqrt{7}$ ft^3, and $4\sqrt{112}$ ft^3 of concrete. How much concrete will be needed to build the three ramps? _____

A. 27 ft^3

B. $27\sqrt{147}$ ft^3

C. $49\sqrt{7}$ ft^3

D. $117\sqrt{7}$ ft^3

Learn Multiplying Radical Expressions

Multiplying radical expressions is similar to multiplying monomials. Let $x \geq 0$.

Monomials

$$5x(3x) = 5 \cdot 3 \cdot x \cdot x$$

$$= 15x^2$$

Radical Expressions

$$5\sqrt{x}(3\sqrt{x}) = 5 \cdot 3 \cdot \sqrt{x} \cdot \sqrt{x}$$

$$= 15x$$

You can apply the Distributive Property to radical expressions. You can also multiply radical expressions with more than one term in each factor. This is similar to multiplying two binomials.

Binomials

$$(2x + 3x)(4x + 5x) = 2x(4x) + 2x(5x) + 3x(4x) + 3x(5x)$$

$$= 8x^2 + 10x^2 + 12x^2 + 15x^2$$

$$= 45x^2$$

Radical Expressions

$$(2\sqrt{2} + 3\sqrt{2})(4\sqrt{2} + 5\sqrt{2})$$

$$= 2\sqrt{2}(4\sqrt{2}) + 2\sqrt{2}(5\sqrt{2}) + 3\sqrt{2}(4\sqrt{2}) + 3\sqrt{2}(5\sqrt{2})$$

$$= 8\sqrt{2^2} + 10\sqrt{2^2} + 12\sqrt{2^2} + 15\sqrt{2^2}$$

$$= 45\sqrt{2^2}$$

$$= 90$$

Notice that when multiplying radical expressions, you multiply the radicands.

Go Online You can complete an Extra Example online.

Example 4 Multiply Radical Expressions

Simplify $5\sqrt{3} \cdot 4\sqrt{6}$.

$5\sqrt{3} \cdot 4\sqrt{6} = (5 \cdot 4)(\sqrt{3} \cdot \sqrt{6})$

$\qquad\qquad = \underline{\hspace{2cm}}$

$\qquad\qquad = 20(3\sqrt{2})$

$\qquad\qquad = \underline{\hspace{2cm}}$

Check

Simplify $3\sqrt{10} \cdot (-9\sqrt{6})$.

_____

Example 5 Multiply Radical Expressions by Using the Distributive Property

Simplify $4\sqrt{7}(2\sqrt{8} + 3\sqrt{7})$.

$4\sqrt{7}(2\sqrt{8} + 3\sqrt{7}) = (\underline{\hspace{1cm}} \cdot 2\sqrt{8}) + (\underline{\hspace{1cm}} \cdot 3\sqrt{7})$

$\qquad\qquad = [(\underline{\hspace{1cm}})(\sqrt{7} \cdot \sqrt{8})] + [(4 \cdot 3)(\underline{\hspace{1cm}})]$

$\qquad\qquad = \underline{\hspace{0.5cm}}\sqrt{56} + \underline{\hspace{0.5cm}}\sqrt{49}$

$\qquad\qquad = 16\sqrt{14} + 12(7)$

$\qquad\qquad = \underline{\hspace{2cm}}$

Check

Simplify $-4\sqrt{6}(7\sqrt{12} - 4\sqrt{8})$.

_____

Go Online
to learn how to find the sums and products of rational and irrational numbers in Expand 8-6.

Pause and Reflect

Did you struggle with anything in this lesson? If so, how did you deal with it?

Record your observations here.

Go Online You can complete an Extra Example online.

Practice

📡 **Go Online** You can complete your homework online.

Examples 1 and 2

Simplify.

1. $7\sqrt{7} - 2\sqrt{7}$

2. $3\sqrt{13} + 7\sqrt{13}$

3. $7\sqrt{5} + 4\sqrt{5}$

4. $2\sqrt{6} + 9\sqrt{6}$

5. $12\sqrt{r} - 9\sqrt{r}$

6. $9\sqrt{6a} - 11\sqrt{6a} + 4\sqrt{6a}$

7. $3\sqrt{5} - 2\sqrt{20}$

8. $3\sqrt{50} - 3\sqrt{32}$

9. $2\sqrt{13} + 4\sqrt{2} - 5\sqrt{13} + \sqrt{2}$

10. $5\sqrt{8} + 2\sqrt{20} - \sqrt{8}$

11. $7\sqrt{3} - 2\sqrt{2} + 3\sqrt{2} + 5\sqrt{3}$

12. $8\sqrt{12} - \sqrt{5} - 4\sqrt{3}$

Example 3

13. **ARCHITECTURE** The Pentagon is the building that houses the U.S. Department of Defense. If the building is a regular pentagon with each side measuring $23\sqrt{149}$ meters, find the perimeter. Leave your answer as a radical expression.

$23\sqrt{149}$ m

14. **BIKING** Iker rode his bike on three trails this week. The first was $2\sqrt{3}$ kilometers, the second was $4\sqrt{3}$ kilometers, and the third was $3\sqrt{3}$ kilometers. How long did Iker ride this week? Give your answer as a radical expression.

15. **RAMP** A ramp at a dog park is made of three sections. The incline and decline pieces are the same length, $13\sqrt{37}$ inches. The center is $10\sqrt{41}$ inches long. What is the total length of the ramp? Give your answer as a radical expression.

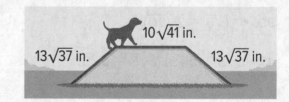

$10\sqrt{41}$ in.

$13\sqrt{37}$ in. $13\sqrt{37}$ in.

16. **DISTANCE** The diagonal length of a football field, including both end zones, is $40\sqrt{97}$ feet. The diagonal length of a practice field is $29\sqrt{97}$ feet. How much longer is the diagonal length of a football field than the diagonal length of the practice field? Give your answer as a radical expression.

17. **CIRCLES** A large circle has an area of 400 cm² and a small circle has an area of 200 cm². The radius of the large circle is $\sqrt{\frac{400}{\pi}}$ cm. The radius of the small circle is $\sqrt{\frac{200}{\pi}}$ cm. Find the difference of the radius of the large circle and the radius of the small circle.

18. GARDEN Camila put a frame around a triangular garden in her yard. The two legs of the triangle are 8 feet long each, and the third side is equal to the length of a leg times $\sqrt{2}$. What is the perimeter of the triangle? Give your answer as a simplified radical expression.

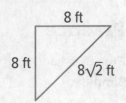

8 ft

8 ft

$8\sqrt{2}$ ft

Examples 4 and 5

Simplify.

19. $2\sqrt{3} \cdot 3\sqrt{15}$

20. $5\sqrt{3} \cdot 2\sqrt{21}$

21. $6\sqrt{7} \cdot 2\sqrt{8}$

22. $7\sqrt{10} \cdot 4\sqrt{10}$

23. $11\sqrt{6} \cdot 3\sqrt{12}$

24. $10\sqrt{5} \cdot 5\sqrt{11}$

25. $\sqrt{2}(\sqrt{8} + \sqrt{6})$

26. $\sqrt{5}(\sqrt{10} - \sqrt{3})$

27. $\sqrt{5}(\sqrt{2} + 4\sqrt{2})$

28. $\sqrt{6}(2\sqrt{10} + 3\sqrt{2})$

29. $4\sqrt{5}(3\sqrt{5} + 8\sqrt{2})$

30. $5\sqrt{3}(6\sqrt{10} - 6\sqrt{3})$

Mixed Exercises

Simplify.

31. $\sqrt{\frac{1}{25}} - \sqrt{5}$

32. $\sqrt{\frac{2}{9}} + \sqrt{6}$

33. $2 + 2\sqrt{2} - \sqrt{8}$

34. $8\sqrt{\frac{5}{4}} + 3\sqrt{20} - \sqrt{45}$

35. $\sqrt{24} - 5\sqrt{12} + 4\sqrt{2} - 3\sqrt{3}$

36. $\frac{1}{2}\sqrt{8} - \sqrt{\frac{2}{9}} + 9\sqrt{2}$

37. $5\sqrt{3} \cdot 3\sqrt{5}$

38. $8\sqrt{7} \cdot \frac{2}{3}\sqrt{7}$

39. $\sqrt{\frac{3}{4}}(\sqrt{6} + 2\sqrt{20})$

40. $6\sqrt{6}(9\sqrt{2} - 5\sqrt{12})$

41. $(a\sqrt{3} + b\sqrt{5})(c\sqrt{8} + d\sqrt{5})$

42. $(6\sqrt{2} + 2\sqrt{3})(\sqrt{10} - 4\sqrt{7})$

43. GEOMETRY The area of a trapezoid is found by multiplying its height by the average length of its bases. Find the area of the deck attached to Mr. Wilson's house. Give your answer as a simplified radical expression.

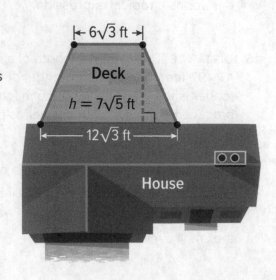

$\leftarrow 6\sqrt{3}$ ft $\rightarrow$

Deck

$h = 7\sqrt{5}$ ft

$\leftarrow$ $12\sqrt{3}$ ft $\rightarrow$

House

44. TRAVEL Lucia used the straight-line distance between Lincoln, Nebraska, and Houston, Texas, and the straight-line distance between Houston and Tallahassee, Florida, to estimate the straight-line distance between Lincoln and Tallahassee, as shown on the map. Lucia traveled from Lincoln to Tallahassee for vacation. A week after returning home from vacation, Lucia traveled from Lincoln to Houston to visit a relative. About how far did Lucia travel after returning home from Houston? Give your answer as a radical expression.

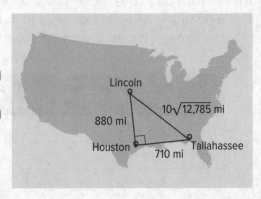

45. FREE FALL A ball is dropped from a building window 800 feet above the ground. Another ball is dropped from a lower window 288 feet high. Both balls are released at the same time. Assume air resistance is not a factor, and use the formula $t = \frac{1}{4}\sqrt{h}$ to find how many seconds t it will take a ball to fall h feet.

a. How much longer after the first ball hits the ground does the second ball land?

b. Find a decimal approximation of the answer for part **a**. Round your answer to the nearest tenth.

46. REASONING Deja is putting a fence around her yard. The long side of the yard is $12\sqrt{2}$ meters, the two shorter sides are each $5\sqrt{2}$ meters, and there is a $\sqrt{2}$-meter section that runs up to the house. When Deja buys the fencing, she approximates the length of fencing she needs. Write and solve an equation to find the total length of fencing f needed for the yard. Give your answer as a radical expression. Then approximate the amount of fencing Deja should buy to the nearest meter.

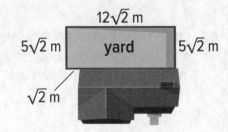

47. SHADOWS The distance from the top of Gino's head to the end of his head's shadow is the hypotenuse of a 45°-45°-90° triangle, or Gino's height times $\sqrt{2}$. Gino is 4.5 feet tall. The distance from a nearby tree to the end of its shadow is 2.5 times as long as the distance from the top of Gino's head to the end of his shadow. How far is it from the top of the tree to the end of its shadow? Give your answer as a radical expression.

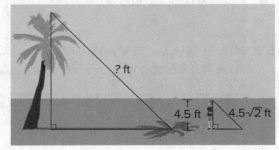

48. STRUCTURE The table shows several radical expressions. Complete the table by writing *yes* or *no* in each row to indicate whether the expression in column A and the expression in column B have the same radicand when simplified.

Expression A	Expression B	Simplify to have same radicand? (Write Yes or No)
$4\sqrt{24}$	$6\sqrt{28}$	
$3\sqrt{18}$	$2\sqrt{32}$	
$5\sqrt{15}$	$12\sqrt{30}$	
$10\sqrt{20}$	$7\sqrt{45}$	

49. SCALING Austin has a model car that uses a 1:24 scale. The real car's hood is $20\sqrt{5}$ inches long. How long is the model car's hood? Give your answer as a radical expression.

50. FIND THE ERROR Jhumpa said $3\sqrt{32}$ cannot be subtracted from $8\sqrt{8}$ because they have different radicands. Kenyon said they can be subtracted. Is either correct? Explain your reasoning.

51. PERSEVERE Determine whether the following statement is *true* or *false*. Provide a proof or a counterexample to support your answer.

$(x + y)^2 > \left(\sqrt{x^2 + y^2}\right)^2$ when $x > 0$ and $y > 0$

52. CONSTRUCT ARGUMENTS Make a conjecture about the sum of a rational number and an irrational number. Is the sum *rational* or *irrational*? Is the product of a nonzero rational number and an irrational number *rational* or *irrational*? Explain your reasoning.

53. CREATE Write an equation that shows a sum of two radicals with different radicands. Explain how you could combine these terms.

54. WRITE A radical expression is an exact value. Rounding a radical gives an approximate value. Give examples of when an exact answer is preferred and when an approximate answer is acceptable.

55. WHICH ONE DOESN'T BELONG? Consider the expressions. Identify which one does not belong and explain your reasoning.

$10\sqrt{5}$	$2\sqrt{20}$	$6\sqrt{12}$	$8\sqrt{45}$

Exponential Equations

Explore Solving Exponential Equations

Online Activity Use an interactive tool to complete the Explore.

☒

@ INQUIRY How can you solve an equation where the variable is in the exponent?

Learn Exponential Equations

In an **exponential equation**, the independent variable is an exponent. The Power Property of Equality and the other properties of exponents can be used to solve exponential equations.

Key Concept • Power Property of Equality	
Words	For any real number $b > 0$ and $b \neq 1$, $b^x = b^y$ if and only if $x = y$.
Examples	If $4^x = 4^5$, then $x = 5$. If $n = \frac{2}{5}$, $5^n = 5^{\frac{2}{5}}$. If $x^{\frac{1}{2}} = 3^{\frac{1}{2}}$, then $x = 3$.

Example 1 Solve a One-Step Exponential Equation

Solve $5^x = 625$.

$$5^x = 625 \qquad \text{Original equation}$$

$$5^x = \underline{\hspace{1cm}} \qquad \text{Rewrite 625 as } 5^4.$$

$$x = \underline{\hspace{1cm}} \qquad \text{Simplify.}$$

Check

Solve $2^x = 512$

$x = \underline{\hspace{1cm}}$

Go Online You can complete an Extra Example online.

Copyright © McGraw-Hill Education

Today's Goals
• Solve exponential equations.

Today's Vocabulary
exponential equation

Go Online You may want to complete the Concept Check to check your understanding.

Think About It!

Can you use the Power Property of Equality if the two bases are not equal? Justify your argument.

Need Help?

In order to apply the Power Property of Equality in this example, the two bases must be raised to the same exponent. Thus, you rewrite 8^2 as $(8^4)^{\frac{1}{2}}$ so that you can set the bases equal to each other.

Example 2 Solve a Multi-Step Exponential Equation

Solve $27^{x-1} = 3$.

$27^{x-1} = 3$	Original equation
$(3^3)^{x-1} = 3$	Rewrite 27 as 3^3.
$\underline{\hspace{3em}} = 3$	Power of a Power Property, Distributive Property
$3^{3x-3} = 3^1$	Rewrite 3 as 3^1.
$\underline{\hspace{3em}} = \underline{\hspace{2em}}$	Power Property of Equality
$3x = \underline{\hspace{2em}}$	Add 3 to each side.
$x = \underline{\hspace{2em}}$	Divide each side by 3.

🌐 Example 3 Solve a Real-World Exponential Equation

AEROSPACE In the 1950s, scientists proposed a space station that could house a crew of approximately 80 people. The station could produce artificial gravity by rotating at a speed of $w = \sqrt{gr}$, where g is 32 feet per second squared, and r is the radius of the station. If the station design required a rotating speed of approximately 64 feet per second to simulate gravity on Earth, what would the radius need to be?

$w = \sqrt{gr}$	Original equation
$64 = \sqrt{32r}$	$w = 64, g = 32$
$64 = \underline{\hspace{3em}}$	$\sqrt[n]{b} = b^{\frac{1}{n}}$
$8^2 = (32r)^{\frac{1}{2}}$	$64 = 8^2$
$(8^4)^{\frac{1}{2}} = (32r)^{\frac{1}{2}}$	$8^2 = (8^4)^{\frac{1}{2}}$
$8^4 = 32r$	Power Property of Equality
$4096 = \underline{\hspace{2em}}$	$8^4 = 4096$
$\underline{\hspace{2em}} = r$	Divide each side by 32.

The radius of the space station would need to be approximately $\underline{\hspace{3em}}$ feet.

Check

SUNSCREEN The degree to which sunscreen protects your skin is measured by its sun protection factor (SPF). For a sunscreen with SPF f, the percentage of UV-B rays absorbed p is $p = 50f^{\frac{1}{5}}$. What SPF absorbs 100% of UV-B rays?

$f = \underline{\hspace{3em}}$

🅑 **Go Online** You can complete an Extra Example online.

Practice

Go Online You can complete your homework online.

Examples 1 and 2

Solve each equation.

1. $2^x = 512$

2. $3^x = 243$

3. $6^x = 46{,}656$

4. $5^x = 125$

5. $3^x = 6561$

6. $16^x = 4$

7. $3^{x-3} = 243$

8. $4^{x-1} = 1024$

9. $6^{x-1} = 1296$

10. $4^{2x+1} = 1024$

11. $2^{4x+3} = 2048$

12. $3^{3x+3} = 6561$

Example 3

13. ELECTRICITY The relationship of the current, power, and resistance of an appliance can be modeled by $I\sqrt{R} = \sqrt{P}$, where I is the current in amperes, P is the power in watts, and R is the resistance in ohms. Find the resistance that an appliance is using if the current is 2.5 amps and the power is 100 watts.

14. VIDEO Felipe uploaded a funny video of his dog. The relationship between the elapsed time in days, d, since the video was first uploaded and the total number of views, v, that the video received is modeled by $v = 4^{1.25d}$. Find the number of days it took Felipe's video to get 1024 views.

15. CONSTRUCTION A large plot of land has been purchased by developers. They roll out a schedule of construction. The relationship between the area of the undeveloped land in hectares, A, and the elapsed time in months, t, since the construction began is modeled by the function $A = 6250 \cdot 10^{-0.1t}$. How many months of construction will there be before the area of the undeveloped land decreases to 62.5 hectares?

Mixed Exercises

Solve each equation.

16. $2^{5x} = 8^{2x-4}$

17. $81^{2x-3} = 9^{x+3}$

18. $2^{4x} = 32^{x+1}$

19. $16^x = \frac{1}{2}$

20. $25^x = \frac{1}{125}$

21. $6^{8-x} = \frac{1}{216}$

22. $5^x = 125$

23. $2^{5x-4} = 64$

24. $4^{x+1} = 256$

25. $3^{4x-2} = 729$

26. **USE A MODEL** Without advertising, a Web site had 96 total visits. Today, the owners of the site are starting a new promotion, which is expected to double the total number of visits to their Web site every 5 days.

a. Write an equation that relates the total number of visits, v, to the number of days the promotion has been running, d.

b. Use your equation from part **a** to find how many days the promotion should be run in order to increase the traffic to the Web site to 12,288 total visits.

27. **PHYSICS** The velocity v of an object dropped from a tall building is given by the formula $v = \sqrt{64d}$, where d is the distance the object has dropped. What distance was the object dropped from if it has a velocity of 49 feet per second? Round your answer to the nearest hundredth.

28. **FENCING** Representatives from the neighborhood have requested that the city install a fence around a newly-built playground. The equation $f = 4\sqrt{A}$ represents the amount of fence f needed based on the area A of the playground. If the playground has 324 feet of fencing, find the area of the playground.

29. **CREATE** Write an equation equivalent to $8^{(2 + x)} = 16^{(2.5 - 0.5x)}$ with a base of 4. Then solve for x. Justify your answer.

30. **FIND THE ERROR** Zari and Jenell are solving $128^x = 4$. Is either correct? Explain your reasoning.

Zari	Jenell
$128^x = 4$	$128^x = 4$
$(2^7)^x = 2^2$	$(2^7)^x = 4$
$2^{7x} = 2^2$	$2^{7x} = 4^1$
$7^x = 2$	$7^x = 1$
$x = \frac{2}{7}$	$x = \frac{1}{7}$

31. **CONSTRUCT ARGUMENTS** Make a conjecture about why the equation $2^b = 5^{b-1}$ cannot be solved by the methods used in this section.

32. **CREATE** Write and solve an exponential equation where both bases need to be changed.

33. **PERSEVERE** Find the solutions to the equation $32^{(x^2 + 4x)} = 16^{(x^2 + 4x)}$.

34. **CREATE** Write an exponential equation that has a solution of $x = -1$.

35. **WHICH ONE DOESN'T BELONG?** Which exponential equation does not belong? Justify your conclusion.

$3^{2x + 1} = 243$	$4^{2x - 1} = 1024$	$2^{3x + 2} = 2048$	$5^{x + 2} = 3125$

36. **WRITE** Explain how to solve the exponential equation $9^{4n - 3} = 3^6$. Include the solution in your explanation.

ⓔ Essential Question

How do you perform operations and represent real-world situations with exponents?

Module Summary

Lessons 8-1 and 8-2

Properties of Exponents

- To multiply two powers with the same base, add their exponents.
- To find the power of a power, multiply the exponents.
- To find the power of a product, find the power of each factor and multiply.
- To divide two powers with the same base, subtract their exponents.
- To find the power of a quotient, find the power of the numerator and the power of the denominator.

Lessons 8-3 and 8-4

Negative and Rational Exponents

- Any nonzero number raised to the zero power is 1.
- A negative exponent is the reciprocal of a number raised to the positive exponent.
- A rational exponent is an exponent that is a rational number.
- If $a^2 = b$, then a is the square root of b.
- If $a^3 = b$, then a is the cube root of b.
- If $a^n = b$ for any positive integer n, then a is an nth root of b.
- The nth root of b can be written as $b^{\frac{1}{n}}$ or $\sqrt[n]{b}$.
- For any positive real number b and any integers m and $n > 1$, $b^{\frac{m}{n}} = \left(\sqrt[n]{b}\right)^m$ or $\sqrt[n]{b^m}$.

Lessons 8-5 and 8-6

Radical Expressions

- For any nonnegative real numbers a and b, $\sqrt{ab} = \sqrt{a} \cdot \sqrt{b}$.
- For any nonnegative real numbers a and b, $\sqrt{\dfrac{a}{b}} = \dfrac{\sqrt{a}}{\sqrt{b}}$, if $b \neq 0$.
- A square root is in simplest form if it contains no perfect square factors other than 1, it contains no fractions, and there are no square roots in the denominator of a fraction.

Lesson 8-7

Exponential Equations

- In an exponential equation, the independent variable is an exponent.
- The Power Property of Equality and the other properties of exponents can be used to solve exponential equations.
- If you square both sides of an equation, the resulting equation is still true.

Study Organizer

⬛ Foldables

Use your Foldable to review this module. Working with a partner can be helpful. Ask for clarification of concepts as needed.

Name _____ Period _____ Date _____

Test Practice

1. MULTI-SELECT Select all expressions equivalent to $(5xy^3)(3xy^3)$. (Lesson 8-1)

  Ⓐ $(5 \cdot 3)(xy^3 \cdot xy^3)$

  Ⓑ $15x^2y^6$

  Ⓒ $15xy^6$

  Ⓓ $(5 + 3)(xy^3 + xy^3)$

  Ⓔ $15xy^9$

2. OPEN RESPONSE Earth's mass is about 5,973,600,000,000,000,000,000,000 kg. Write this mass in scientific notation. (Lesson 8-1)

  [   ]

3. MULTIPLE CHOICE The volume of a cube is $V = s^3$, where s is the length of one side. Which expression represents the volume of the cube? (Lesson 8-1)

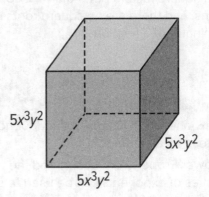

$5x^3y^2$

$5x^3y^2$

$5x^3y^2$

  Ⓐ $5x^9y^6$

  Ⓑ $125x^9y^6$

  Ⓒ $25x^9y^6$

  Ⓓ $15x^3y^2$

4. MULTIPLE CHOICE Simplify $(w^3)^5$. (Lesson 8-1)

  Ⓐ w^8

  Ⓑ w^{15}

  Ⓒ w^{27}

  Ⓓ w^{243}

5. OPEN RESPONSE A manufacturer sells square trivets in various sizes. In the corner of each trivet sold, they engrave the company's square logo. The diagram shows the ratio of the size of each trivet to the size of the company's logo.

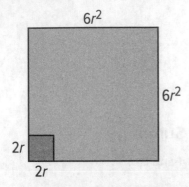

$6r^2$

$6r^2$

$2r$

$2r$

What is the ratio of the area of the trivet to the area of the logo in simplified form? (Lesson 8-2)

Ratio of area of trivet to area of logo:

____r^2 to 1.

  [   ]

6. OPEN RESPONSE Simplify $\dfrac{k^6r^3t^5}{kr^2t^4}$. (Lesson 8-2)

  [   ]

7. MULTIPLE CHOICE Simplify $\left(\dfrac{2x^2y^3}{3x^5}\right)^0$. (Lesson 8-3)

  Ⓐ 0

  Ⓑ $\dfrac{2}{3}$

  Ⓒ 1

  Ⓓ $3x^5$

8. MULTIPLE CHOICE Simplify $\left(\dfrac{2m^2j^3}{3p^5}\right)^4$. (Lesson 8-2)

(A) $\dfrac{8m^8j^{12}}{12p^{20}}$

(B) $\dfrac{16m^8j^{12}}{81p^{20}}$

(C) $\dfrac{16m^8j^{12}}{3p^5}$

(D) $\dfrac{16m^6j^7}{81p^9}$

9. OPEN RESPONSE The maximum speed of an object made of material A when pushed by wind is about 0.001 foot per minute, so the order of magnitude is about 10^{-3}.

The maximum speed of an object made of material B when pushed by the same wind is about 1000 feet per minute, so the order of magnitude is about 10^3 feet per minute.

What is the ratio of the maximum speed of the object made of material A to the maximum speed of the object made of material B? Write your answer in decimal form. (Lesson 8-3)

Ratio of A to B: _____

10. OPEN RESPONSE Explain whether the following equation is true or false. If false, explain how to make it true. (Lesson 8-3)

$$\dfrac{49x^{-3}y^4z^{-1}}{4x^2y^{-2}} = \dfrac{7y^6}{2x^5yz}$$

11. OPEN RESPONSE Explain how to simplify the following expression in your own words. (Lesson 8-4)

$729^{\frac{1}{6}}$

12. MULTIPLE CHOICE Evaluate $\sqrt[4]{81}$. (Lesson 8-4)

(A) 4

(B) $\dfrac{1}{4}$

(C) 81

(D) 3

13. MULTIPLE CHOICE Evaluate $\sqrt[6]{4096}$. (Lesson 8-4)

(A) 2

(B) 4

(C) 16

(D) 64

14. MULTIPLE CHOICE Simplify $25^{-\frac{3}{2}}$. (Lesson 8-4)

(A) -125

(B) -37.5

(C) $\dfrac{1}{125}$

(D) $\dfrac{1}{15}$

15. **OPEN RESPONSE** Evaluate $81^{\frac{3}{4}}$. (Lesson 8-4)

16. **OPEN RESPONSE** Simplify $\sqrt{98}$. (Lesson 8-5)

17. **MULTIPLE CHOICE** Simplify $\sqrt{121a^5b^3c}$. (Lesson 8-5)

Ⓐ $11a^2b\sqrt{abc}$

Ⓑ $11a^2b^4$

Ⓒ $11ab\sqrt{ab^3c}$

Ⓓ $11b\sqrt{a^3b^3c}$

18. **MULTIPLE CHOICE** Simplify $\sqrt[3]{\frac{125a^6}{27b^3}}$. (Lesson 8-5)

Ⓐ $\frac{a^2}{b}$

Ⓑ $\frac{5a^6}{3b^3}$

Ⓒ $\frac{\sqrt[3]{a^2}}{\sqrt[3]{b}}$

Ⓓ $\frac{5a^2}{3b}$

19. **OPEN RESPONSE** Simplify

$8\sqrt{5} + 2\sqrt{7} - 6\sqrt{5} + 4\sqrt{7}$. (Lesson 8-6)

20. **OPEN RESPONSE** Simplify $4\sqrt{3}(5\sqrt{8} + 2\sqrt{3})$.
(Lesson 8-6)

21. **MULTIPLE CHOICE** Solve
$7 = 343^{x-2}$. (Lesson 8-7)

Ⓐ 5

Ⓑ $\frac{7}{3}$

Ⓒ 2

Ⓓ 1

22. **MULTIPLE CHOICE** Solve
$4 = 16^{x+2}$. (Lesson 8-7)

Ⓐ -4

Ⓑ $-\frac{3}{2}$

Ⓒ $\frac{2}{3}$

Ⓓ 4

23. **OPEN RESPONSE** The volume of a pyramid is $V = \frac{1}{3}Bh$, where B is the area of the base and h is the height. What is the value of x if the volume of the pyramid, in simplest form, is equivalent to $6^{\frac{x}{3}}$ cm³?

(Lesson 8-7)

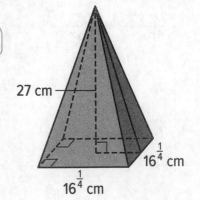

27 cm

$16^{\frac{1}{4}}$ cm

$16^{\frac{1}{4}}$ cm

Exponential Functions

e Essential Question
When and how can exponential functions represent real-world situations?

What Will You Learn?
Place a check mark (✓) in each row that corresponds with how much you already know about each topic **before** starting this module.

KEY

👎 — I don't know.   👍 — I've heard of it.   👍 — I know it!

	Before			After		
	👎	👍	👍	👎	👍	👍
graph exponential growth functions						
graph exponential decay functions						
translate exponential functions						
dilate exponential functions						
reflect exponential functions						
solve problems involving exponential growth and decay						
transform exponential expressions						
generate geometric sequences						
write recursive formulas						
translate between recursive and explicit formulas						

📖 **Foldables** Make this Foldable to help you organize your notes about exponential functions. Begin with a sheet of 11″ × 17″ paper and six index cards.

1. **Fold** lengthwise about 3″ from the bottom.

2. **Fold** the paper in thirds.

3. **Open** and staple the edges on either side to form three pockets.

4. **Label** the pockets as shown. Place two index cards in each pocket.

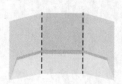
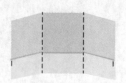
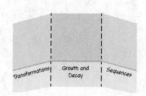

What Vocabulary Will You Learn?

Check the box next to each vocabulary term that you may already know.

- ☐ asymptote
- ☐ common ratio
- ☐ compound interest
- ☐ explicit formula
- ☐ exponential decay functions

- ☐ exponential function
- ☐ exponential growth function
- ☐ geometric sequence
- ☐ recursive formula

Are You Ready?

Complete the Quick Review to see if you are ready to start this module.
Then complete the Quick Check.

Quick Review

Example 1

Evaluate $2x^3$ for $x = 5$.

$2x^3$	Original expression
$= 2(5)^3$	Substitute 5 for x.
$= 2(125)$	Evaluate the exponent.
$= 250$	Multiply.

Example 2

Divide $\frac{5}{6} \div \frac{1}{3}$.

$\frac{5}{6} \div \frac{1}{3} = \frac{5}{6} \cdot \frac{3}{1}$	Multiply by the reciprocal.
$= \frac{15}{6}$	Multiply the numerators and multiply the denominators.
$= \frac{5}{2}$	Find the simplest form.

Quick Check

Evaluate each expression for the given value.

1. $-4x^2$ for $x = 7$

2. $3x^2$ for $x = 3$

3. $0.25x^4$ for $x = 1$

4. x^5 for $x = 3$

Divide.

5. $128 \div 4$

6. $\frac{1}{3} \div 2$

7. $-9 \div 3$

8. $\frac{1}{8} \div \frac{1}{2}$

How did you do?

Which exercises did you answer correctly in the Quick Check? Shade those exercise numbers below.

       ⑧

Exponential Functions

Today's Goals
- Recognize situations modeled by linear or exponential functions.
- Graph exponential functions, showing intercepts and end behavior.

Today's Vocabulary
exponential function

exponential growth function

exponential decay function

asymptote

Explore Exponential Behavior

Online Activity Use a table to complete the Explore.

> ⊘ **INQUIRY** How does exponential behavior differ from linear behavior?

Learn Identifying Exponential Behavior

An **exponential function** is a function of the form $y = ab^x$, where $a \neq 0$, $b > 0$, and $b \neq 1$. Some examples of exponential functions are $y = 2(3)^x$, $y = 4^x$, and $y = \left(\frac{1}{2}\right)^x$.

Linear

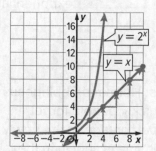

the rate of change remains constant

Exponential

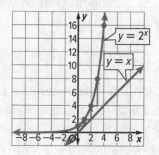

the rate of change increases by the same factor

Note that in the linear function, the rate of change remains constant. In the exponential function, the rate of change increases by the same factor.

🌥 **Think About It!**

Is a function that relates an independent variable to a change in order of magnitude always exponential? Explain your reasoning.

🌥 **Think About It!**

Can you confirm that the entire data set displays exponential behavior by looking at the first two intervals?

🖱 **Go Online**

An alternate method is available for this example.

🌐 **Example 1** Identify Exponential Behavior

EARTHQUAKES The Richter Scale measures the energy that an earthquake releases and assigns a magnitude to it. These orders of magnitude can be approximated by comparing them to the explosive power of TNT. Determine whether the set of data displays exponential behavior.

Magnitude	TNT (tons)
1	0.6
2	6
3	60
4	600
5	6000
6	60,000
7	600,000
8	6,000,000

Magnitudes 1 and 2

As the order of magnitude increases from ___ to ___, the amount of TNT that is approximately equal in magnitude increases from ___ tons to ___ tons. That is an increase by a factor of ___.

Magnitudes 2 and 3

As the order of magnitude increases from ___ to ___, the amount of TNT that is approximately equal in magnitude increases from ___ tons to ___ tons. That is an increase by a factor of ___.

Because the amount of TNT increases by the _____ given an _____ in magnitude, the data set _____ exponential behavior.

Check

MEMORY In the 19th century, psychologist Hermann Ebbinghaus created a formula to approximate how quickly people forget information over time. The approximate percentage of the newly learned information a person retains over time is shown in the table. Determine whether the data displays exponential behavior.

Time (days)	% Retained
0	100
1	80
2	64
3	51.2
4	40.96
5	32.768

Source: Indiana University

The data set _____ display exponential behavior.

Explore Restrictions on Exponential Functions

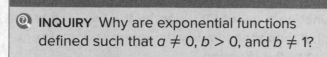

🏃 **Online Activity** Use graphing technology to complete the Explore.

> ❓ **INQUIRY** Why are exponential functions defined such that $a \neq 0$, $b > 0$, and $b \neq 1$?

Learn Graphing Exponential Functions

Functions of the form $f(x) = ab^x$, where $a > 0$ and $b > 1$, are called **exponential growth functions**. Functions of the form $f(x) = ab^x$, where $a > 0$ and $0 < b < 1$, are called **exponential decay functions**.

The graphs of exponential functions have an **asymptote**. An asymptote is a line that a graph approaches.

🏃 **Go Online**
You can watch a video to see how to graph exponential functions.

Key Concept • Types of Exponential Functions

Exponential Growth Functions	Exponential Decay Functions
Equation	
$f(x) = ab^x$, $a > 0$, $b > 1$	$f(x) = ab^x$, $a > 0$, $0 < b < 1$
Domain, Range	
D = all real numbers; R = $\{y \mid y > 0\}$	D = all real numbers; R = $\{y \mid y > 0\}$
Intercepts	
one y-intercept, no x-intercepts	one y-intercept, no x-intercepts
End Behavior	
as x increases, $f(x)$ increases; as x decreases, $f(x)$ approaches 0	as x increases, $f(x)$ approaches 0; as x decreases, $f(x)$ increases
Graph	
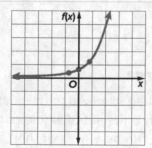	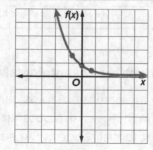

Talk About It!

How can you determine the *y*-intercept without substituting into the original equation? (Hint: Consider what $x = 0$ and the *y*-intercept mean in the context of the situation.)

Problem-Solving Tip

Make an Organized List Making an organized list of *x*-values and corresponding *y*-values is helpful in graphing the function. It can also help you identify patterns in the data.

Go Online

You can watch a video to see how to use a graphing calculator with this example.

Example 2 Exponential Growth Function

FOLDING **Each time you fold a piece of paper in half, it doubles in thickness. If a piece of paper is 0.05 millimeter thick, then you can determine the thickness *y* of a piece of paper given the number of folds *x* with the function $y = 0.05(2)^x$. Identify the key features of the function, graph it, and then identify the relevant domain and range in the context of the situation.**

Part A Identify key features.

Because $a > 0$ and $b > 1$, $y = 0.05(2)^x$ is an exponential _____ function. The domain is _____ and the range is _____.

The *y*-intercept is the value of *y* when $x = 0$.

$$y = 0.05(2)^0$$

$$y = 0.05(\_\_\_)$$

$$= 0.05$$

The *y*-intercept is _____.

Because $y = 0.05(2)^x$ is an exponential growth function, as *x* increases, *y* increases, and as *x* decreases, *y* approaches 0.

Part B Graph the function.

Make a table of values. Then, plot the points and draw a curve to approximate the function.

x	$y = 0.05(2)^x$	y
−2	$y = 0.05(2)^{-2}$	
−1	$y = 0.05(2)^{-1}$	
0	$y = 0.05(2)^0$	
1	$y = 0.05(2)^1$	
2	$y = 0.05(2)^2$	
3	$y = 0.05(2)^3$	
4	$y = 0.05(2)^4$	

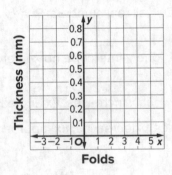

Part C Identify relevant domain and range.

Because the number of folds cannot be negative and folds must be counted in integers, the potential domain is the set of whole numbers and the potential range is the set of real numbers greater than or equal to _____. However, because the paper cannot be folded indefinitely, the thickness of the paper cannot continue to grow to infinity. So the domain will be restricted to the greatest possible number of _____, and the range will be restricted to the greatest _____ of the paper.

Go Online You can complete an Extra Example online.

Check

Consider $y = 3^x$.

Part A

List the key features that apply to $y = 3^x$. Include the domain, range, y-intercept, and end behavior of the function.

_____

_____

FILM The function $y = 3^x$ can be used to model a real-world situation. Sarah wants to crowdfund a film project. To spread the word, she shares the page with 3 friends, and requests that each friend share it with 3 more friends. The function that models the number of new people given the link to the crowdfunding page y given the number of cycles x is defined by the function $y = 3^x$.

Part B

Select the correct graph of $y = 3^x$. _____

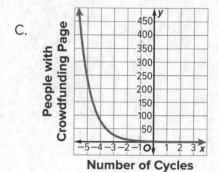

A.

B.

C.

D.

Part C

Describe the relevant range in the context of the situation.

_____

Math History Minute

While a junior in high school, **Britney Gallivan (1985-)** showed that a single length of toilet paper 4000 feet long can be folded in half a maximum of twelve times. Britney's proof involved the exponential equation

$L = \frac{\pi t}{6}(2^n + 4)(2^n - 1)$,

where L is the length of the paper, t is the thickness of the material to be folded, and n represents the number of folds desired.

Copyright © McGraw-Hill Education

Use a Source

Research the amount of caffeine in another caffeinated drink. Write a function that models the amount of caffeine left in your system after x hours, and identify the key features of the function.

🌐 Example 3 Exponential Decay Function

CAFFEINE The half-life of a substance describes how long it takes for the substance to deplete by half. The half-life of caffeine in the body of a healthy adult is approximately 5 hours, meaning that it takes 5 hours for the body to break down half of the caffeine. Suppose an energy drink contains 160 milligrams of caffeine. The amount of caffeine y left in your system after x hours is modeled by the function $y = 160\left(\frac{1}{2}\right)^{\frac{x}{5}}$. Identify the key features of the function, graph it, and then identify the relevant domain and range in the context of the situation.

Part A Identify key features.

Because $a > 0$ and $0 < b < 1$, $y = 160\left(\frac{1}{2}\right)^{\frac{x}{5}}$ is an exponential _____ function. The domain is _____ and the range is _____.

The y-intercept is the value of y when $x = 0$.

$$y = 160\left(\frac{1}{2}\right)^{\frac{0}{5}}$$

$$= 160(\underline{}) = 160.$$

The y-intercept is _____.

Part B Graph the function.

x	$160\left(\frac{1}{2}\right)^{\frac{x}{5}}$	y
−1	$160\left(\frac{1}{2}\right)^{\frac{-1}{5}}$	
0	$160\left(\frac{1}{2}\right)^{\frac{0}{5}}$	
1	$160\left(\frac{1}{2}\right)^{\frac{1}{5}}$	
2	$160\left(\frac{1}{2}\right)^{\frac{2}{5}}$	
3	$160\left(\frac{1}{2}\right)^{\frac{3}{5}}$	
4	$160\left(\frac{1}{2}\right)^{\frac{4}{5}}$	
5	$160\left(\frac{1}{2}\right)^{\frac{5}{5}}$	

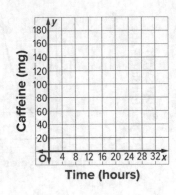

Part C Identify relevant domain and range.

Because time cannot be negative, the relevant domain is {_____}. Because the amount of caffeine cannot be negative and the amount of caffeine when $x = 0$ is 160 mg, the relevant range is {_____}.

🌐 **Go Online** You can complete an Extra Example online.

Practice

Go Online You can complete your homework online.

Example 1

Determine whether the set of data displays exponential behavior. Write *yes* or *no*. Explain why or why not.

1.

x	−3	−2	−1	0
y	9	12	15	18

2.

x	0	5	10	15
y	20	10	5	2.5

3.

x	4	8	12	16
y	20	40	80	160

4.

x	50	30	10	−10
y	90	70	50	30

5. PICTURE FRAMES Because a picture frame includes a border, the picture must be smaller in area than the entire frame. The table shows the relationship between picture area and frame length for a particular line of frames. Is this an exponential relationship? Explain.

Frame Length (in.)	Picture Area (in²)
5	6
6	12
7	20
8	30
9	42

Examples 2 and 3

6. WASTE Suppose the waste generated by nonrecycled paper and cardboard products in tons y after x days can be approximated by the function $y = 1000(2)^{0.3x}$.

   a. Identify key features.

   b. Graph the function.

   c. Identify the relevant domain and range.

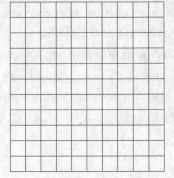

7. IODINE Iodine 131 is a radioisotope that is related to nuclear energy, medical diagnostic and treatment procedures, and natural gas production. A scientist is testing 50 milligrams of Iodine 131. The scientist knows that the half-life of Iodine 131 is about 8.02 days. The function $y = 50\left(\frac{1}{2}\right)^{\frac{x}{8.02}}$ represents the amount of Iodine 131 remaining in milligrams y after x days.

   a. Identify key features.

   b. Graph the function.

   c. Identify the relevant domain and range.

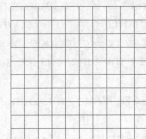

8. DEPRECIATION Suppose a company's computer equipment is decreasing in value according to the function $y = 4000(0.87)^x$. In the equation, x represents the number of years that have elapsed since the equipment was purchased and y represents the value in dollars. What was the value 5 years after the computer equipment was purchased? Round your answer to the nearest dollar.

Mixed Exercises

Graph each function. Find the *y*-intercept and state the domain, range, and the equation of the asymptote.

9. $y = 2\left(\frac{1}{6}\right)^x$

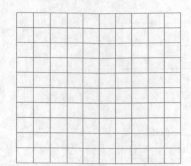

10. $y = \left(\frac{1}{12}\right)^x$

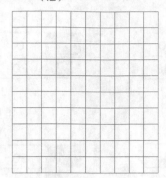

11. $y = -3(9^x)$

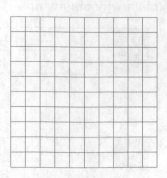

12. $y = -4(10^x)$

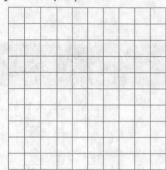

13. $y = 3(11^x)$

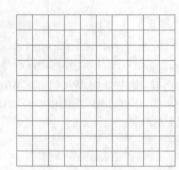

14. $y = 4^x + 3$

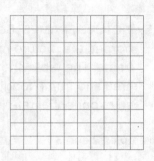

15. METEOROLOGY The atmospheric pressure in millibars at altitude *x* meters above sea level can be approximated by the function $f(x) = 1038(1.000134)^{-x}$ when *x* is between 0 and 10,000.

   a. What is the atmospheric pressure at sea level?

   b. The McDonald Observatory in Texas is at an altitude of 2000 meters. What is the approximate atmospheric pressure there?

   c. As altitude increases, what happens to atmospheric pressure?

16. PERSEVERE Use tables and graphs to compare and contrast an exponential function $f(x) = ab^x + c$, where $a \neq 0$, $b > 0$, and $b \neq 1$, and a linear function $g(x) = ax + c$. Include intercepts, symmetry, end behavior, extrema, and intervals where the functions are increasing, decreasing, positive, or negative.

17. CREATE Write an exponential function that passes through (0, 3) and (1, 6).

18. ANALYZE Determine whether the graph of $y = ab^x$, where $a \neq 0$, $b > 0$, and $b \neq 1$, *sometimes, always,* or *never* has an *x*-intercept. Justify your argument.

19. WRITE Find an exponential function that represents a real-world situation, and graph the function. Analyze the graph, and explain why the situation is modeled by an exponential function rather than a linear function.

Transformations of Exponential Functions

Explore Translating Exponential Functions

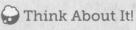 **Online Activity** Use graphing technology to complete the Explore.

> **? INQUIRY** What effect does adding to or subtracting from a function before or after it has been evaluated have on the function?

Learn Translations of Exponential Functions

Key Concept • Vertical Translations of Exponential Functions

- The graph of $g(x) = b^x + k$ is the graph of $f(x) = b^x$ translated vertically.
- If $k > 0$, the graph of $f(x)$ is translated k units up.
- If $k < 0$, the graph of $f(x)$ is translated $|k|$ units down.

Key Concept • Horizontal Translations of Exponential Functions

- The graph of $g(x) = b^{x-h}$ is the graph of $f(x) = b^x$ translated horizontally.
- If $h > 0$, the graph of $f(x)$ is translated h units right.
- If $h < 0$, the graph of $f(x)$ is translated $|h|$ units left.

Example 1 Vertical Translations of Exponential Functions

Describe the translation in $g(x) = 2^x + 3$ as it relates to the graph of the parent function $f(x) = 2^x$.

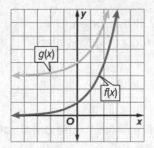

The constant k is added to the function after it has been evaluated, so k affects the _____ values.

The value of k is _____ than 0, so the graph of $f(x) = 2^x$ is translated _____ units _____.

Today's Goals
- Apply translations to exponential functions.
- Apply dilations to exponential functions.
- Apply reflections to exponential functions.
- Use transformations to identify exponential functions from graphs and write equations of exponential functions.

Go Online You can watch a video to see how to describe translations of functions.

Think About It! For $h < 0$, why must you move $|h|$ units left instead of h units?

Think About It! What do you notice about the asymptote of a vertically translated exponential function compared to the asymptote of the parent function?

💭 **Think About It!**

What do you notice about the asymptote of a horizontally translated exponential function compared to the asymptote of the parent function?

Example 2 Horizontal Translations of Exponential Functions

Describe the translation in $g(x) = 3^{x+1}$ as it relates to the graph of the parent function $f(x) = 3^x$.

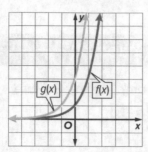

The constant h is subtracted from x before the function is performed, so h affects the _____ values.

The value of h is _____ than 0, so the graph of $f(x) = 3^x$ is translated _____ unit _____.

💭 **Think About It!**

How can the placement of the constant tell you if the resulting transformation will be a vertical or horizontal translation?

Example 3 Multiple Translations of Exponential Functions

Describe the translation in $g(x) = \left(\frac{1}{2}\right)^{x-2} - 4$ as it relates to the graph of the parent function $f(x) = \left(\frac{1}{2}\right)^x$.

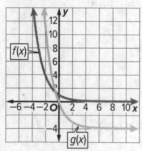

The value of h is subtracted from x _____ the function is performed and is _____, so the graph of $f(x) = \left(\frac{1}{2}\right)^x$ is translated _____.

The value of k is added to the function _____ it has been evaluated and is _____, so the graph of $f(x) = \left(\frac{1}{2}\right)^x$ is also translated _____.

Check

Describe the translation in $g(x) = 2^{x+1} - 8$ as it relates to the graph of the parent function $f(x) = 2^x$.

The graph of $g(x) = 2^{x+1} - 8$ is the translation of the graph of the parent function 1 unit _____ and 8 units _____.

▶ **Go Online** You can complete an Extra Example online.

Example 4 Identify Exponential Functions from Graphs (Vertical Translations)

The given graph is a translation of the parent function $f(x) = \left(\frac{1}{4}\right)^x$.

Use the graph of the function to write its equation.

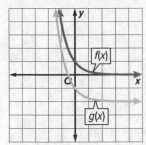

$g(x) = $ _____

The horizontal asymptote of $g(x)$ is different from the horizontal asymptote of $f(x)$, implying a vertical translation of the form $g(x) = \left(\frac{1}{4}\right)^x + k$. The parent graph has a y-intercept at $(0, 1)$. The translated graph has a y-intercept at $(0, \underline{\hspace{1cm}})$. The y-intercept is shifted ____ units down, so $k = $ ____.

Study Tip

Vertical Translations
Any exponential parent function has a y-intercept at $(0, 1)$ and an asymptote at $y = 0$. By examining how far these features are shifted up or down, you can easily determine the value of k when identifying exponential functions.

Example 5 Identify Exponential Functions from Graphs (Horizontal Translations)

The given graph is a translation of the parent function $f(x) = 2^x$. Use the graph of the function to write its equation.

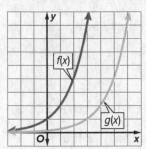

$g(x) = $ ____

The horizontal asymptote of $g(x)$ is the same as the horizontal asymptote of $f(x)$, implying a horizontal translation of the form $g(x) = 2^{x-h}$. The parent graph passes through $(0, 1)$. The translated graph has a y-value of 1 at $(\underline{\hspace{0.8cm}}, 1)$. The graph is shifted 3 units right, so $h = $ ____.

Study Tip

Horizontal Translations
Any exponential parent function has a y-intercept at $(0, 1)$. By examining how far this point is shifted right or left, you can easily determine the value of h when identifying exponential functions.

Check

The given graph is a translation of $f(x) = 5^x$. Which is the equation for the function shown in the graph? ____

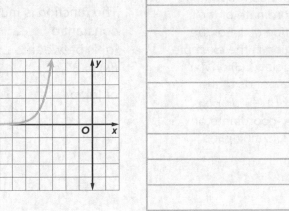

A. $g(x) = 5^x + 4$

B. $g(x) = 5^{x-4}$

C. $g(x) = 5^x - 4$

D. $g(x) = 5^{x+4}$

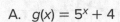 **Go Online** You can complete an Extra Example online.

Copyright © McGraw-Hill Education

Explore Dilating Exponential Functions

🔎 **Online Activity** Use graphing technology to complete the Explore.

> ⊘
>
> ❓ **INQUIRY** What effect does multiplying a function by a value before or after it has been evaluated have on the function?

Learn Dilations of Exponential Functions

Key Concept • Vertical Dilations of Exponential Functions

- The graph of $g(x) = ab^x$ is the graph of $f(x) = b^x$ stretched or compressed vertically by a factor of $|a|$.

- If $|a| > 1$, the graph of $f(x)$ is stretched vertically away from the x-axis.

- If $0 < |a| < 1$, the graph of $f(x)$ is compressed vertically toward the x-axis.

Key Concept • Horizontal Dilations of Exponential Functions

- The graph of $g(x) = b^{ax}$ is the graph of $f(x) = b^x$ stretched or compressed horizontally by a factor of $\frac{1}{|a|}$.

- If $|a| > 1$, the graph of $f(x)$ is compressed horizontally toward the y-axis.

- If $0 < |a| < 1$, the graph of $f(x)$ is stretched horizontally away from the y-axis.

🖱 **Go Online**
You can watch a video to see how to describe dilations of functions.

Study Tip

Vertical Dilations For a vertical dilation, if you multiply each y-coordinate of the function $f(x)$ by a, you'll get the corresponding y-coordinate of the function $g(x)$. For the function in the example the point $(1, 3)$ on $f(x)$ corresponds to the point $\left(1, \frac{3}{4}\right)$ on $g(x)$. The y-coordinate of $f(x)$, 3, is multiplied by a.

Example 6 Vertical Dilations of Exponential Functions

Describe the dilation in $g(x) = \frac{1}{4}(3)^x$ as it relates to the graph of the parent function $f(x) = 3^x$.

Since $f(x) = 3x$, $g(x) = a \cdot f(x)$, where $a = \frac{1}{4}$.

$g(x) = \frac{1}{4}(3)^x \rightarrow g(x) = \frac{1}{4}f(x)$

The function is multiplied by the positive constant a _____ it has been evaluated, and $|a|$ is between ___ and ___, so the graph of $f(x) = 3^x$ is _____ by a factor of $|a|$, or _____.

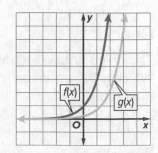

🖱 **Go Online** You can complete an Extra Example online.

Example 7 Horizontal Dilations of Exponential Functions

Describe the dilation in $g(x) = \left(\frac{5}{3}\right)^{2x}$ as it relates to the graph of the parent function $f(x) = \left(\frac{5}{3}\right)^{x}$.

x is multiplied by the positive constant a _____ it has been evaluated, and $|a|$ is greater than ___, so the graph of $f(x) = \left(\frac{5}{3}\right)^{x}$ is _____ by a factor of $\frac{1}{|a|}$, or _____.

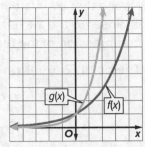

Check

Identify the dilation in each function as it relates to the parent function $f(x) = 4^x$ by writing the type of dilation and dilation factor next to each equation.

$g(x) = 4^{\frac{2}{3}x}$	$\longleftrightarrow$	
$h(x) = 3(4)^x$	$\longleftrightarrow$	
$k(x) = \frac{6}{7}(4)^x$	$\longleftrightarrow$	
$g(x) = 4^{\frac{5}{2}x}$	$\longleftrightarrow$	

🌎 Example 8 Describe Dilations of Exponential Functions

ENERGY Since 2000, solar PV capacity in the world has been growing exponentially. It can be approximated by the function $c(x) = 0.897(1.46)^x$, where $c(x)$ is the solar PV capacity in gigawatts, x is the number of years since 2000, and 0.897 is the initial capacity. Describe the dilation in $c(x) = 0.897(1.46)^x$ as it is related to the parent function $f(x) = (1.46)^x$.

The parent function is $f(x) = (1.46)^x$.

Then $c(x) = af(x)$, where $a =$ _____.

$c(x) = 0.897(1.46)^x \rightarrow c(x) =$ _____ $f(x)$

The function is multiplied by the positive constant a _____ it has been evaluated and $|a|$ is _____ 0 and 1, so the graph of $f(x) = (1.46)^x$ is _____ by a factor of $|a|$, or _____.

PV Capacity

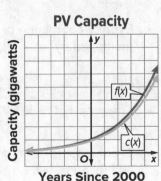

Years Since 2000

Copyright © McGraw-Hill Education

💬 **Talk About It!**

Make a conjecture about the *y*-intercept of *g*(*x*) and the value of *a*. Will this hold true for any vertical dilation of an exponential function? Explain your reasoning.

Example 9 Identify Exponential Functions from Graphs (Dilations)

The given graph is a dilation of the parent function $f(x) = 2^x$. Use the graph of the function to write its equation.

Notice that every point on the graph of *g*(*x*) is closer to the *x*-axis, implying a vertical compression of the form $g(x) = a(2)^x$.

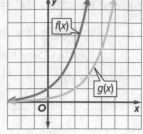

The point (2, 1) lies on the graph. Solve for *a*.

$$\underline{\quad} = a(2)^2$$

$$\underline{\quad} = \underline{\quad}a$$

$$\underline{\quad} = a$$

$$g(x) = \underline{\quad\quad}$$

Check

The given graph is a dilation of $f(x) = 3^x$. Which is the equation for the function shown in the graph? _____

A. $g(x) = 0.007(3)^x$

B. $g(x) = \frac{2}{3}(3)^x$

C. $g(x) = 3^x + \frac{3}{2}$

D. $g(x) = \frac{3}{2}(3)^x$

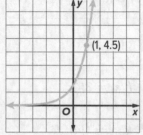

(1, 4.5)

Explore Reflecting Exponential Functions

 Online Activity Use graphing technology to complete the Explore.

> @ **INQUIRY** What effect does multiplying a function by −1 before or after it has been evaluated have on the function?

🤔 **Think About It!**

What does the general form of an exponential function look like once it has been reflected across the *x*-axis?

Learn Reflections of Exponential Functions

Key Concept • Reflections of Exponential Functions Across the *x*-axis

- The graph of −*f*(*x*) is the reflection of the graph of $f(x) = b^x$ across the *x*-axis.

- Every *y*-coordinate of −*f*(*x*) is the corresponding *y*-coordinate of *f*(*x*) multiplied by −1.

▶ **Go Online**

You can watch a video to see how to describe reflections of functions.

Key Concept • Reflections of Exponential Functions Across the *y*-axis

- The graph of *f*(−*x*) is the reflection of the graph of $f(x) = b^x$ across the *y*-axis.

- Every *x*-coordinate of *f*(−*x*) is the corresponding *x*-coordinate of *f*(*x*) multiplied by −1.

Example 10 Vertical Reflections of Exponential Functions

Describe how the graph of $g(x) = -3(2)^x$ is related to the graph of the parent function $f(x) = 2^x$.

The function is multiplied by −1 and the positive constant a _____ it has been evaluated and $|a|$ is _____ 1, so the graph of $f(x) = 2^x$ is _____ and reflected across the _____.

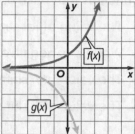

Example 11 Horizontal Reflections of Exponential Functions

Describe how the graph of $g(x) = (3)^{-2x}$ is related to the graph of the parent function $f(x) = 3^x$.

The function is multiplied by −1 and the constant a _____ it is evaluated and $|a|$ is _____ 1, so the graph of $f(x) = 3^x$ is _____ and reflected across the _____.

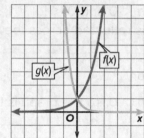

Check

Match each function with its graph.

1. ___ $g(x) = -3^x$
2. ___ $h(x) = \left(\frac{1}{3}\right)^{-x}$
3. ___ $k(x) = -\left(\frac{1}{3}\right)^x$
4. ___ $g(x) = 3^{-x}$

A.

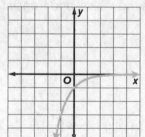

B.

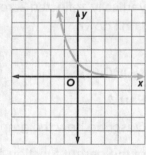

C.

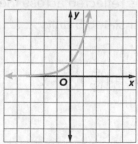

D.

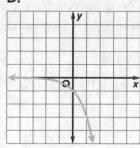

Think About It!

The example shows the reflection of an exponential function of the form $f(x) = ab^x$ over the y-axis for the case where $b > 1$. Examine the following cases and describe the effect a reflection across the y-axis would have on the end behavior of the parent function $f(x) = ab^x$.

Case 1: $g(x) = ab^{-x}$ where $b > 1$

Case 2: $g(x) = ab^{-x}$ where $0 < b < 1$

Go Online You can complete an Extra Example online.

Go Online
You can watch a video
to see how to graph
transformations of
exponential functions
using a graphing
calculator.

Think About It!
Write an exponential
function that is the
parent function $f(x) = 2^x$
stretched vertically and
translated 3 units left
and 6 units up.

Learn Transformations of Exponential Functions

The general form of an exponential function is $f(x) = ab^{x-h} + k$, where a, h, and k are parameters that dilate, reflect, or translate a parent function with base b.

- The value of $|a|$ stretches or compresses (dilates) the parent graph.
- When a is negative, the graph is reflected across the x-axis.
- The value of h shifts (translates) the parent graph right or left.
- The value of k shifts (translates) the parent graph up or down.

Example 12 Multiple Transformations of Exponential Functions

Describe how the graph of $g(x) = -\frac{1}{2}(3)^{x-2} - 1$ is related to the graph of the parent function $f(x) = 3^x$.

$a < 0$ and $0 < |a| < 1$, so the graph of $f(x) = 3^x$ is reflected across the _____ and _____ by a factor of $|a|$, or _____.

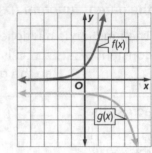

$h > 0$, so the graph is then translated h units _____, or _____.

$k < 0$, so the graph is then translated $|k|$ units _____, or _____.

$g(x) = -\frac{1}{2}(3)^{(x-2)} - 1$ is the graph of the parent function compressed vertically, reflected across the x-axis, and translated 2 units right and 1 unit down.

Check

Part A

Describe how the graph of $g(x) = -4\left(\frac{1}{2}\right)^{x+5} - 2$ is related to the graph of the parent function $f(x) = \left(\frac{1}{2}\right)^x$.

The graph of $f(x) = \left(\frac{1}{2}\right)^x$ is reflected across the _____ and _____ vertically. The graph is translated 5 units _____ and 2 units _____.

Part B

Sketch the graph of $g(x) = -4\left(\frac{1}{2}\right)^{x+5} - 2$.

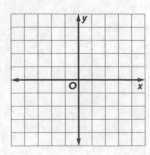

Go Online You can complete an Extra Example online.

Practice

🡒 **Go Online** You can complete your homework online.

Examples 1–3, 6–7, 10–12

Describe the transformation of $g(x)$ as it relates to the parent function $f(x)$.

1. $f(x) = 6^x$; $g(x) = 6^x + 8$

2. $f(x) = 5^x$; $g(x) = -5^x$

3. $f(x) = 3^x + 1$; $g(x) = 3^{2x} + 1$

4. $f(x) = 4^x - 3$; $g(x) = 4^{0.5x} - 3$

5. $f(x) = 2.3^x$; $g(x) = -2.3^{x-1}$

6. $f(x) = 2^x$; $g(x) = 2^{-x} + 1$

7. $f(x) = 5^x + 2$; $g(x) = 5^{-x} + 6$

8. $f(x) = 1.4^x - 1$; $g(x) = -1.4^x + 6$

9. $f(x) = 3^x + 1$; $g(x) = 2(3^x + 1)$

10. $f(x) = -4^x$; $g(x) = \frac{1}{3}(-4^x)$

11. $f(x) = 4^x$; $g(x) = 4^{x-3}$

12. $f(x) = \left(\frac{1}{2}\right)^x + 5$; $g(x) = \left(\frac{1}{2}\right)^x$

Examples 4–5, 9

Each graph is a transformation of the parent function $y = 2^x$. Use the graph of the function to write its equation.

13.

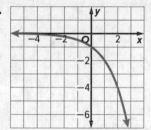

14.

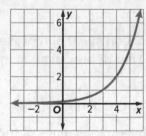

15.

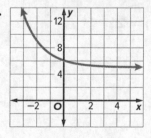

16.

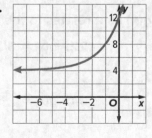

Example 8

17. **SAVING** Celia invests $2000 in a savings account that earns 1.25% interest per year compounded annually. The amount of money in her bank account after x years can be modeled by $g(x) = 2000(1.0125)^x$. Describe the dilation in $g(x)$ as it relates to the parent function $f(x) = 1.0125^x$.

18. **CAFFEINE** Suppose an 8-ounce cup of coffee contains 100 milligrams of caffeine. The rate at which caffeine is eliminated from an adult's body is 11% per hour. The function $f(x) = 100(0.89)^x$ can be used to model the amount of caffeine left in a person's bloodstream after x hours of consuming the cup of coffee. Suppose the function $g(x) = 25(0.89)^x$ represents the amount of caffeine left in a person's bloodstream after x hours of consuming an 8-ounce cup of green tea. Describe $g(x)$ as a transformation of $f(x)$.

19. **VISITORS** The number of visitors to a new skateboarding park can be modeled by the exponential function $g(x) = 20(2^x)$, where x represents the number of months since the park's grand opening. Explain how the number of visitors during that first month is a dilation of the parent function $f(x) = 2^x$.

20. **DEPRECIATION** Depreciation is the decrease in the value of an item resulting from its age or wear. When an item loses about the same percent of its value each year, an exponential function can be used to model its decreasing value over time. The function $g(x) = 12{,}000(0.85)^x$ can be used to model the value of a $12,000 car as it depreciates at an annual rate of 15% over x years. $g(x)$ is a dilation of the parent function $f(x) = 0.85^x$. The graph shows the function $g(x)$.

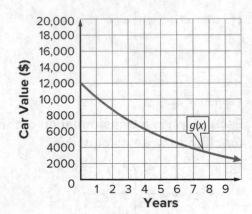

   a. Write the equation of a function $h(x)$ that represents the depreciation of a $20,000 car depreciating at the same rate over x years.

   b. Describe $h(x)$ as it relates to the parent function.

   c. What is the difference between the values of the $12,000 car and the $20,000 car after 5 years?

Mixed Exercises

Describe the transformation of $g(x)$ as it relates to the parent function $f(x) = 2^x$.

21. $g(x) = 2^x + 6$

22. $g(x) = 3(2)^x$

23. $g(x) = -\frac{1}{4}(2)^x$

24. $g(x) = -3 + 2^x$

25. $g(x) = 2^{-x}$

26. $g(x) = -5(2)^x$

Write a function $g(x)$ to represent the transformation of the parent function $f(x)$.

27. $f(x) = 2^x$ translated 3 units up

28. $f(x) = 8^x$ translated 1 unit down

29. $f(x) = 5^x$ translated 2 units right

30. $f(x) = 3^x$ translated 4 units left

31. $f(x) = 6^x + 7$ translated 2 units down

32. $f(x) = -2^x + 3$ translated 5 units right

33. $f(x) = 4^x$ is compressed vertically by a factor of $\frac{1}{2}$

34. $f(x) = 3^x$ is stretched vertically by a factor of 5

35. $f(x) = 2^x$ is compressed horizontally by a factor of $\frac{1}{3}$

36. $f(x) = 5^x$ is stretched horizontally by a factor of 4

Each graph is a transformation of the parent function $f(x) = 5^x$. Use the graph of the function to write its equation.

37.
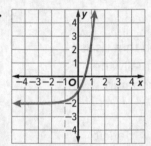

38.
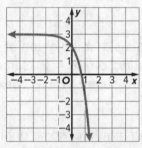

39.
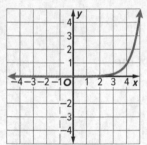

40.
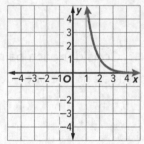

41. ASSETS Thomas and Rebecca each put $1000 into a bank account that earns 1.5% interest per year compounded annually. Thomas also has an antique toy automobile. The graph shows the amount of their assets over time.

a. Describe the graph of Thomas' assets as a transformation of Rebecca's assets.

b. Use the graph to extrapolate the value of Thomas' antique toy automobile.

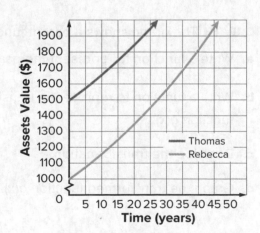

The graph of g(x) is a transformation of the parent function f(x). Graph g(x) and describe the transformation in each function as it relates to the parent function.

42. $f(x) = 3^x$

$g(x) = 5 \cdot 3^{x+2} - 4$

43. $f(x) = 2^x$

$g(x) = -3 \cdot 2^{-x} + 1$

44. CONSTRUCT ARGUMENTS Name the coordinates of the point at which the graphs of $g(x) = 2^x + 3$ and $h(x) = 5^x + 3$ intersect. Explain your reasoning.

45. STRUCTURE Describe the similarities between the graph of $f(x) = 4^{x+2}$ and the graph of $g(x) = 16 \cdot 4^x$. Use the properties of exponents to justify your answer.

46. ANALYZE What would happen to the shape of the graph of an exponential function if the function is multiplied by a number between 0 and −1? What would happen to its shape if the exponent is multiplied by a number between 0 and −1? Justify your argument.

47. FIND THE ERROR Jennifer claims that the graph of $g(x) = 2(2^x)$ is a graph that rises more rapidly than its parent function $f(x) = 2^x$. James claims that it is actually the parent graph shifted to the left 2 units. Who is correct? Explain your reasoning.

48. WRITE A deficit is a negative amount of some quantity, such as money. A deficit that is growing exponentially can be modeled by $y = ab^{c(x-h)} + k$. Describe the constraints on a, b, and c.

49. WHICH ONE DOESN'T BELONG? Consider each pair of transformations of the function f(x) to g(x). Which one does not belong? Justify your conclusion.

$f(x) = 3^{x+2}$ $g(x) = 3^{x-1} + 2$	$f(x) = 2^x$ $g(x) = 2^{x+3} + 2$	$f(x) = 4^{x+1}$ $g(x) = 4^{x+4} + 2$

50. CREATE The graph shows a parent function f(x).

a. Write a function to represent the parent function f(x).

b. Write a function to represent a transformation of the parent function g(x).

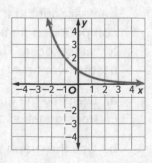

c. Describe the transformation.

d. Graph the transformed function g(x).

Writing Exponential Functions

Copyright © McGraw-Hill Education

Explore Writing an Exponential Function to Model Population Growth

 Online Activity Use a real-world situation to complete the Explore.

> ×
>
> ② **INQUIRY** How can you find an equation that models the population growth of a colony of organisms that grows exponentially?

Today's Goals
- Construct exponential functions by using a graph, a description, or two points.
- Create equations and solve problems involving exponential growth.
- Create equations and solve problems involving exponential decay.

Today's Vocabulary
compound interest

Learn Constructing Exponential Functions

If you are given two points, a graph, or a description of an exponential function, you can write an exponential function to model the data.

Given Two Points

Substitute the values of x and y into the equation $y = ab^x$. The result will be a system of equations in two variables, each with a and b as unknowns. You can then solve the system using substitution.

Given a Graph

Use any two points to write an equation in the form $y = ab^x$. Substitute the values of x and y into the equation, and solve the resulting system of equations for a and b using substitution.

Given a Description

Use the information to create a table or a graph. Then look for a pattern between the input and output values. Keep an eye out for words such as *exponential*, *linear*, *multiple*, *constant*, and *factor*, which can help you determine whether the function is exponential.

Study Tip

Graphs If you write an exponential equation based on a graph that includes the y-intercept, you can substitute the y-value of the intercept for a in $y = ab^x$. Since $b^0 = 1$, $y = a$. You can then use another point to find the common ratio b.

Example 1 Write an Exponential Function Given Two Points

Write an exponential function for the graph that passes through (1, 6) and (3, 24).

Substitute x and y into $y = ab^x$ to get a system of two equations.

$y = ab^x$	General form	$y = ab^x$	General form
$\underline{\quad} = ab^1$	$y = 6$ and $x = 1$	$\underline{\quad} = ab^3$	$y = 24$ and $x = 3$

(continued on the next page)

Solve the system of equations using substitution.

Solve the first equation for a.

$\underline{\quad} = ab^1$ First equation

$\underline{\quad} = a$ Division Property of Equality

Substitute $\frac{6}{b}$ for a in the second equation to find b.

$24 = ab^3$ Second equation

$24 = \underline{\quad}b^3$ Substitute $\frac{6}{b}$ for a.

$24 = \underline{\quad}$ Multiply.

$24 = \underline{\quad}$ Quotient of Powers

$\underline{\quad} = b^2$ Division Property of Equality; then simplify

$\underline{\quad} = b$ Definition of exponent

Substitute 2 for b in either equation to find a.

$6 = ab^1$ Second equation

$6 = a(\underline{\quad})^1$ Substitute 2 for b.

$\underline{\quad} = a$ Simplify.

Write the equation.

$y = \underline{\quad} \times \underline{\quad}^x$

Check

Write an exponential function that passes through $(-1, 20)$ and $(1, 5)$.

$y = \underline{\quad} \times \underline{\quad}^x$

Example 2 Write an Exponential Function Given a Graph

Write an exponential function for the graph.

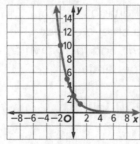

x	y
−2	
−1	
0	
1	

You can use any two points to write a system of two equations using

$y = ab^x$.

$2.5 = ab^0$

$1.25 = ab^1$

Since there is a zero exponent in the first equation, solve it for a.

$a = \underline{\quad}$

Then substitute this value into the second equation to find b.

$b = \underline{\quad}$

The graph can be modeled by the function $y = \underline{\qquad\qquad}$.

 Go Online You can complete an Extra Example online.

Study Tip

Assumptions The graphs of multiple functions pass through the points (1, 6) and (3, 24). For example, the linear equation $y = 9x - 3$ also passes through these points. However, because you are told to find the exponential function that passes through those points, you can assume that the exponential relationship is the correct one.

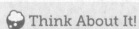 Think About It!

Use the graph to estimate the value of y when $x = 2$. Then use the function to find y when $x = 2$.

Check

Part A Use the graph to estimate the value of y when $x = 4$. _____

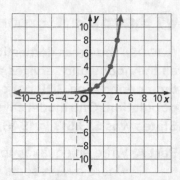

Part B Write an exponential function that models the graph. _____

Part C Use the function in Part B to find y when $x = 4$. _____

 Example 3 Write an Exponential Function Given a Description

CONTEST A radio station is giving away $1000 to the first listener who answers a question correctly. If the question goes unanswered for one hour, the prize increases by 10% until it is answered correctly. Write a function to describe this situation.

Hour (h)	Prize Total (P)
0	1000
1	1100
2	1210
3	1331

This is an example of exponential growth, so $b > 1$. Divide each value of P by the preceding term to find a common ratio of 1.1.

The value of a can be found by identifying the y-intercept.

$$a = _____$$

Which function best models the situation? _____

A. $P = 0.1(1000)^h$

B. $P = 1000(0.1)^h$

C. $P = 1000(0.9)^h$

D. $P = 1000(1.1)^h$

Learn Solving Problems Involving Exponential Growth

Key Concept • Equation for Exponential Growth

$$y = a(1 + r)^t$$

y is the final amount.

t is time.

a is the initial amount.

r is the rate of growth expressed as a decimal, $r > 0$.

Key Concept • Equation for Compound Interest

$$A = P\left(1 + \frac{r}{n}\right)^{nt}$$

A is the current amount.

r is the annual interest rate expressed as a decimal, $r > 0$.

n is the number of times the interest is compounded each year.

t is time in years.

P is the principal, or the initial amount.

 Talk About It!

Why is the common ratio 1.1 and not 0.1? Explain.

 Think About It!

Why is the constant 1 in the exponential growth formula? What does it represent?

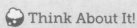
🌐 Example 4 Exponential Growth

GOODS The gross domestic product (GDP) is the monetary value of all the goods and services produced within a country in a specific time period. The GDP per capita is this value divided by the population. One model says that the GDP per capita in the United States was $13,513 in 1946, and it has increased by 2% every year.

Part A Write an equation to represent the GDP per capita after t years, where a represents the initial GDP in 1946, and r represents the rate of growth each year.

If the initial year was 1946, the initial GDP was $13,513. So $a =$ _____. The GDP grew 2% each year, so $r =$ _____ or _____.

$y = a(1 + r)^t$ Equation for exponential growth

$y =$ _____ $(1 +$ _____ $)^t$ $a = 13{,}513$ and $r = 2\%$ or 0.02

$y =$ _____   Simplify.

In this equation, y is the GDP per capita, and t is the time in years since 1946.

Part B If this trend continued, calculate the GDP per capita in 2016.

$t =$ _____

Using $y = 13{,}513(1.02)^t$, the GDP per capita in 2016 was _____.

Check

POPULATION From 2013 to 2014, the city of Austin, Texas, saw one of the highest population growth rates in the country at 2.9%. The population of Austin in 2014 was estimated to be about 912,000.

Part A If the trend were to continue, which equation represents the estimated population t years after 2014? _____

A. $y = 912{,}000(0.029)^t$

B. $y = 912{,}000(3.9)^t$

C. $y = 1.029(912{,}000)^t$

D. $y = 912{,}000(1.029)^t$

Part B To the nearest person, predict the population of Austin in 2019.

_____ people

🌐 **Go Online** You can complete an Extra Example online.

🌐 Apply Example 5 Compound Interest

COLLEGE PLANNING **Maria invests $5500 into a college savings account that pays 3.25% compounded quarterly. How much money will there be in the account after 5 years?**

1 What is the task?

Describe the task in your own words. Then list any questions that you may have. How can you find answers to your questions?

_____

_____

2 How will you approach the task? What have you learned that you can use to help you complete the task?

_____

_____

_____

3 What is your solution?

Use your strategy to solve the problem.

Write an equation to represent the amount of money in Maria's account after t years.

_____

How much money will be in Maria's account after 5 years?

_____

4 How can you know that your solution is reasonable?

✏️ **Write About It!** Write an argument that can be used to defend your solution.

_____

_____

_____

Check

BANKING Twin brothers Amare and Jermaine each received $1000 for graduation. Amare invests his money in an account that pays 2.25% compounded daily. Jermaine invests his money in an account that pays 2.25% compounded annually.

Part A Which brother will have more money at the end of 10 years?

_____

    A. Amare

    B. Jermaine

    C. The accounts will be equal.

Part B To the nearest cent, how much more money? $_____

🌐 **Go Online** You can complete an Extra Example online.

Think About It!

What major difference do you notice between the equation for exponential growth and the equation for exponential decay?

Learn Solving Problems Involving Exponential Decay

Key Concept • Equation for Exponential Decay

$$y = a(1 - r)^t$$

y is the final amount.

t is time.

a is the initial amount.

r is the rate of decay expressed as a decimal, $0 < r < 1$.

🌐 Example 6 Exponential Decay

BANKS In banking, a dormant account is one that has not been used in over a year. A bank charges a monthly fee on dormant accounts of 0.8% of the account balance. One dormant account initially had a balance of $1609.

Part A Write an equation to represent the balance in the account after t months.

$y = a(1 - r)^t$ Equation for exponential decay

$y = \underline{\hspace{1cm}} (1 - \underline{\hspace{1cm}})^t$ $a = 1609$ and $r = 0.8\%$ or 0.008

$y = \underline{\hspace{1cm}}(\underline{\hspace{1cm}})^t$ Simplify.

The equation is _____, where y is the balance in the account after t months.

Think About It!

Why is the rate subtracted from 1 in the equation for exponential decay, but added to 1 in the equation for exponential growth?

Part B Estimate the balance in the account after a year.

$t = \underline{\hspace{0.8cm}}$

$y = 1609(0.992)^t = \underline{\hspace{0.8cm}}$

Check

CITY PLANNING A city has been experiencing a slight population loss over the last few years. In 2014, the population was 1.8503 million, representing a 0.18% decrease from the previous year.

Part A If the trend were to continue, which equation represents the estimated population in millions after t years? ____

    A. $y = 1.8503(0.18)^t$

    B. $y = 1.8503(0.9982)^t$

    C. $y = 1.8503(1.0018)^t$

    D. $y = 1.8503(1.9982)^t$

Part B To the nearest ten thousandth, predict the population in 2029. _____ million people

🌐 **Go Online** You can complete an Extra Example online.

🡆 Go Online

to practice what you've learned about exponential growth and decay in the Put It All Together over Lessons 9-1 through 9-3.

Practice

🔄 **Go Online** You can complete your homework online.

Example 1

Write an exponential function for a graph that passes through the points.

1. (2, 16) and (3, 32)

2. (1, 1) and (3, 0.25)

3. (2, 90) and (4, 810)

4. (−2, 4) and (1, 0.5)

5. (1, 12) and (3, 192)

6. (1, 18) and (3, 72)

Example 2

Write an exponential function for the graph.

7.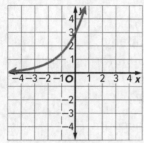

8.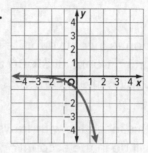

9.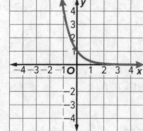

10.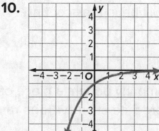

Example 3

11. BIOLOGY A certain species of bacteria in a laboratory culture begins with 50 cells and doubles in number every 30 minutes. Write a function to model the situation.

12. DEPRECIATION Amrita bought a new delivery van for $32,500. The value of this van depreciates at a rate of 12% each year. Write a function to model the value of the van after *x* years of ownership.

13. COMMUNICATION Cell phone usage grew about 23% each year from 2010 to 2016. If cell phone usage in 2010 was 43,000,000, write a function to model U.S. cell phone usage over that time period.

14. INVESTING Robyn invests $1500 at 4.85% compounded quarterly. Write an equation to represent the amount of money she will have in t years.

15. POPULATION The population of New York City increased from 8,192,426 in 2010 to 8,550,405 in 2015. The annual rate of population increase for the period was about 0.9%.

    a. Write an equation for the population, P, t years after 2010.

    b. Use the equation to predict the population of New York City in 2025.

16. SAVINGS A company has a bonus incentive for its employees. The company pays employees an initial signing bonus of $1000 and invests that amount for the employees. Suppose the investment earns 8% interest compounded quarterly.

    a. If an employee receiving this incentive withdraws the balance of the account after 5 years, how much will be in the account?

    b. If an employee receiving this incentive withdraws the balance of the account after 35 years, how much will be in the account?

17. MANUFACTURING A textile company bought a piece of weaving equipment for $60,000. It is expected to depreciate at an average rate of 10% per year.

    a. Write an equation for the value of the piece of equipment Z after t years.

    b. Find the value of the piece of equipment after 6 years.

18. HIGHER EDUCATION The table lists the average annual costs of attending a four-year college in the United States during a recent year.

College Sector	Tuition and Fees	Room and Board
Four-year Public	$9,410	$10,138
Four-year Private	$32,410	$11,516

Source: College Board

Rayelle's parents plan to invest $15,000 in a mutual fund earning an average of 4.5 percent interest, compounded monthly. After 15 years, for how many years will this investment be able to cover the tuition, fees, room, and board for Rayelle at a public college if costs stay the same? Round your answer to the nearest month.

19. DEPRECIATION The value of a home theater system depreciates by about 7% each year. Aeryn purchases a home theater system for $3000. What is its value 4 years after purchase? Round your answer to the nearest hundred.

20. MONEY Hans opens a savings account by depositing $1200. The account earns 0.2 percent interest compounded weekly. How much will be in the account in 10 years if he makes no more deposits? Assume that there are exactly 52 weeks in a year, and round your answer to the nearest cent.

21. POPULATION In 2016 the U.S. Census Bureau estimated the population of the United States at 322 million. If the annual rate of growth was about 0.81%, find the expected population at the time of the 2030 census. Round your answer to the nearest ten million.

Mixed Exercises

Write an exponential function for a graph that passes through the points.

22. (2, 1.4) and (4, 5.6)   **23.** (1, 10.4) and (4, 665.6)   **24.** (1, 42) and (3, 2688)

25. POPULATION The population of Camden, New Jersey, has been decreasing by 0.12% a year on average. If this trend continues, and the population was 79,318 in 2006, estimate Camden's population in 2025.

26. MEDICINE When doctors prescribe medication, they have to consider the rate at which the body filters a drug from the bloodstream. Suppose it takes the human body 6 days to filter out half of a certain vaccine. The amount of the vaccine remaining in the bloodstream x days after an injection is given by the equation $y = y_0(0.5)^{\frac{x}{6}}$, where y_0 is the initial amount. Suppose a doctor injects a patient with 20 μg (micrograms) of the vaccine.

   a. How much of the vaccine will remain after 1 day? Round your answer to the nearest tenth, if necessary.

   b. How much of the vaccine will remain after 12 days? Round your answer to the nearest tenth, if necessary.

   c. After how many days will the amount of vaccine be less than 1 μg?

27. USE TOOLS Graham invested money to save for a car. After x years, the value of Graham's investment can be modeled by the equation $y = 2400(0.95)^x$. How much did Graham originally invest? Is the value of his investment increasing or decreasing? Explain your reasoning. Use technology to find when the investment will be worth half of its starting value.

28. USE A MODEL There is a leak in a container that holds a certain nontoxic gas. Each hour, it loses 10% of its volume.

a. Write an equation that models the amount of gas left in the container after x hours, assuming there were 300 cubic centimeters in the container before the leak. Then use your equation to determine the amount of gas left in the container after 11 hours. Round your answer to the nearest tenth.

b. Dewanda believes a graph of this function should be a scatter plot instead of a continuous curve. Do you agree? Explain how this relates to the domain of the function.

29. STRUCTURE A wildlife researcher is studying the population of deer in a forest.

Years of Study	0	1	2	3
Population	128	160	200	250

a. The table shows the estimated number of deer in the forest over a 3-year period. Write an exponential function that fits this data and can be used to predict the deer population in future years.

b. The average rate of change is the change in the value of the dependent variable divided by the change in the value of the independent variable. What was the average rate of change in population during those three years?

c. If the population growth follows the model from **part a**, do you expect the deer population to continue to increase by the value you came up with in **part b**? Explain.

d. Use the values in the table to show how you know the function is exponential, not linear.

30. ANALYZE Determine the growth rate (as a percent) of a population that quadruples every year. Justify your argument.

31. PERSEVERE Santos invested $1200 into an account with an interest rate of 8% compounded monthly. Use a calculator to approximate how long it will take for Santos's investment to reach $2500.

32. ANALYZE The amount of water in a container doubles every minute. After 8 minutes, the container is full. After how many minutes was the container half-full? Justify your argument.

33. WRITE What should you consider when using exponential models to make decisions?

34. WRITE Compare and contrast the exponential growth formula and the exponential decay formula.

35. CREATE Honovi purchased a new car for $25,000 and has $5000 left to invest.

a. Choose an interest rate between 4% and 7% for Honovi's investment, and find the length of time it would take for the investment to double.

b. Choose an annual depreciation rate from 8% to 10% for the new car that Honovi purchased, and find the length of time it would take for the car's value to be equal to one-half of the purchase price.

c. Using the rates from part **a** and part **b**, find the length of time it would take for the investment to be equal to the value of the car. What is the value at that time?

Transforming Exponential Expressions

🌐 Example 1 Write Equivalent Exponential Expressions

BANKING Savewell Bank offers a savings account with 0.15% interest compounded monthly, and Second Local Bank offers a savings account with 2% interest compounded annually.

To compare the two accounts, we need to compare rates with the same compounding frequency. One way to do this is to compare the approximate monthly interest rates offered by each bank, which is also called the *effective* monthly interest rate.

Part A Compare monthly rates.

Write a function to represent the balance A of a savings account after t years at Second Local Bank. Then write an equivalent function that represents monthly compounding.

For convenience, let the initial amount of the investment be \$1.

$y = a(1 + r)^t$ Equation for exponential growth

$A(t) = \underline{\quad}(1 + \underline{\quad})^t$ $y = A(t), a = 1, r = 2\%$ or 0.02

$A(t) = \underline{\quad}$ Simplify.

Then write a function that represents 12 compoundings per year, a power of $12t$, instead of 1 compounding per year, a power of $1t$.

$A(t) = \underline{\quad}$ Original function

$A(t) = 1.02^{\left(\frac{1 \cdot 12}{12}\right)t}$ $1 \text{ year} = \frac{1 \text{ year}}{12 \text{ months}} \cdot 12 \text{ months}$

$A(t) = \left(1.02^{\frac{1}{12}}\right)^{12t}$ Power of a Power

$A(t) \approx \underline{\quad\quad}$ $(1.02)^{\frac{1}{12}} = \sqrt[12]{1.02}$ or about 1.00165

The effective monthly interest rate offered by Second Local Bank is about _____ or about _____ per month. It is slightly more than the 0.15% offered by Savewell Bank. So, Second Local Bank is a better choice.

Part B Compare annual rates.

Write a function to represent the balance A of a savings account after t months at Savewell Bank. Then write an equivalent function that represents annual compounding.

$y = a(1 + r)^t$ Equation for exponential growth

$A(t) = \underline{\quad}(1 + \underline{\quad})^t$ $y = A(t), a = 1, r = 0.15\%$ or 0.0015

$A(t) = \underline{\quad}$ Simplify.

(continued on the next page)

Today's Goal
- Use the properties of exponents to transform expressions for exponential functions.

🌐 **Go Online** You can watch a video to see how to compare savings accounts.

Talk About It

Does the result in **Part B** make sense compared to the result of **Part A**? Explain.

$A(t) = 1.0015^t$ represents the balance of a savings account at Savewell Bank after t months.

Write an equivalent function that represents 1 compounding per year. Since there are 12 months in a year, the exponent should be $\frac{1}{12}t$.

$A(t) = \underline{\hspace{1cm}}$	Original function
$A(t) = 1.0015^{\left(12 \cdot \frac{1}{12}\right)t}$	$1 \text{ year} = 12 \text{ months} \cdot \frac{1 \text{ year}}{12 \text{ months}}$
$A(t) = \left(1.0015^{12}\right)^{\frac{1}{12}t}$	Power of a Power
$A(t) = \underline{\hspace{2cm}}$	$1.0015^{12} \approx 1.0181$

From this expression, we can determine that the effective annual interest rate of Savewell Bank is about _____, or about _____, which is less than the 2% interest rate offered by Second Local Bank.

Check

SAVINGS Tareq is planning to invest money into a savings account. Oak Hills Financial offers 3.1% interest compounded annually. First City Bank has savings accounts with a quarterly compounded interest rate of 0.7%.

Part A Write a function $A(t)$ that represents the balance of a savings account after t years through Oak Hills Financial if interest were compounded quarterly.

$A(t) \approx \underline{\hspace{1.5cm}}$

What is the effective quarterly interest rate of Oak Hills Financial, rounded to the nearest hundredth? _____

Part B Write a function $A(t)$ that represents the balance of a savings account after t quarters through First City Bank if interest were compounded annually.

$A(t) \approx \underline{\hspace{1.5cm}}$

What is the effective annual rate of First City Bank, rounded to the nearest hundredth? _____

_____ is the better bank for Tareq's savings account.

 Go Online You can complete an Extra Example online.

Practice

🡢 **Go Online** You can complete your homework online.

Example 1

1. **INVESTING** Kimiyo is planning to invest money in a savings account. She is comparing the interest rates of savings accounts at two banks. Bank A offers a savings account with 2.1% interest compounded annually. Bank B offers a savings account with a quarterly compounded interest rate of 0.8%.

    a. Write a function to represent the balance A of Kimiyo's account after t years through Bank A, assuming an initial investment of $1. Then write an equivalent function that represents quarterly compounding.

    b. Which is the better plan? Explain.

    c. What is the approximate effective annual interest rate at Bank B? How does your result relate to your answer to part **b**?

2. **COLLECTIONS** Keandra is comparing the growth rates in the value of two items in a collection. The value of a necklace increases by 3.2% per year. The value of a ring increases by 0.33% per month.

    a. Write a function to represent the value A of the necklace after t years, assuming an initial value of $1. Then write an equivalent function that represents monthly compounding.

    b. Which item is increasing in value at a faster rate? Explain.

    c. What is the approximate annual rate of growth of the ring? How does your result relate to your answer to part **b**?

3. **SAVINGS** Amir is trying to decide between two savings account plans at two different banks. He finds that Bank A offers a quarterly compounded interest rate of 0.95%, while Bank B offers 3.75% interest compounded annually. Which is the better plan? Explain.

4. **BACTERIA** The scientist found that Bacteria A has a growth rate of 0.99% per minute, while Bacteria B has a growth rate of 0.018% per second. Determine which bacterium has a faster growth rate. Explain.

5. **POPULATION** The population of Species A is decreasing at a rate of about 0.25% per quarter. The population of Species B is decreasing at a rate of about 1.34% per year. Determine which species has a population that is decreasing at a faster rate. Explain.

Mixed Exercises

6. **POPULATION** The table shows the population of two small towns that experience increases in population.

    a. Write a function that can be used to estimate the population $P(t)$ of Town A t years after 2012.

    b. Write a function that can be used to estimate the population $P(t)$ of Town B t years after 2012.

Year	Population Town A	Population Town B
2012	8,000	9,500
2013	8,480	9,975
2014	8,989	10,474
2015	9,528	10,997
2016	10,100	11,547

    c. Use your equations and properties of exponents to find the approximate effective monthly increase in the populations of Town A and Town B.

7. **ACCOUNTS** Dominic is trying to decide between two checking account plans. Plan A offers a monthly compounded interest rate of 0.05%, while Plan B offers 0.5% interest compounded annually. Which is the better plan?

8. **CAR DEPRECIATION** Juana is deciding between two cars to purchase. Car A depreciates annually at a rate of 3.5%, while Car B depreciates monthly at a rate of 0.32%. Which car has a better effective rate of depreciation?

9. **INVESTMENT** As a wedding gift, Dotty and Brad received $10,000 cash from Dotty's grandparents. The couple is trying to decide where to invest the money. Account A offers 2.3% interest compounded semi-annually. Account B offers 4.2% interest compounded annually. Which account has the better rate? Explain.

10. **SAVINGS** Hernando is deciding between two certificate of deposit accounts. Account Y offers 4.5% interest compounded annually. Account Z offers 1.13% interest compounded quarterly. Which is the better deal? Explain.

11. **FINANCE** Gita is deciding between two retirement accounts. Account A offers 0.5% interest compounded monthly. Account B offers 2.5% interest compounded annually. Which is the better deal? Explain.

12. **WILDLIFE** The table shows that the population of hawks in two different nature preserves has been decreasing.

Year	Hawk Population (Nature Preserve A)	Hawk Population (Nature Preserve B)
2013	114	120
2014	111	115
2015	108	110
2016	105	106

   a. Write a function that can be used to estimate the population $P(t)$ of the hawks in Nature Preserve A t years after 2013.

   b. Write a function that can be used to estimate the population $P(t)$ of the hawks in Nature Preserve B t years after 2013.

   c. Use your equations and properties of exponents to find the approximate effective quarterly decrease in population of hawks in Nature Preserve A and Nature Preserve B.

Higher-Order Thinking Skills

13. **PERSEVERE** The rate at which an object cools is related to the temperature of the surrounding environment. At the time of an experiment, Mrs. Haubner's lab temperature was 72°F. The approximate temperature of the water at time t in minutes in Mrs. Haubner's lab is predicted by the function $T(t) = 72 + (212 - 72)2.72^{-0.4t}$, where −0.4° per minute is defined as the rate of cooling. Rewrite this function so that the coefficient of t in the exponent is 1.

14. **WRITE** Explain why it is important for a consumer to compare rates in the same unit before making a purchase.

15. **CREATE** Write a scenario that compares two accounts with interest rates compounded at different rate units. Then determine which account has the better rate.

16. **FIND THE ERROR** Marsha is opening a savings account. Eagle Savings Bank is offering her an account with a 0.13% monthly interest rate, while Admiral Savings Bank is offering an account with a 1% annual interest rate. Marsha believes the account at Admiral Savings bank is better because 1% is a greater interest rate than 0.13%. Why is Marsha incorrect? Explain your reasoning.

Geometric Sequences

Explore Modeling Geometric Sequences

🧭 **Online Activity** Use a real-world situation to complete the Explore.

> ❓ **INQUIRY** How can you create a formula to predict how a ball bounces? ✕

Learn Geometric Sequences

A **geometric sequence** is a pattern of numbers that begins with a nonzero term and each term after is found by multiplying the previous term by a nonzero constant r. This constant is called the **common ratio**. Dividing a term by the previous term results in the common ratio.

To find the common ratio, divide each term by the previous term. Then, write the ratio in simplest form.

Example 1 Geometric Sequences

Determine whether the sequence −432, 144, −48, 16, … is geometric. Explain.

$$\frac{144}{-432} = \underline{\quad} \qquad \frac{-48}{144} = \underline{\quad} \qquad \frac{16}{-48} = -\frac{1}{3}$$

Since the ratio is the same for all of the terms, ____, the sequence ____ geometric.

Example 2 Identify Geometric Sequences

Determine whether the sequence 16, 12, 8, 4, … is geometric. Explain.

$$\frac{12}{16} = \underline{\quad} \qquad \frac{8}{12} = \underline{\quad} \qquad \frac{4}{8} = \underline{\quad}$$

The ratios _____ the same, so the sequence _____ geometric.

🧭 **Go Online** You can complete an Extra Example online.

Today's Goals
- Identify and generate geometric sequences.
- Construct and use exponential functions for geometric sequences.

Today's Vocabulary
geometric sequence
common ratio

💬 **Talk About It!**

Why must neither the first term nor the common ratio of a geometric sequence be zero?

Watch Out!

Find the Common Ratio Be sure to write the term as the numerator and the previous term as the denominator when finding the common ratio. Otherwise, you will be calculating the reciprocal of the common ratio.

Check

Determine whether each sequence is geometric. If so, determine its common ratio.

a.

n	1	2	3	4	...
a_n	8	20	50	125	...

This sequence _____ a geometric sequence. It has a common ratio of _____.

b. $-0.7, 0.07, -0.007, 0.0007, ...$

This sequence _____ a geometric sequence. It has a common ratio of _____.

Example 3 Find Terms of Geometric Sequences

Find the next three terms in each geometric sequence.

a. 64, 16, 4, 1, ...

Step 1 Find the common ratio.

$$\frac{16}{64} = \underline{\quad} \qquad \frac{4}{16} = \frac{1}{4} \qquad \frac{1}{4} = \underline{\quad}$$

The common ratio is _____.

Step 2 Multiply by the common ratio.

$$1 \times \frac{1}{4} = \underline{\quad} \qquad \frac{1}{4} \times \frac{1}{4} = \underline{\quad} \qquad \frac{1}{16} \times \frac{1}{4} = \underline{\quad}$$

The next three terms are _____.

b.

n	1	2	3	4	...
a_n	8	12	18	27	...

Step 1 Find the common ratio.

$$\frac{12}{8} = \underline{\quad} \qquad \frac{18}{12} = 1.5 \qquad \frac{27}{18} = \underline{\quad}$$

The common ratio is ___.

Step 2 Multiply by the common ratio.

$$27 \times 1.5 = \underline{\quad} \quad 40.5 \times 1.5 = \underline{\quad} \quad 60.75 \times 1.5 = \underline{\quad}$$

The next three terms are _____.

Check

Find the next three terms in each geometric sequence.

a. 729, 243, 81, ___, ___, ___, ...

b. $-4, -44, -484,$ _____, _____, _____, ...

🧭 **Go Online** You can complete an Extra Example online.

> 💭 **Think About It!**
> How could you determine a term prior to any given term of a geometric sequence?

Learn Geometric Sequences as Exponential Functions

Key Concept • *n*th Term of a Geometric Sequence

The *n*th term a_n of a geometric sequence with first term a_1 and common ratio r is given by the following formula, where n is any positive integer and $a_1, r \neq 0$.

$$a_n = a_1 r^{n-1}$$

Think About It!

When finding the *n*th term of a geometric sequence, why is r raised to the $n - 1$ power instead of to the *n*th power?

Example 4 Find the *n*th Term of a Geometric Sequence

Use an explicit formula to find the 11th term of each geometric sequence.

512, 256, 128, 64, ...

The first term of the sequence is 512. So, $a_1 = 512$.

Find the common ratio.

$$\frac{256}{512} = \underline{\quad} \qquad \frac{128}{256} = \frac{1}{2} \qquad \frac{64}{128} = \underline{\quad}$$

The common ratio is $\frac{1}{2}$.

Use the common ratio to find the 11th term of the sequence.

$a_n = a_1 r^{n-1}$	Formula for the *n*th term
$a_n = 512\left(\frac{1}{2}\right)^{n-1}$	$a_1 = 512$ and $r = \frac{1}{2}$
$a_{11} = 512\left(\frac{1}{2}\right)^{11-1}$	To find the eleventh term, $n = 11$.
$= 512\left(\frac{1}{2}\right)^{10}$	Simplify.
$= 512\left(\frac{1}{1024}\right)$	$\left(\frac{1}{2}\right)^{10} = \left(\frac{1}{1024}\right)$
$= \frac{1}{2}$	Simplify.

Watch Out!

Exponents Remember that the base, which is the common ratio, is raised to $n - 1$ instead of *n*.

Check

Write the equation for the *n*th term of the geometric sequence.

n	1	2	3	4	...
a_n	729	243	81	27	...

$$a_n = \underline{\qquad}$$

Find the 8th term of the sequence.

_____

Go Online You can complete an Extra Example online.

Copyright © McGraw-Hill Education

⊕ Example 5 Use a Geometric Sequence

🐸 **Think About It!**

What assumption is made when calculating the population of North Dakota in 2030?

POPULATION **North Dakota's population is increasing more quickly than any other state's population. In 2011, the population was 685,242, and it has been increasing by an average of 2.5% each year. If this trend continues, determine the estimated population in 2030.**

Since the population is growing exponentially, we can apply the equation for exponential growth, $y = a(1 + r)^t$ to determine the common ratio. An increase of 2.5% means that the population is being multiplied by $1 + 0.025$, or 1.025, each year. So, $r = 1.025$. Since $a_1 = 685,242$ in 2011, the population in 2030 is represented by the twentieth term, a_{20}.

$a_n = a_1 r^{n-1}$	Formula for the nth term
$a_{20} = 685,242(1.025)^{20-1}$	$a_1 = 685,242, r = 1.025, n = 20$
$a_{20} = 685,242(1.025)^{19}$	Simplify.
$a_{20} = 1,095,462$	Use a calculator.

In 2030, the estimated population of North Dakota will be 1,095,462.

Check

BOUNCES A rubber bouncy ball is dropped from a height of 5 feet. Each time the ball bounces back to 85% of the height from which it fell. Determine the height of the ball after 6 bounces and after 10 bounces. Round to the nearest hundredth.

$a_6 = $ _____ ft

$a_{10} = $ _____ ft

🔎 **Go Online** You can complete an Extra Example online.

Practice

🅡 **Go Online** You can complete your homework online.

Examples 1 and 2

Determine whether each sequence is geometric. Explain.

1. 4, 1, 2, ...

2. 10, 20, 30, 40, ...

3. 4, 20, 100, ...

4. 212, 106, 53, ...

5. −10, −8, −6, −4, ...

6. 5, −10, 20, 40, ...

7. −96, −48, −24, −12, ...

8. 7, 13, 19, 25, ...

9. 3, 9, 81, 6561, ...

10. 108, 66, 141, 99, ...

11. $\frac{3}{8}, -\frac{1}{8}, -\frac{5}{8}, -\frac{9}{8}, ...$

12. $\frac{7}{3}$, 14, 84, 504, ...

Example 3

Find the next three terms in each geometric sequence.

13. 2, −10, 50, ...

14. 36, 12, 4, ...

15. 4, 12, 36, ...

16. 400, 100, 25, ...

17. −6, −42, −294, ...

18. 1024, −128, 16, ...

19. 2, 6, 18, ...

20. 2500, 500, 100, ...

21. $\frac{4}{5}, \frac{2}{5}, \frac{1}{5}, ...$

22. −4, 24, −144, ...

23. 72, 12, 2, ...

24. −3, −12, −48, ...

Example 4

Use an explicit formula to find the 10th term of each geometric sequence.

25. 1, 9, 81, 729, ...

26. 2, 8, 32, 128, ...

27. −9, 27, −81, 243, ...

28. 6, −24, 96, −384, ...

Example 5

29. MUSEUMS The table shows the annual visitors to a museum in millions. Write an equation for the projected number of visitors after n years.

Year	Visitors (millions)
1	4
2	6
3	9
4	$13\frac{1}{2}$
n	?

30. WORLD POPULATION The CIA estimates that the world population is growing at a rate of 1.167% each year. The world population in 2015 was about 7.3 billion.

   a. Write an equation for the world population after n years.

   b. Find the estimated world population in 2025.

31. DEPRECIATION Te'Andra has a computer system that she bought for $5000. Each year, the computer system loses one-fifth of its then-current value. How much money will the computer system be worth after 6 years?

Mixed Exercises

32. POPULATION The table shows the projected population of the United States through 2060. Does this table show an *arithmetic sequence*, a *geometric sequence*, or neither? Explain.

Year	Population
2020	334,503,000
2030	359,402,000
2040	380,219,000
2050	398,328,000
2060	416,795,000

Source: U.S. Census Bureau

33. SAVINGS ACCOUNTS A bank offers a savings account that earns 0.5% interest each month.

   a. Write an equation for the balance of the savings account after n months.

   b. Given an initial deposit of $500, what will the account balance be after 15 months?

34. Write an equation for the nth term of the geometric sequence 3, −24, 192, ... Then find the 9th term of this sequence.

35. Write an equation for the nth term of the geometric sequence $\frac{9}{16}, \frac{3}{8}, \frac{1}{4}, ...$. Then find the 7th term of this sequence.

36. Write an equation for the nth term of the geometric sequence 1000, 200, 40, ...Then find the 5th term of this sequence.

37. Write an equation for the nth term of the geometric sequence $-8, -2, -\frac{1}{2}, \ldots$. Find the 8th term of this sequence.

38. Write an equation for the nth term of the geometric sequence $32, 48, 72, \ldots$. Find the 6th term of this sequence.

39. USE A SOURCE Research the average annual salary for a 25-year-old and the average rate of increase in salary per year. Then write an equation for the nth year of employment. Find the 20th term of this sequence, and explain what it means.

40. STRUCTURE For each of the geometric sequences below, fill in the missing terms, write the corresponding exponential equation, and use the exponential equation to determine the 10th term in the sequence.

    a. $0.5, 6,$ _____; $f(x) =$ _____; 10th term: _____

    b. __, $10,$ __, $40,$ __; $g(x) =$ _____; 10th term: _____

41. REASONING Find the previous three terms of the geometric sequence $-192, -768, -3072, \ldots$.

42. STATE YOUR ASSUMPTION Consider two different geometric sequences. Each starts with the same constant. The common ratio producing subsequent terms in the first is positive and is the reciprocal of the common ratio producing subsequent terms in the second. How would the graphs of the two sequences compare? Think about intercepts, asymptotes, and end behavior. Then graph an example of the situation.

43. REASONING You have just been offered a part-time job. The employer offers two different methods of payment. They are shown in the table.

Month	Method 1 Payment	Method 2 Payment
1	$100.00	$0.01
2	$108.00	$0.02
3	$116.00	$0.04
4	$124.00	$0.08

    a. Describe the two different methods of payment being offered.

    b. What kind of mathematical equations can you use to model each situation? How do you know? Write each equation.

    c. You are planning to work at this job for two years. Your manager promises to raise your salary the way it is described in the table, as long as you meet the minimum performance rating each month. Which payment plan would you choose? Explain your reasoning.

44. CONSTRUCT ARGUMENTS The terms of a geometric sequence are defined by the equation $a_n = 512(0.5)^x$. A second sequence contains the terms $b_3 = 7168$ and $b_7 = 28$.

    a. Determine which sequence has the greater common ratio.

    b. What is the initial term of each sequence? Explain your reasoning.

45. REGULARITY The sum of the interior angles of a triangle is 180°. The interior angles of a pentagon add to 540°. Is the relationship between the number of sides in a polygon and the sum of interior angles a geometric sequence? Use the sum of the measures of the interior angles of a square to justify your answer.

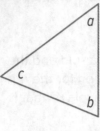 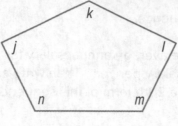

$$a + b + c = 180° \qquad j + k + l + m + n = 540°$$

46. PERSEVERE Write a sequence that is both geometric and arithmetic. Explain your answer.

47. FIND THE ERROR Haro and Matthew are finding the ninth term of the geometric sequence −5, 10, −20, … . Is either of them correct? Explain your reasoning.

Haro	Matthew
$r = \frac{10}{-5}$ or −2	$r = \frac{10}{-5}$ or −2
$a_9 = -5(-2)^{9-1}$	$a_9 = -5 \cdot (-2)^{9-1}$
$= -5(512)$	$= -5 \cdot -256$
$= -2560$	$= 1280$

48. ANALYZE Write a sequence of numbers that form a pattern but are neither arithmetic nor geometric. Justify your argument.

49. WRITE How are graphs of geometric sequences and exponential functions similar? How are they different?

50. WRITE Summarize how to find a specific term of a geometric sequence.

51. CREATE Give a counterexample for the following statement:
As n increases in a geometric sequence, the value of a_n will move farther away from zero.

52. CREATE Write a geometric sequence. Then explain why your sequence is geometric.

Recursive Formulas

Today's Goals
• Calculate terms in sequences by using recursive formulas.

• Write arithmetic and geometric sequences recursively and use them to model situations.

Today's Vocabulary
explicit formula
recursive formula

Learn Using Recursive Formulas

An **explicit formula** allows you to find any term of a sequence by using a formula written in terms of n. A **recursive formula** allows you to find the nth term of a sequence by performing operations to one or more of the preceding terms.

Example 1 Recursive Formula for an Arithmetic Sequence

Find the first five terms of the sequence $a_1 = 7$ and $a_n = a_{n-1} - 9$ if $n \geq 2$.

Use $a_1 = 7$ and the recursive formula to find the next four terms.

$a_2 = a_{2-1} - 9$ $n = 2$ $a_3 =$ _____ $n = 3$
$\quad =$ _____   Simplify. $\quad = a_2 - 9$ Simplify.
$\quad = 7 - 9$ $a_1 = 7$ $\quad =$ _____ $a_2 = -2$
$\quad =$ _____   Simplify. $\quad = -11$ Simplify.
$a_4 = a_{4-1} - 9$ $n = 4$ $a_5 =$ _____ $n = 5$
$\quad = a_3 - 9$ Simplify. $\quad =$ _____   Simplify.
$\quad =$ _____ $a_3 = -11$ $\quad = -20 - 9$ $a_4 = -20$
$\quad =$ ____   Simplify. $\quad = -29$ Simplify.

The first five terms of the sequence are ___, ___, ___, ___, and ___.

Example 2 Recursive Formula for a Geometric Sequence

Find the first five terms of the sequence $a_1 = 5$ and $a_n = 3a_{n-1}$ if $n \geq 2$.

n	$a_n = 3a_{n-1}$	a_n
1	—	5
2	$a_n = 3(5)$	
3	$a_n = 3(15)$	
4	$a_n = 3a_{4-1}$	
5	$a_n = 3a_{5-1}$	

The first five terms of the sequence are _____, _____, _____, _____, and _____.

Go Online You can complete an Extra Example online.

Study Tip

Recursive and Explicit Formulas Recursive formulas are used for generating sequences of numbers. They are not as useful for finding, for example, the fiftieth term of a sequence since you would first have to find terms one through forty-nine. For this type of calculation, it is better to use an explicit formula.

Explore Writing Recursive Formulas from Sequences

Online Activity Use an interactive tool to complete the Explore.

×

@ INQUIRY How can you write a formula that relates the numbers in a geometric sequence?

Learn Writing Recursive Formulas

Key Concept • Writing Recursive Formulas

Step 1 Determine whether the sequence is arithmetic or geometric by finding a common difference or a common ratio.

Step 2 Write a recursive formula.

Arithmetic Sequence

$a_n = a_{n-1} + d$, where d is the common difference

Geometric Sequence

$a_n = r \cdot a_{n-1}$, where r is the common ratio

Step 3 State the first term and domain for n.

Example 3 Write a Recursive Formula Using a List

Write a recursive formula for 16, 48, 144, 432, ...

Step 1 Determine whether a common difference or ratio exists.

Subtract each term from the term that follows it to check for a common difference.

$48 - 16 =$ _____ $144 - 48 =$ _____ $432 - 144 =$ _____

There is no common difference.

Check for a common ratio by dividing each term by the term that precedes it.

$\frac{48}{16} =$ _____ $\frac{144}{48} = 3$ $\frac{432}{144} =$ _____

The common ratio is _____. The sequence is _____.

Step 2 Write a recursive formula.

$a_n = r \cdot a_{n-1}$ Recursive formula for geometric sequence

$a_n =$ _____ $\cdot a_{n-1}$ $r = 3$

Step 3 State the first term and domain for n.

The first term a_1 is 16, and the domain of the function is $n \geq 1$.

A recursive formula for the sequence is $a_1 =$ _____,
$a_n =$ _____ $a_{n-1}, n \geq 2$.

Notice that n must be greater than or equal to 2 in the recursive formula.

Check

SOCIAL MEDIA The table shows the total number of views at the end of each day for a video.

Write a recursive formula for the sequence.

$a_1 =$ _____

$a_n = a_{n-1}$ _____

Day	Views
1	100
2	9000
3	17,900
4	26,800

Example 4 Write a Recursive Formula Using a Graph

Write a recursive formula for the graph.

Step 1 Find a common difference or common ratio, or determine that neither exists.

$84 - 109 =$ _____

$59 - 84 =$ _____

$34 - 59 =$ _____

The common difference is _____. The sequence is _____.

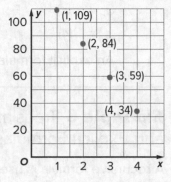

Step 2 Write a recursive formula.

$a_n = a_{n-1} + d$ Recursive formula for arithmetic sequence

$a_n = a_{n-1} +$ _____ $d = -25$

Step 3 State the first term and domain for n.

The first term a_1 is 109, and $n \geq 1$.

A recursive formula for the sequence is $a_1 =$ _____, $a_n = a_{n-1} -$ _____, $n \geq 2$.

Example 5 Write Recursive and Explicit Formulas

MOVIES **The premise of a movie is that a new virus is spreading, turning infected persons into zombie-like creatures. The table outlines the total number of infected persons at the end of each day.**

Day	Infected Persons
1	3
2	12
3	48
4	192
5	768

a. Write a recursive formula for the sequence.

Step 1 Find a common difference or common ratio.

$12 - 3 =$ _____ $48 - 12 =$ _____ $192 - 48 =$ _____

There is no common difference. Check for a common ratio by dividing each term by the term that precedes it.

$\frac{12}{3} =$ _____ $\frac{48}{12} =$ _____ $\frac{192}{48} =$ _____ $\frac{768}{192} =$ _____

There is a common ratio of 4. The sequence is geometric.

(continued on the next page)

Think About It!

How can you make sure that your recursive formula is correct?

Math History Minute

Hungarian mathematician **Rózsa Péter** (1905–1977) was the first Hungarian female mathematician to become an Academic Doctor of Mathematics. She helped to establish the modern field of recursive function theory, and she was the author of *Playing with Infinity: Mathematical Explorations and Excursions.*

Step 2 Write a recursive formula.

$$a_n = r \cdot a_{n-1} \qquad \text{Recursive formula for geometric sequence}$$
$$a_n = 4a_{n-1} \qquad r = 4$$

Step 3 State the first term and domain for n.

The first term a_1 is 3, and $n \geq 1$. A recursive formula for the sequence is $a_1 = 3$, $a_n = 4a_{n-1}$, $n \geq 2$.

b. Write an explicit formula for the sequence.

Steps 1 and 2 The common ratio is 4.

Step 3 Use the formula for the nth term of a geometric sequence.

$$a_n = a_1 r^{n-1} \qquad \text{Formula for the } n\text{th term}$$
$$= 3(4)^{n-1} \qquad a_1 = 3 \text{ and } r = 4$$

An explicit formula for the sequence is $a_n = 3(4)^{n-1}$.

Think About It!

For the sequence in part **b**, find a_2.

Study Tip

Geometric Sequences
Recall that the formula for the nth term of a geometric sequence is $a_n = a_1 r^{n-1}$.

Example 6 Translate Between Recursive and Explicit Formulas

A recursive formula is useful when finding a number of successive terms in a sequence. An explicit formula is useful when finding the nth term of a sequence. Therefore, it may be necessary to translate between the two forms.

a. Write a recursive formula for $a_n = 0.5n + 2$.

$a_n = 0.5n + 2$ is an explicit formula for an arithmetic sequence with $d = 0.5$ and $a_1 = 0.5(1) + 2$ or 2.5.

Therefore, a recursive formula for a_n is $a_1 = \underline{\quad}$, $a_n = a_{n-1} \underline{\quad}$, $n \geq 2$.

b. Write an explicit formula for $a_1 = 1011$, $a_n = 1.25a_{n-1}$, $n \geq 2$.

$a_n = 1.25a_{n-1}$ is a recursive formula for a $\underline{\quad\quad}$ sequence with $a_1 = 1011$ and $r = 1.25$.

Therefore, an explicit formula for a_n is $a_n = \underline{\quad} \cdot (\underline{\quad})^{n-1}$.

Check

Part A Write a recursive formula for $a_n = \frac{1}{2} + (n-1)10$.

$$a_1 = \underline{\quad}$$

$$a_n = a_{n-1} \underline{\quad}$$

Part B Write an explicit formula for $a_1 = -60$, $a_n = 1.5a_{n-1}$, $n \geq 2$.

$$a_n = \underline{\quad}(\underline{\quad})^{n-1}$$

 Go Online You can complete an Extra Example online.

Practice

🔍 **Go Online** You can complete your homework online.

Examples 1 and 2

Find the first five terms of each sequence.

1. $a_1 = 23$, $a_n = a_{n-1} + 7$, $n \geq 2$

2. $a_1 = 48$, $a_n = -0.5a_{n-1} + 8$, $n \geq 2$

3. $a_1 = 8$, $a_n = 2.5a_{n-1}$, $n \geq 2$

4. $a_1 = 12$, $a_n = 3a_{n-1} - 21$, $n \geq 2$

5. $a_1 = 13$, $a_n = -2a_{n-1} - 3$, $n \geq 2$

6. $a_1 = \frac{1}{2}$, $a_n = a_{n-1} + \frac{3}{2}$, $n \geq 2$

Example 3

Write a recursive formula for each sequence.

7. 12, −1, −14, −27, ...

8. 27, 41, 55, 69, ...

9. 2, 11, 20, 29, ...

10. 100, 80, 64, 51.2, ...

11. 40, −60, 90, −135, ...

12. 81, 27, 9, 3, ...

Example 4

Write a recursive formula for each graph.

13.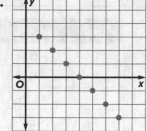

14.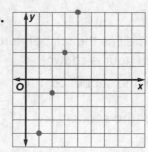

15.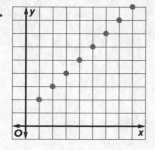

16.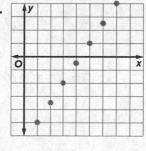

Write a recursive formula for each graph.

17.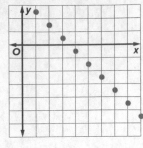

18.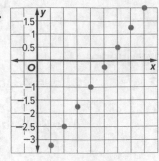

Example 5

19. VIRAL VIDEOS A viral video got 175 views in one hour, 350 views in two hours, 525 views in three hours, 700 views in four hours, and so on.

   a. Find the next 5 terms in the sequence.

   b. Write a recursive formula for the sequence.

   c. Write an explicit formula for the sequence.

20. PAPER A piece of paper is folded several times. The number of sections into which the piece of paper is divided after each fold is shown.

   a. Write a recursive formula for the sequence.

   b. Write an explicit formula for the sequence.

Number of Folds	Sections
1	2
2	4
3	8
4	16
5	32

21. SNOW A snowman begins to melt as the temperature rises. The height of the snowman in feet after each hour is shown.

   a. Write a recursive formula for the sequence.

   b. Write an explicit formula for the sequence.

Hour	Height (ft)
1	6.0
2	5.4
3	4.86
4	4.374

Example 6

For each recursive formula, write an explicit formula. For each explicit formula, write a recursive formula.

22. $a_n = 3(4)^{n-1}$

23. $a_1 = -2, a_n = a_{n-1} - 12, n \geq 2$

24. $a_1 = 38, a_n = \frac{1}{2}a_{n-1}, n \geq 2$

25. $a_n = -7n + 52$

26. $a_1 = 38, a_n = a_{n-1} - 17, n \geq 2$

27. $a_n = 5n - 16$

28. $a_n = 50(0.75)^{n-1}$

29. $a_1 = 16, a_n = 4a_{n-1}, n \geq 2$

Mixed Exercises

30. CLEANING An equation for the cost a_n in dollars that a carpet cleaning company charges for cleaning n rooms is $a_n = 50 + 25(n - 1)$. Write a recursive formula to represent the cost a_n.

31. SAVINGS A recursive formula for the balance of a savings account a_n in dollars at the beginning of year n is $a_1 = 500$, $a_n = 1.05a_{n-1}$, $n \geq 2$. Write an explicit formula to represent the balance of the savings account a_n.

32. USE TOOLS In 2010, County A had a population of 1.3 million people. The largest factory in the area produced 1700 million widgets per year. The population of County A is projected to grow at 1.2% per year, and the number of widgets produced is expected to grow by 10 million per year.

   a. Develop explicit formulas for the population and annual widget production, in millions, as functions of the number of years n after 2010.

   b. The graph of $y = \dfrac{1700 + 10x}{1.3(1.012)^x}$ represents the annual widget production per person for County A from 2010 to 2020, where x is the number of years after 2010. The next-highest widget-producing county produces widgets at a constant rate of 1200 widgets per person. Use a graphing calculator to extend the graph and find the year when County A will no longer be the leader in widget production. Explain your results.

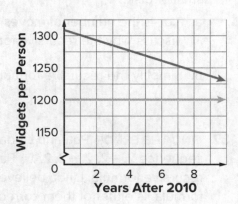

33. USE A MODEL Ramon has been tracing his family tree with his parents. He claims that he has over 250 great- great- great- great- great- great-grandparents. Is this possible? Write both an explicit and recursive formula for this situation.

34. REASONING Carl Friedrich Gauss, a German mathematician of the 1700s, was asked as a young boy for the sum of the integers from 1 to 100, and he unhesitatingly replied with the correct answer.

   a. Identify the type of the sequence 1, 2, 3, ... 100, and explore a way to find its sum based on grouping pairs of numbers from each end of the sequence. Then explain how Gauss was able to find the sum so quickly.

   b. Find an explicit formula for the sum S of n terms of an arithmetic sequence in which n is an even number, the first term is a_1 and the nth, or last term is a_n.

35. REGULARITY The first ten numbers in the Fibonacci sequence can be defined by $a_{n+1} = a_n + a_{n-1}$, and each ratio $\frac{a_n}{a_{n-1}}$ can be computed using a spreadsheet (see column C).

Fibonacci sequence

	A	B	C
1	1	1	
2	2	1	1
3	3	2	2
4	4	3	1.5
5	5	5	1.666667
6	6	8	1.6
7	7	13	1.625
8	8	21	1.615385
9	9	34	1.619048
10	10	55	1.617647

Sheet 1   Sheet 2   Sheet 3

  a. Which spreadsheet formulas could have been used to calculate the entries in cells B3 and C2?

  b. Compute the ratio $\frac{a_n}{a_{n-1}}$ up to $n = 50$. What do you observe?

36. STRUCTURE There is a famous puzzle called the "Tower of Hanoi." There are three pegs, and a certain number of disks of varying sizes can be set on each peg. The puzzle starts with the disks in a stack on the left-most peg, with the largest disk on the bottom and the disks getting smaller as they are stacked. The goal is to move the disks from the left-most peg to the right-most peg while obeying three rules. First, only one disk can be moved at a time. Second, only the top disk on any peg can be moved. Third, at no time can a larger disk be placed on a smaller disk.

  a. If a_n is the number of moves it takes to solve a puzzle consisting of n disks, discuss why the recursive formula $a_n = a_{n-1} + 1 + a_{n-1}$ makes sense.

  b. Simplify the recursive formula. What is a_1? Why?

37. FIND THE ERROR Pati and Linda are working on a math problem that involves the sequence 2, −2, 2, −2, 2, ... Pati thinks that the sequence can be written as a recursive formula. Linda believes that the sequence can be written as an explicit formula. Is either of them correct? Explain your reasoning.

38. PERSEVERE Find a_1 for the sequence in which $a_4 = 1104$ and $a_n = 4a_{n-1} + 16$.

39. ANALYZE Determine whether the following statement is *true* or *false*. Justify your argument. *There is only one recursive formula for every sequence.*

40. PERSEVERE Find a recursive formula for 4, 9, 19, 39, 79, ...

41. WRITE Explain the difference between an explicit formula and a recursive formula.

42. CREATE Give a counterexample for the following statement: In a recursive sequence, if $a_1 = a_2$, then $a_2 = a_3$, and so on.

@ Essential Question

When and how can exponential functions represent real-world situations?

Module Summary

Lesson 9-1

Exponential Functions

- Functions of the form $y = ab^x$, where $a \neq 0$ and $b > 1$, are exponential growth functions.
- Functions of the form $y = ab^x$, where $a \neq 0$ and $0 < b < 1$, are exponential decay functions.
- The graphs of exponential functions have an asymptote.

Lessons 9-2 through 9-4

Transforming and Writing Exponential Functions

- The graph $f(x) = b^x$ is a parent graph of an exponential function.
- The graph of $g(x) = b^x + k$ is the graph of $f(x) = b^x$ translated vertically.
- The graph of $g(x) = b^{x-h}$ is the graph of $f(x) = b^x$ translated horizontally.
- The graph $g(x) = ab^x$ is the graph of $f(x) = b^x$ stretched or compressed vertically by a factor of $|a|$.
- The graph $g(x) = b^{ax}$ is the graph of $f(x) = b^x$ stretched or compressed horizontally by a factor of $\frac{1}{|a|}$.
- When an exponential function $f(x)$ is multiplied by -1, the result is a reflection across the x- or y-axis.
- In the equation $y = a(1 + r)^t$, y is the final amount, a is the initial amount, r is the rate of change expressed as a decimal, and t is time.

Lesson 9-5

Geometric Sequences

- A geometric sequence is a pattern of numbers that begins with a nonzero term and each term after is found by multiplying the previous term by a nonzero constant r.
- The nth term a_n of a geometric sequence with first term a_1 and common ratio r is given by the formula $a_n = a_1 r^{n-1}$, where n is any positive integer, $a_1 \neq 0$, and $r \neq 0$.

Lesson 9-6

Recursive Functions

- An explicit formula allows you to find any term a_n of a sequence by using a formula written in terms of n.
- To write a recursive formula for an arithmetic or geometric sequence, determine whether the sequence is arithmetic or geometric by finding a common difference or a common ratio.

Study Organizer

📖 Foldables

Use your Foldable to review this module. Working with a partner can be helpful. Ask for clarification of concepts as needed.

Test Practice

1. GRAPH The table shows the function $y = 2^x - 1$. (Lesson 9-1)

x	y
0	0
1	1
2	3
3	7

Graph the function.

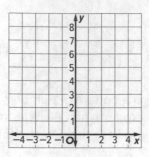

2. MULTIPLE CHOICE The table shows the number of text messages Ernesto sent each month. (Lesson 9-1)

Month	Text Messages
April	2
May	6
June	18
July	54

What type of behavior is shown in the table?

Ⓐ linear

Ⓑ piece-wise

Ⓒ exponential

Ⓓ none of the above

3. OPEN RESPONSE Describe the end behavior of the graph of the exponential function shown on the graph. (Lesson 9-1)

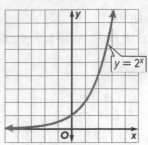

4. MULTIPLE CHOICE Consider the graph. Which function represents the reflection of the parent function $f(x) = 3^x$ across the y-axis? (Lesson 9-2)

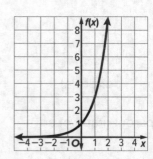

Ⓐ $f(x) = -3^x$

Ⓑ $f(x) = 3^{-x}$

Ⓒ $f(x) = 3^{-2x}$

Ⓓ $f(x) = -2(3)^x$

5. **MULTIPLE CHOICE** Describe the translation in $h(x) = 2^x + 5$ as it relates to the parent function $h(x) = 2^x$. (Lesson 9-2)

Ⓐ Up 5 units

Ⓑ Down 5 units

Ⓒ Right 5 units

Ⓓ Left 5 units

6. **OPEN RESPONSE** Horticulturists can estimate the number of hybrid plants of a certain type they will sell based on the parent function $b(x) = 2.5^x$. Suppose a new facility starts with 4 of these plants to hybridize, which can be modeled with the function $b(x) = 4(2.5)^x$. Describe the effect on the graph as it relates to the parent function. (Lesson 9-2)

7. **MULTIPLE CHOICE** Which exponential function models the graph? (Lesson 9-3)

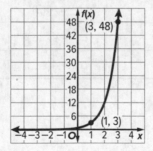

Ⓐ $y = \frac{3}{4}(4)^x$

Ⓑ $y = 4\left(\frac{3}{4}\right)^x$

Ⓒ $y = \frac{3}{4}(x)^4$

Ⓓ $y = 4(x)^x$

8. **OPEN RESPONSE** A population, $f(x)$, after x years may be modeled with $f(x) = 2(3)^x$. What is the initial amount, growth rate, domain and range? (Lesson 9-3)

Use the table below for Exercises 9–11.
Joey wants to invest money in a savings account. The table compares two banks he is considering. Joey needs to decide which is the better deal for investing his money.

	Interest Rate	Compound Frequency
First & Loan	0.6%	monthly
Local Credit Union	9%	annually

9. **MULTIPLE CHOICE** What is the effective monthly interest rate offered by Local Credit Union? (Lesson 9-4)

Ⓐ 5.5%

Ⓑ 2.2%

Ⓒ 0.75%

Ⓓ 0.72%

10. **MULTIPLE CHOICE** What is the effective annual interest rate offered by First & Loan? (Lesson 9-4)

Ⓐ 7.4%

Ⓑ 7.2%

Ⓒ 1.006%

Ⓓ 0.6%

11. **OPEN RESPONSE** Which bank gives Joey the better savings plan? Justify your answer. (Lesson 9-4)

Name _____ Period _____ Date _____

12. MULTIPLE CHOICE Whitney invests $3000 in an account earning 4.5% interest that is compounded annually. How much money will be in Whitney's account after 10 years? (Lesson 9-3)

Ⓐ $1893.02

Ⓑ $4658.91

Ⓒ $4700.98

Ⓓ $123,254.07

13. OPEN RESPONSE Attendance for local baseball games has been increasing by an average of 10% per year for the last few years. In 2018, the average attendance was 100 people.

Predict the average number of people attending local baseball games in 2022 if this trend continues. Round to the nearest whole number. (Lesson 9-5)

14. MULTIPLE CHOICE What equation can be written for the nth term of this geometric sequence? (Lesson 9-5)

n	1	2	3	4
a_n	100	−50	25	−12.5

Ⓐ $a_n = 100(2)^{n-1}$

Ⓑ $a_n = 100(-2)^{n-1}$

Ⓒ $a_n = 100\left(-\frac{1}{2}\right)^{n-1}$

Ⓓ $a_n = 100\left(\frac{1}{2}\right)^{n-1}$

15. OPEN RESPONSE The table shows the number of pages Aaron read in his book each day. Write a recursive formula for the sequence. (Lesson 9-6)

Day	1	2	3	4
Pages Read	20	35	50	65

16. OPEN RESPONSE What are the first five terms of the sequence for $a_1 = -2$ and $a_n = 2a_{n-1} + 5$ if $n \geq 2$. (Lesson 9-6)

17. TABLE ITEM Complete the table for the geometric sequence. (Lesson 9-6)
$a_1 = 3$ and $a_n = 4a_{n-1}$, if $n \geq 2$

n	formula	a_n
1	—	3
2	$a_n = 4(3)$	—
3	$a_n = 4(\_\_)$	—
4	$a_n = 4(\_\_)$	—

Polynomials

e Essential Question

How can you perform operations on polynomials and use them to represent real-world situations?

What Will You Learn?

Place a check mark (✓) in each row that corresponds with how much you already know about each topic **before** starting this module.

KEY

👎 — I don't know.   👍(side) — I've heard of it.   👍 — I know it!

	Before			After		
	👎	👍	👍	👎	👍	👍
write polynomials in standard form						
add polynomials						
subtract polynomials						
multiply polynomials by a monomial						
solve equations with polynomial expressions						
multiply binomials						
multiply polynomials						
factor polynomials using the Distributive Property						
factor quadratic trinomials by grouping						
factor polynomials that are the result of special products						

📖 **Foldables** Make this Foldable to help you organize your notes about polynomials. Begin with four sheets of grid paper.

1. **Fold** in half along the width. On the first two sheets, cut 5 centimeters along the fold at the ends. On the second two sheets cut in the center, stopping 5 centimeters from the ends.

2. **Insert** the first sheets through the second sheets and align the folds. Label the front Module 10, Polynomials. Label the pages with lesson numbers and the last page vocabulary.

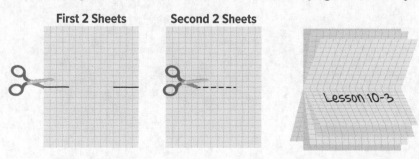

First 2 Sheets   Second 2 Sheets

Lesson 10-3

What Vocabulary Will You Learn?

Check the box next to each vocabulary term that you may already know.

☐ binomial

☐ degree of a monomial

☐ degree of a polynomial

☐ difference of two squares

☐ factoring

☐ factoring by grouping

☐ leading coefficient

☐ perfect square trinomials

☐ polynomial

☐ prime polynomial

☐ quadratic expression

☐ standard form of a polynomial

☐ trinomial

Are You Ready?

Complete the Quick Review to see if you are ready to start this module.
Then complete the Quick Check.

Quick Review

Example 1

Rewrite $6x(-3x - 5x - 5x^2 + x^3)$ using the Distributive Property. Then simplify.

$6x(-3x - 5x - 5x^2 + x^3)$

$\quad = 6x(-3x) + 6x(-5x) + 6x(-5x^2) + 6x(x^3)$

$\quad = -18x^2 + (-30x^2) - 30x^3 + 6x^4$

$\quad = -48x^2 - 30x^3 + 6x^4$

Example 2

Simplify $8c + 6 - 4c + 2c^2$.

$8c + 6 - 4c + 2c^2$	Original expression
$= 2c^2 + 8c - 4c + 6$	Rewrite in descending order.
$= 2c^2 + (8 - 4)c + 6$	Use the Distributive Property.
$= 2c^2 + 4c + 6$	Combine like terms.

Quick Check

Rewrite each expression using the Distributive Property. Then simplify.

1. $a(a + 5)$

2. $2(3 + x)$

3. $n(n - 3n^2 + 2)$

4. $-6(x^2 - 5x + 4)$

Simplify each expression. If not possible, write *simplified*.

5. $3x + 10x$

6. $4w^2 + w + 15w^2$

7. $6m^2 - 8m$

8. $2x^2 + 5 + 11x^2 + 7$

How did you do?

Which exercises did you answer correctly in the Quick Check? Shade those exercise numbers below.

Adding and Subtracting Polynomials

Learn Types of Polynomials

A **polynomial** is a monomial or the sum of two or more monomials. Some polynomials have special names.

- A monomial is a number, a variable, or a product of a number and one or more variables.
- A **binomial** is the sum of *two* monomials.
- A **trinomial** is the sum of *three* monomials.

The **degree of a monomial** is the sum of the exponents of all its variables. A nonzero constant term has degree 0, and zero has no degree.

The **degree of a polynomial** is the greatest degree of any term in the polynomial. You can find the degree of a polynomial by finding the degree of each term. Polynomials are named by their degree.

Degree	Name
0	constant
1	linear
2	quadratic
3	cubic
4	quartic
5	quintic
6 or more	6th degree, 7th degree, . . .

Addition is commutative, and therefore the terms of a polynomial can be written in any order. However, the **standard form of a polynomial** has the terms written in order from greatest degree to least degree. When a polynomial is in standard form, the coefficient of the first term is called the **leading coefficient**.

Example 1 Identify Polynomials

Determine whether each expression is a polynomial. If it is a polynomial, find the degree and determine whether it is a monomial, binomial, or trinomial.

a. **8ab − 2c**

8ab − 2c is the sum of two monomials, 8ab and −2c, so this __ a polynomial. Degree: __; binomial.

Today's Goals
- Identify and write polynomials by using the standard form.
- Add polynomials.
- Subtract polynomials.

Today's Vocabulary
polynomial
binomial
trinomial
degree of a monomial
degree of a polynomial
standard form of a polynomial
leading coefficient

 Think About It!

Is $4x - 2x^2 + 9$ written in standard form? Justify your argument.

 Talk About It!

Explain why 8ab − 2c is a 2nd degree polynomial and not a 1st degree polynomial.

b. −11.25

−11.25 is a real number, so this ___ a polynomial. Degree: ___;
_____.

c. $2x^{-2} + 3xy$

$2x^{-2} = \frac{2}{x^2}$, which _____ a monomial, so this _____ a polynomial.

d. $9x^3 - 8x + 5x - 27$

The simplified form is $9x^3 - 3x - 27$, which is the sum of _____ monomials, so this ___ a polynomial. Degree: ___; _____.

e. $2m^2 + 2mn - n^2$

$2m^2 + 2mn - n^2$ is the sum of _____ monomials, so this ___ a polynomial. Degree: ___; _____.

Check

Determine whether each expression is a polynomial. If it is a polynomial, find the degree and determine whether it is a *monomial*, *binomial*, or *trinomial*.

Expression	Is it a polynomial ?	Degree	Classification
a. $3z^{-2}$			
b. $2x^3 + x - 12$			
c. $9b$			
d. $9x - 2$			

Example 2 Standard Form of a Polynomial

Write $4x + 12 + 2x^3 - 3x^2$ in standard form. Identify the leading coefficient.

To write the polynomial in standard form, rewrite the terms in order from greatest degree, 3, to least degree, 0. The polynomial can be rewritten as _____ with a leading coefficient of ___.

Check

Part A Write $5b - 10b^2 + 35 - b^3$ in standard form.

_____

Part B Identify the leading coefficient in $5b - 10b^2 + 35 - b^3$.

The leading coefficient of the polynomial is ___.

🫧 **Think About It!**

In standard form, why is the constant term at the end of the polynomial rather than the beginning?

🐺 **Go Online** You can complete an Extra Example online.

Explore Using Algebra Tiles to Add and Subtract Polynomials

🔊 **Online Activity** Use algebra tiles to complete the Explore.

@ **INQUIRY** How are the processes for adding and subtracting polynomials similar?

×

Learn Adding Polynomials

Adding polynomials involves adding like terms. When adding polynomials, you can group like terms by using a horizontal or vertical format.

Method 1 Horizontal Method

Group and combine like terms.

$(3x^2 + 9x + 27) + (2x^2 + 4x - 12)$

$= [3x^2 + 2x^2] + [9x + 4x] + [27 + (-12)]$ Group like terms.

$= 5x^2 + 13x + 15$ Combine like terms.

Method 2 Vertical Method

Align like terms in columns and combine.

$3x^2 + 9x + 27$

$\underline{(+) \quad 2x^2 + 4x - 12}$ Align like terms.

$5x^2 + 13x + 15$ Combine like terms.

Example 3 Add Polynomials

Find each sum.

a. $(3x^2 - 4) + (x^2 - 9)$

$(3x^2 - 4) + (x^2 - 9) = [3x^2 + x^2] + [-4 + (-9)]$ Group like terms.

$= 4x^2 - 13$ Combine like terms.

b. $(8 - x^2) + (4x + 2x^2 - 9)$

$-x^2 + 8 \qquad \rightarrow \qquad -x^2 + 0x + 8$ Insert a placeholder to align the terms.

$4x + 2x^2 - 9 \quad \rightarrow \quad \underline{(+) \quad 2x^2 + 4x - 9}$ Align and combine like terms.

🔊 **Go Online** You can complete an Extra Example online.

💭 **Think About It!**

How can writing polynomials in standard form be helpful when adding?

Study Tip

Placeholders When adding polynomials, it may be necessary to insert a placeholder to help align the terms. For example, if one of the polynomials does not have an x^2 term, add $0x^2$ to keep the terms aligned.

Check

Find each sum. Write your answer in standard form.

$(12y + 20y^2 - 2) + (-13y^2 + y - 10)$ _____

$(-4b - b^2 + 2) + 2(b^2 + 2b - 1)$ _____

$(-f + 5f^2 + 5) + (3f^3 - f + f^2)$ _____

Learn Subtracting Polynomials

You can subtract a polynomial by adding its additive inverse. To find the additive inverse of a polynomial, write the opposite of each term.

Select a method to find $(11x - 13 - 7x^3 - 8x^2) - (2x + 8x^2 + 20)$.

Method 1 Horizontal method

Subtract $2x + 8x^2 + 20$ by adding its additive inverse.

$(11x - 13 - 7x^3 - 8x^2) - (2x + 8x^2 + 20)$

$\quad = (11x - 13 - 7x^3 - 8x^2) + (-2x - 8x^2 - 20)$ The additive inverse of $2x + 8x^2 + 20$ is $-2x - 8x^2 - 20$.

$\quad = -7x^3 + [-8x^2 + (-8x^2)] + [11x + (-2x)] + [-13 + (-20)]$

$\quad = -7x^3 - 16x^2 + 9x - 33$

Method 2 Vertical method

Align like terms in columns and subtract by adding the additive inverse.

$$
\begin{array}{r}
-7x^3 - 8x^2 + 11x - 13 \\
(-)\ 0x^3 + 8x^2 + 2x + 20 \\
\hline
\end{array}
\quad \rightarrow \quad
\begin{array}{r}
-7x^3 - 8x^2 + 11x - 13 \\
(+)\ -0x^3 - 8x^2 - 2x - 20 \\
\hline
-7x^3 - 16x^2 + 9x - 33
\end{array}
$$

Adding or subtracting integers results in an integer, so the set of integers is closed under addition and subtraction. Similarly, when you add or subtract polynomials, you are combining like terms. This results in a polynomial with the same variables and exponents as the original polynomials, but possibly different coefficients. Thus, the sum or difference of two polynomials is always a polynomial, and the set of polynomials is closed under addition and subtraction.

Example 4 Subtract Polynomials Horizontally

Find $(6x - 11) - (2x - 19)$.

Subtract $(2x - 19)$ by adding its additive inverse.

$(6x - 11) - (2x - 19) = (6x - 11) + (-2x + 19)$

$\qquad\qquad\qquad\qquad = [6x + (-2x)] + [-11 + 19]$ Group like terms.

$\qquad\qquad\qquad\qquad = \underline{\quad} + \underline{\quad}$ Combine like terms.

 Go Online You can complete an Extra Example online.

💭 **Think About It!**

In the example, why is the term $0x^3$ introduced when subtracting the polynomials?

Example 5 Subtract Polynomials Vertically

Find $(x + 2) - (7x - 3x^2 + 14)$.

Align like terms in columns and subtract by adding the additive inverse.

$$0x^2 + x + 2 \qquad \rightarrow \qquad 0x^2 + x + 2$$
$$(-) \; -3x^2 + 7x + 14 \qquad\qquad (+)3x^2 - 7x - 14$$

Check

Find $(z^2 + 2z - 5) - (9z - 3z^2)$. Write your answer in standard form.

_____

Find $(8r - 14 + 7r^2) - (-16r^2 - 7r - 3)$. Write your answer in standard form.

_____

Find $(h - 2h - h^2) - (5h^2 - 2 + 8h)$. Write your answer in standard form.

_____

 ## Example 6 Add and Subtract Polynomials

ALBUM SALES Today's recording artists can sell hard copies H and digital copies D of their albums. The equations $H = 9w + 53$ and $D = 13w + 126$ represent the number of albums (in thousands) one artist sold in w weeks. Write an equation that shows how many more digital albums were sold than hard copies S. Then predict how many more digital albums are sold than hard copies in 52 weeks.

To write an equation that represents how many more digital albums were sold than hard copies S, subtract the equation for the number of hard copies H sold from the equation for the number of digital albums D sold.

$$S = (13w + 126) - (9w + 53)$$

$$= \underline{}w + \underline{}$$

Substitute 52 for w to predict how many more digital albums are sold than hard copies in 52 weeks.

There will be _____ more digital albums sold than hard copies in 52 weeks.

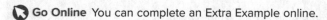 **Go Online** You can complete an Extra Example online.

 Think About It!
What is the first step for finding the difference of polynomials?

_____
_____
_____
_____
_____
_____
_____

Study Tip
Units Pay attention to the language in the question. The equations represent the number of hard copy and digital albums sold in the thousands, so your answer should represent album sales in the thousands.

Think About It!
What assumption did you make about the trend of sales over the 52 weeks? Can you determine the sales of hard copy and digital albums for a specific week? Explain.

_____
_____
_____
_____

Check

COLLEGE LIVING The total number of students T who attend a college consists of two groups: students who live in dorm rooms on campus D and students who live in apartments off campus A. The number (in hundreds) of students who live in dorm rooms and the total number of students enrolled in the college can be modeled by the following equations, where n is the number of years since 2001.

$$T = 17n + 23$$
$$D = 11n + 8$$

Part A Write an equation that models the number of students who live in apartments.

_____

Part B Predict the number of students who will live in apartments in 2020.

_____

Part C What do you need to assume in order to predict the number of students who will live on campus in 2020? ____

A. The total number of students does not include students who commute.

B. Students do not share dorm rooms.

C. The number of students enrolled in the college remains the same.

D. Many students live at home during the summer.

Pause and Reflect

Did you struggle with anything in this lesson? If so, how did you deal with it?

Record your observations here.

🔊 **Go Online** You can complete an Extra Example online.

Name _____ Period _____ Date _____

Practice

🅡 **Go Online** You can complete your homework online.

Example 1

Determine whether each expression is a polynomial. If it is a polynomial, find the
degree and determine whether it is a *monomial*, *binomial*, or *trinomial*.

1. $\frac{5y^3}{x^2} + 4x$

2. 21

3. $c^4 - 2c^2 + 1$

4. $d + 3d^c$

5. $a - a^2$

6. $5n^3 + nq^3$

Example 2

Write each polynomial in standard form. Identify the leading coefficient.

7. $5x^2 - 2 + 3x$

8. $8y + 7y^3$

9. $4 - 3c - 5c^2$

10. $-y^3 + 3y - 3y^2 + 2$

11. $11t + 2t^2 - 3 + t^5$

12. $2 + r - r^3$

13. $\frac{1}{2}x - 3x^4 + 7$

14. $-9b^2 + 10b - b^6$

Examples 3–5

Find each sum or difference.

15. $(2x + 3y) + (4x + 9y)$

16. $(6s + 5t) + (4t + 8s)$

17. $(5a + 9b) - (2a + 4b)$

18. $(11m - 7n) - (2m + 6n)$

19. $(m^2 - m) + (2m + m^2)$

20. $(x^2 - 3x) - (2x^2 + 5x)$

21. $(d^2 - d + 5) - (2d + 5)$

22. $(2h^2 - 5h) + (7h - 3h^2)$

23. $(5f + g - 2) + (-2f + 3)$

24. $(6k^2 + 2k + 9) + (4k^2 - 5k)$

25. $(2c^2 + 6c + 4) + (5c^2 - 7)$

26. $(2x + 3x^2) - (7 - 8x^2)$

Lesson 10-1 • Adding and Subtracting Polynomials **551**

Copyright © McGraw-Hill Education

Find each sum or difference.

27. $(3c^3 - c + 11) - (c^2 + 2c + 8)$

28. $(z^2 + z) + (z^2 - 11)$

29. $(2x - 2y + 1) - (3y + 4x)$

30. $(4a - 5b^2 + 3) + (6 - 2a + 3b^2)$

31. $(x^2y - 3x^2 + y) + (3y - 2x^2y)$

32. $(-8xy + 3x^2 - 5y) + (4x^2 - 2y + 6xy)$

33. $(5n - 2p^2 + 2np) - (4p^2 + 4n)$

34. $(4rxt - 8r^2x + x^2) - (6rx^2 + 5rxt - 2x^2)$

Example 6

35. PROFIT Company A and Company B both started their businesses in the same year. The profit P, in millions, of Company A is given by the equation $P = 3.2x + 12$, where x is the number of years in business. The profit P, in millions, of Company B is given by the equation $P = 2.7x + 10$, where x is the number of years in business.

  a. Write a polynomial equation to give the difference in profit D after x years.

  b. Predict the difference in profit after 10 years.

36. ENVELOPES An office supply company produces yellow document envelopes. The envelopes come in a variety of sizes, but the length is always 4 centimeters more than double the width, x.

  a. Write a polynomial equation to give the perimeter P of any of the envelopes.

  b. Predict the perimeter of an envelope with a width of 6 centimeters.

Mixed Exercises

Classify each polynomial according to its degree and number of terms.

37. $4x - 3x^2 + 5$

38. $11z^3$

39. $9 + y^4$

40. $3x^3 - 7x$

41. $-2x^5 - x^2 + 5x - 8$

42. $10t - 4t^2 + 6t^3$

Find each sum or difference.

43. $(4x + 2y - 6z) + (5y - 2z + 7x) + (-9z - 2x - 3y)$

44. $(5a^2 - 4) + (a^2 - 2a + 12) + (4a^2 - 6a + 8)$

45. $(3c^2 - 7) + (4c + 7) - (c^2 + 5c - 8)$

46. ROCKETS Two toy rockets are launched straight up into the air. The height, in feet, of each rocket at t seconds after launch is given by the polynomial equations shown. Write an equation to find the difference in height of Rocket A and Rocket B. Predict the difference in height after 5 seconds.

Rocket A: $D_1 = -16t^2 + 122t$

Rocket B: $D_2 = -16t^2 + 84t$

47. INDUSTRY Two identical right cylindrical steel drums containing oil need to be covered with a fire-resistant sealant. In order to determine how much sealant to purchase, George must find the surface area of the two drums. The surface area, including the top and bottom bases, is given by the formula $S = 2\pi rh + 2\pi r^2$.

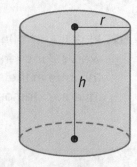

  a. Write a polynomial to represent the total surface area of the two drums.

  b. Find the total surface area, to the nearest tenth of a square meter, if the height of each drum is 2 meters and the radius of each is 0.5 meter.

48. MANUFACTURING A company delivers their product in cubic boxes that have volume x^3. When the company begins to manufacture a second product, manufacturing designs a new shipping box that is 3 inches longer in one dimension and 1 inch shorter in another dimension. The volume of the new box is $x^3 + 2x^2 - 3x$.

  a. Write an expression to represent the total volume of 4 of each kind of box.

  b. Find an expression that shows the difference in volume between the two boxes.

49. FIND THE ERROR Claudio says that polynomials are not closed under addition and gives this counterexample, which he says is not a polynomial: $(x^2 - 2x) + (-x^2 + 2x) = 0$. Is Claudio correct? Explain your reasoning.

50. VOLUME The volume of a sphere with radius x is $\frac{4}{3}\pi x^3$ units^3, and the volume of a cube with side length x is x^3 units^3.

  a. Find the total combined volume of the sphere and cube.

  b. How much more volume does the sphere contain?

51. STRUCTURE Compute the following differences.

  a. $(5x^2 - 3x + 7) - (18x^2 - 2x - 3)$

  b. $(18x^2 - 2x - 3) - (5x^2 - 3x + 7)$

  c. What do you notice about **part a** compared to **part b**? How does this relate to the structure of the integers?

52. REGULARITY In the set of integers, every integer has an additive inverse. In other words, for every integer n, there is another integer $-n$ such that $n + (-n) = 0$. Is this true in the set of polynomials? Does every polynomial have an additive inverse? Demonstrate with an example.

53. FIND THE ERROR Cheyenne and Nicolas are finding $(2x^2 - x) - (3x + 3x^2 - 2)$. Is either correct? Explain your reasoning.

Cheyenne	Nicolas
$(2x^2 - x) - (3x + 3x^2 - 2)$	$(2x^2 - x) - (3x + 3x^2 - 2)$
$= (2x^2 - x) + (-3x + 3x^2 - 2)$	$= (2x^2 - x) + (-3x - 3x^2 - 2)$
$= 5x^2 - 4x - 2$	$= -x^2 - 4x - 2$

54. ANALYZE Determine whether each of the following statements is *true* or *false*. Justify your argument.

  a. A binomial can have a degree of zero.

  b. The order in which polynomials are subtracted does not matter.

55. PERSEVERE Write a polynomial that represents the sum of $2n + 1$ and the next two consecutive odd integers.

56. WRITE Why would you add or subtract equations that represent real-world situations? Explain.

57. WRITE Describe how to add and subtract polynomials using both the vertical and horizontal methods.

58. CREATE Write two polynomials that can be added to have a sum of $-11x^3 - x^2 + 5x + 6$.

Multiplying Polynomials by Monomials

Explore Using Algebra Tiles to Find Products of Polynomials and Monomials

 Online Activity Use algebra tiles to complete the Explore.

> ⊙ **INQUIRY** How can you use the Distributive Property to find the product of a polynomial and a monomial?

Learn Multiplying a Polynomial by a Monomial

To find the product of a monomial and a binomial, you can use the Distributive Property. You can also use the Distributive Property to find the product of a monomial and a longer polynomial. When polynomials are multiplied, the product is also a polynomial. Therefore, the set of polynomials is closed under multiplication. This is similar to the system of integers, which is also closed under multiplication.

Example 1 Multiply a Polynomial by a Monomial

Simplify $-2x(4x^2 + 3x - 5)$.

$-2x(4x^2 + 3x - 5)$ Original expression

$= -2x(4x^2) + (-2x)(3x) - (-2x)(5)$ Distributive Property

$= -8x^3 + (-6x^2) - (-10x)$ Multiply.

$= \underline{\hspace{3cm}}$ Simplify.

Go Online
An alternate method is available for this example.

Example 2 Simplify Expressions

Simplify $3n(6n^3 - 4n) - 2(9n^4 - 11)$.

$3n(6n^3 - 4n) - 2(9n^4 - 11)$

$= 3n(6n^3) - 3n(4n) + (-2)(9n^4) + (-2)(-11)$ Distributive Property

$= 18n^4 - 12n^2 - 18n^4 + 22$ Multiply.

$= (18n^4 - 18n^4) - 12n^2 + 22$ Commutative and Associative properties

$= \underline{\hspace{3cm}}$ Combine like terms.

Watch Out!

Negatives If the monomial has a negative coefficient, remember to distribute the negative sign to each term in the polynomial.

 Go Online You can complete an Extra Example online.

Study Tip

Check To ensure that your answer is correct, substitute it into the original equation and verify that the simplified expressions are equal.

🌐 **Example 3** Write and Evaluate a Polynomial Expression

ARCHITECTURE The world's largest basket is a building. Each face of the building is in the shape of a trapezoid, with the largest face having a height of h and two base lengths, $h + 90$ and $2h + 84$. Write and simplify an expression to represent the area of one side of the building.

Let h = the height of a trapezoid, a = _____ and b = _____.

$$A = \frac{1}{2}h(a + b) \qquad \text{Area of a trapezoid}$$
$$= \frac{1}{2}h[(\text{\_\_\_\_\_}) + (\text{\_\_\_\_\_})] \qquad a = h + 90 \text{ and } b = 2h + 84$$
$$= \frac{1}{2}h(\text{\_\_\_\_\_}) \qquad \text{Add and simplify.}$$
$$= \text{\_\_\_\_\_} \qquad \text{Distributive Property}$$

The area of one side of the building is _____.

Example 4 Solve Equations with Polynomial Expressions

Solve each equation.

a. $-16p = -2(p + 3) + 3(6p - 30)$

$-16p = -2(p + 3) + 3(6p - 30)$	Original equation
$-16p = \text{\_\_\_\_} - 6 + \text{\_\_\_\_} - 90$	Distributive Property
$-16p = \text{_____}$	Combine like terms.
$-32p = -96$	Subtract $16p$ from each side.
$p = \text{\_\_\_\_}$	Divide each side by -32.

b. $-2q(4q - 9) = q(-8q + 15) - 3(-4q - 6)$

$-2q(4q - 9) = q(-8q + 15) - 3(-4q - 6)$	Original equation
$-8q^2 + 18q = -8q^2 + \text{\_\_\_\_} + 12q + \text{\_\_\_\_}$	Distributive Property
$-8q^2 + 18q = -8q^2 + \text{\_\_\_\_} + 18$	Combine like terms.
$\text{\_\_\_\_} = 27q + 18$	Add $8q^2$ to each side.
$-9q = \text{\_\_\_\_}$	Subt. $27q$ from each side.
$q = \text{\_\_\_\_}$	Divide each side by -9.

Check

Solve each equation.

$2p = 3(4p - 10)$

$p = \text{\_\_\_\_}$

$-3q(2q + 5) = 2(-3q^2 + 15) - 5(10q + 6)$

$q = \text{\_\_\_\_}$

🔵 **Go Online** You can complete an Extra Example online.

Practice

🔊 **Go Online** You can complete your homework online.

Example 1

Simplify each expression.

1. $b(b^2 - 12b + 1)$

2. $f(f^2 + 2f + 25)$

3. $-3m^3(2m^3 - 12m^2 + 2m + 25)$

4. $2j^2(5j^3 - 15j^2 + 2j + 2)$

5. $2pr^2(2pr + 5p^2r - 15p)$

6. $4t^3u(2t^2u^2 - 10tu^4 + 2)$

Example 2

Simplify each expression.

7. $-3(5x^2 + 2x + 9) + x(2x - 3)$

8. $a(-8a^2 + 2a + 4) + 3(6a^2 - 4)$

9. $-4d(5d^2 - 12) + 7(d + 5)$

10. $-9g(-2g + g^2) + 3(g^3 + 4)$

11. $2j(7j^2k^2 + jk^2 + 5k) - 9k(-2j^2k^2 + 2k^2 + 3j)$

12. $4n(2n^3p^2 - 3np^2 + 5n) + 4p(6n^2p - 2np^2 + 3p)$

Example 3

13. **NUMBER THEORY** The sum of the first n whole numbers is given by the expression $\frac{1}{2}(n^2 + n)$. Expand the equation by multiplying, then find the sum of the first 12 whole numbers.

14. **COLLEGE** Troy's grandfather gave him $700 to start his college savings account. Troy's grandfather also gives him $40 each month to add to the account. Troy's mother gives him $50 each month, but has been doing so for 4 fewer months than Troy's grandfather. Write a simplified expression for the amount of money Troy has received m months after his mother started giving him money.

15. **MARKET** Sophia went to the farmers' market to purchase some vegetables. She bought peppers and potatoes. The peppers were $0.39 each and the potatoes were $0.29 each. She spent $3.88 on vegetables, and bought 4 more potatoes than peppers. If x = the number of peppers, write and solve an equation to find out how many of each vegetable Sophia bought.

16. **GEOMETRY** The volume of a pyramid can be found by multiplying the area of its base B by one-third of its height. The area of the rectangular base of a pyramid is given by the polynomial equation $B = x^2 - 4x - 12$.

    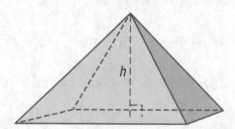

    a. Write a polynomial equation to represent the volume of the pyramid V if its height is 10 meters.

    b. Find the volume of the pyramid if $x = 12$ m.

Example 4

Solve each equation.

17. $7(t^2 + 5t - 9) + t = t(7t - 2) + 13$

18. $w(4w + 6) + 2w = 2(2w^2 + 7w - 3)$

19. $5(4z + 6) - 2(z - 4) = 7z(z + 4) - z(7z - 2) - 48$

20. $9c(c - 11) + 10(5c - 3) = 3c(c + 5) + c(6c - 3) - 30$

21. $2f(5f - 2) - 10(f^2 - 3f + 6) = -8f(f + 4) + 4(2f^2 - 7f)$

22. $2k(-3k + 4) + 6(k^2 + 10) = k(4k + 8) - 2k(2k + 5)$

Mixed Exercises

Simplify each expression.

23. $a(4a + 3)$

24. $-c(11c + 4)$

25. $x(2x - 5)$

26. $2y(y - 4)$

27. $-3n(n^2 + 2n)$

28. $4h(3h - 5)$

29. $3x(5x^2 - x + 4)$

30. $7c(5 - 2c^2 + c^3)$

31. $-4b(1 - 9b - 2b^2)$

32. $6y(-5 - y + 4y^2)$

33. $2m^2(2m^2 + 3m - 5)$

34. $-3n^2(-2n^2 + 3n + 4)$

Simplify each expression.

35. $w(3w + 2) + 5w$

36. $f(5f - 3) - 2f$

37. $-p(2p - 8) - 5p$

38. $y^2(-4y + 5) - 6y^2$

39. $2x(3x^2 + 4) - 3x^3$

40. $4a(5a^2 - 4) + 9a$

41. $4b(-5b - 3) - 2(b^2 - 7b - 4)$

42. $3m(3m + 6) - 3(m^2 + 4m + 1)$

43. $-5q^2w^3(4q + 7w) + 4qw^2(7q^2w + 2q) - 3qw(3q^2w^2 + 9)$

44. $-x^2z(2z^2 + 4xz^3) + xz^2(xz + 5x^3z) + x^2z^3(3x^2z + 4xz)$

Solve each equation.

45. $3(a + 2) + 5 = 2a + 4$

46. $2(4x + 2) - 8 = 4(x + 3)$

47. $5(y + 1) + 2 = 4(y + 2) - 6$

48. $4(b + 6) = 2(b + 5) + 2$

49. $6(m - 2) + 14 = 3(m + 2) - 10$

50. $3(c + 5) - 2 = 2(c + 6) + 2$

51. LANDSCAPING The courtyard on a college campus has a sculpture surrounded by a circle of 50 flags. The university plans to install a new sidewalk 12 feet wide around the perimeter of the outside of the circle of flags. If the outside circumference of the sidewalk is 1.10 times the circumference of the circle of flags, write an equation for the outside circumference of the sidewalk. Solve the equation for the radius of the circle of flags. Recall that the circumference of a circle is $2\pi r$.

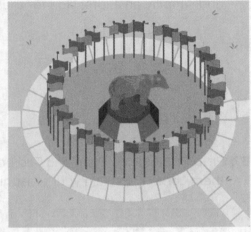

52. STRUCTURE The base lengths of the trapezoid shown are given by polynomial expressions in terms of the trapezoid's height h.

a. Write and simplify an expression for the area of the trapezoid.

b. If the height of the trapezoid is 4 units, what is the area of the trapezoid?

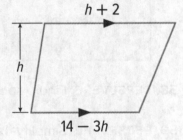

$h + 2$

h

$14 - 3h$

53. FIND THE ERROR Andres simplified the expression $12y^2(3y - 2y^2) - 3y^3(4 - 2y)$. Is he correct? Explain your reasoning.

$12y^2(3y - 2y^2) - 3y^3(4 - 2y)$
$36y^3 - 24y^2 - 12y^3 - 6y^4$
$24y^3 - 24y^2 - 6y^4$

54. STRUCTURE The diagram shows the dimensions of a right rectangular prism.

a. Write and simplify an expression for the volume of the prism.

b. If the height of the rectangular prism is 6 units, what is the volume of the rectangular prism?

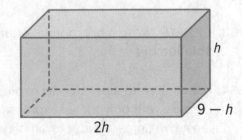

h

$9 - h$

$2h$

55. USE TOOLS Through market research, a company finds that it can expect to sell $45 - 5x$ products if each is priced at $1.25x$ dollars.

   a. Write and simplify an expression for the expected revenue.

   b. Determine the price of each product, to the nearest cent, when $x = 4.5$.

   c. Determine the revenue when the expected number of products are sold, to the nearest cent, when $x = 4.5$.

56. STRUCTURE An area is enclosed by the fence shown at the right, represented by the solid line segments. Write and simplify an expression for the area of the enclosure.

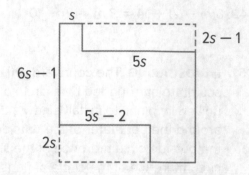

57. FIND THE ERROR Pearl and Ted both simplified $2x^2(3x^2 + 4x + 2)$. Is either of them correct? Explain your reasoning.

Pearl
$2x^2(3x^2 + 4x + 2)$
$6x^4 + 8x^2 + 4x^2$
$6x^4 + 12x^2$

Ted
$2x^2 (3x^2 + 4x + 2)$
$6x^4 + 8x^3 + 4x^2$

58. PERSEVERE Find p such that $3x^p(4x^{2p + 3} + 2x^{3p - 2}) = 12x^{12} + 6x^{10}$.

59. PERSEVERE Simplify $4x^{-3}y^2(2x^5y^{-4} + 6x^{-7}y^6 - 4x^0y^{-2})$.

60. ANALYZE Is there a value of x that makes the statement $(x + 2)^2 = x^2 + 2^2$ true? If so, find a value for x. Justify your argument.

61. CREATE Write a monomial and a polynomial using n as the variable. Find their product.

62. WRITE Describe the steps to multiply a polynomial by a monomial.

63. CREATE Write a polynomial equation, with variables on both sides, that has a solution of $t = 9$.

64. CREATE Write a polynomial expression that can be simplified to $-3c^3 + 74c^2 - 4c$. Be sure your polynomial expression requires the use of the Distributive Property at least two times in order to simplify.

Multiplying Polynomials

Today's Goal
• Multiply binomials by using the Distributive Property and the FOIL Method.

Today's Vocabulary
quadratic expression

Explore Using Algebra Tiles to Find Products of Two Binomials

Online Activity Use algebra tiles to complete the Explore.

INQUIRY How can you use the Distributive Property to find the product of two binomials?

Learn Multiplying Binomials

Binomials can be multiplied horizontally or vertically. Multiply $(x - 2)(x + 6)$.

Method 1 Vertical Method

$$\begin{array}{r} x - 2 \\ (\times) \quad x + 6 \\ \hline 6x - 12 \\ (+) \quad x^2 - 2x \\ \hline x^2 + 4x - 12 \end{array}$$

Multiply by 6.
Multiply by x.
Combine like terms.

Method 2 Horizontal Method

$(x - 2)(x + 6) = x(x + 6) - 2(x + 6)$ Rewrite as the sum of two products.

$= x^2 + 6x - 2x - 12$ Distributive Property

$= x^2 + 4x - 12$ Combine like terms.

You can also use a shortcut version of the Distributive Property, called the FOIL method, to multiply binomials.

Key Concept • FOIL Method

To multiply two binomials, find the sum of the products of **F** the *First terms*, **O** the *Outer terms*, **I** the *Inner terms*, and **L** the *Last terms*.

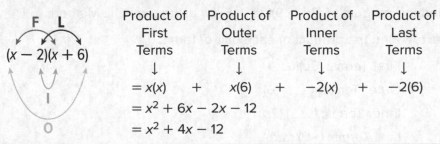

Product of First Terms	Product of Outer Terms	Product of Inner Terms	Product of Last Terms
↓	↓	↓	↓
$= x(x)$ +	$x(6)$ +	$-2(x)$ +	$-2(6)$

$= x^2 + 6x - 2x - 12$

$= x^2 + 4x - 12$

An expression in one variable with a degree of 2 is called a **quadratic expression**.

Go Online You can complete an Extra Example online.

Think About It!

How does the horizontal method using the Distributive Property differ from the FOIL method?

Study Tip

FOIL Method
The FOIL method can be used only when multiplying two binomials. The FOIL method cannot be used when multiplying, for example, two trinomials because nine products are needed.

Your Notes

Example 1 Multiply Binomials by Using the Vertical Method

Find $(x - 1)(x + 7)$ by using the vertical method.

$$
\begin{array}{r}
x \quad - \quad 1 \\
(\times) \quad x \quad + \quad 7 \\
\hline
7x \quad - \quad 7 \\
x^2 \quad - \quad x \\
\hline
\_\_ \quad + \quad \_\_ \quad - \quad \_\_
\end{array}
$$

Multiply by 7.

Multiply by x.

Combine like terms.

Example 2 Multiply Binomials by Using the Horizontal Method

Find $(3x - 4)(4x - 10)$ by using the horizontal method.

$(3x - 4)(4x - 10)$ $= 3x(4x - 10) + -4(4x - 10)$ — Rewrite as sum of two products.

$= \_\_x^2 - \_\_x - \_\_x + 40$ — Distributive Property

$= 12x^2 - \_\_x + 40$ — Combine like terms.

Example 3 Multiply Binomials by Using the FOIL Method

Find $(2a - 12)(5a + 3)$ by using the FOIL method.

$(2a - 12)(5a + 3)$

$= 2a(\_\_) + 2a(\_\_) + (\_\_)(5a) + (\_\_)(3)$

$= \_\_a^2 + \_\_a - \_\_a - \_\_$

$= 10a^2 \_\_\_a - 36$

Check

Find the product of $(3p - 9)(2p + 6)$ by using the FOIL Method.

Part A Find the product of each pair of terms.

First Terms: $(3p)(\_\_)$

Outer Terms: $(\_\_)(6)$

Inner Terms: $(\_\_)(2p)$

Last Terms: $(-9)(\_\_)$

Part B What is the product of $(3p - 9)(2p + 6)$ written in standard form?

_____

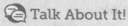 **Go Online** You can complete an Extra Example online.

Study Tip

Signs Notice that in the first step of the solution, you added the product of -4 and $4x$, and in the second step, that turned into a subtraction of $16x$. Remember that $-4(4x) = -16x$.

Study Tip

FOIL Method
The FOIL method is a memory device that can help you remember to find all four products when multiplying two binomials. The order in which the terms are multiplied is not important.

💬 Talk About It!

Is your answer complete after you use the FOIL method to multiply? Explain.

🌐 Apply Example 4 Use the FOIL Method

SNOW REMOVAL A town worker is clearing the snow from a parking lot and the surrounding sidewalk. The sidewalk extends *x* feet on every side of the parking lot. Write an expression for the total area of the parking lot and sidewalk.

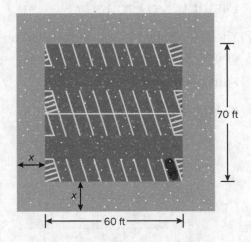

1 What is the task?

Describe the task in your own words. Then list any questions that you may have. How can you find answers to your questions?

_____

_____

2 How will you approach the task? What have you learned that you can use to help you complete the task?

_____

_____

_____

3 What is your solution?

Use your strategy to solve the problem.

Length = _____   Width = _____

The total area of the parking lot and sidewalk is _____.

4 How can you know that your solution is reasonable?

✏️ **Write About It!** Write an argument that can be used to defend your solution.

_____

_____

_____

Check

FRAME Jacinta is framing her newest painting. The dimensions of the frame with the painting are represented by a width of $5y - 5$ and a length of $2y + 6$. Write an expression in standard form that represents the area of the frame with the painting. _____

▶️ **Go Online** You can complete an Extra Example online.

Example 5 Multiply Polynomials by Using the Distributive Property

The Distributive Property can also be used to multiply any two polynomials.

Find each product.

a. $(x + 3)(x^2 - x - 9)$

$(x + 3)(x^2 - x - 9)$

$= \underline{\quad}(x^2 - x - 9) + \underline{\quad}(x^2 - x - 9)$ Distributive Property

$= x^3 - x^2 - 9x + \underline{\quad} - \underline{\quad} - \underline{\quad}$ Multiply.

$= \underline{\hspace{4cm}}$ Combine like terms.

b. $(3t^2 - 5t + 9)(4t^2 - 4t + 14)$

$(3t^2 - 5t + 9)(4t^2 - 4t + 14)$

$= \underline{\quad}(4t^2 - 4t + 14) - \underline{\quad}(4t^2 - 4t + 14) + \underline{\quad}(4t^2 - 4t + 14)$

$= 12t^4 - 12t^3 + 42t^2 \underline{\hspace{3cm}} + 36t^2 - 36t + 126$

$= \underline{\hspace{4cm}}$

Think About It!

When multiplying two trinomials, do you have to apply the Distributive Property in any certain order?

Check

Find each product.

a. $(m - 3)(m^2 + m - 5)$ ____

  A. $m^3 - 2m^2 - 8m + 15$

  B. $m^3 - 4m^2 - 8m + 15$

  C. $m^3 - 2m^2 + 8m + 15$

  D. $m^3 - 2m^2 - 8m - 15$

b. $(6v^2 + 4v - 3)(v^2 - v + 6)$ ____

  A. $6v^4 + 10v^3 + 29v^2 + 27v - 18$

  B. $6v^4 - 2v^3 + 32v^2 + 27v - 18$

  C. $6v^4 - 2v^3 + 29v^2 + 27v - 18$

  D. $6v^4 - 2v^3 + 29v^2 + 24v - 18$

Go Online You can complete an Extra Example online.

Name _____ Period _____ Date _____

Practice

Go Online You can complete your homework online.

Examples 1–3

Find each product.

1. $(3c - 5)(c + 3)$

2. $(g + 10)(2g - 5)$

3. $(6a + 5)(5a + 3)$

4. $(4x + 1)(6x + 3)$

5. $(5y - 4)(3y - 1)$

6. $(6d - 5)(4d - 7)$

7. $(3m + 5)(2m + 3)$

8. $(7n - 6)(7n - 6)$

9. $(12t - 5)(12t + 5)$

10. $(5r + 7)(5r - 7)$

11. $(8w + 4x)(5w - 6x)$

12. $(11z - 5y)(3z + 2y)$

Example 4

13. PLAYGROUND The dimensions of a playground are represented by a width of $9x + 1$ feet and a length of $5x - 2$ feet. Write an expression that represents the area of the playground.

14. THEATER The Loft Theater has a center seating section with $3c + 8$ rows and $4c - 1$ seats in each row. Write an expression for the total number of seats in the center section.

15. CRAFTS Suppose a rectangular quilt made up of squares has a length-to-width ratio of 5 to 4. The length of the quilt is $5x$ inches. The quilt can be made slightly larger by adding a border of 1-inch squares all the way around the perimeter of the quilt. Write a polynomial expression for the area of the larger quilt.

16. FLAG CASE A United States flag is sometimes folded into a triangle shape and displayed in a triangular display case. If a display case has dimensions shown in inches, write a polynomial expression that represents the area of wall space covered by the display case.

17. NUMBER THEORY Think of a whole number. Subtract 2. Write down this number. Take the original number and add 2. Write down this number. Find the product of the numbers you wrote down. Subtract the square of the original number. The result is always -4. Use polynomials to show how this number trick works.

Lesson 10-3 · Multiplying Polynomials 565

Example 5

Find each product.

18. $(2y - 11)(y^2 - 3y + 2)$

19. $(4a + 7)(9a^2 + 2a - 7)$

20. $(m^2 - 5m + 4)(m^2 + 7m - 3)$

21. $(x^2 + 5x - 1)(5x^2 - 6x + 1)$

22. $(3b^3 - 4b - 7)(2b^2 - b - 9)$

23. $(6z^2 - 5z - 2)(3z^3 - 2z - 4)$

Mixed Exercises

Find each product.

24. $(m + 4)(m + 1)$

25. $(x + 2)(x + 2)$

26. $(b + 3)(b + 4)$

27. $(t + 4)(t - 3)$

28. $(r + 1)(r - 2)$

29. $(n - 5)(n + 1)$

30. $(3c + 1)(c - 2)$

31. $(2x - 6)(x + 3)$

32. $(d - 1)(5d - 4)$

33. $(2\ell + 5)(\ell - 4)$

34. $(3n - 7)(n + 3)$

35. $(q + 5)(5q - 1)$

36. $(3b + 3)(3b - 2)$

37. $(2m + 2)(3m - 3)$

38. $(4c + 1)(2c + 1)$

39. $(5a - 2)(2a - 3)$

40. $(4h - 2)(4h - 1)$

41. $(x - y)(2x - y)$

42. $(w + 4)(w^2 + 3w - 6)$

43. $(t + 1)(t^2 + 2t + 4)$

44. $(k - 4)(k^2 + 5k - 2)$

45. $(m + 3)(m^2 + 3m + 5)$

46. $(2x + 1)(x^2 - 3x - 4)$

47. $(3b + 4)(2b^2 - b + 4)$

Simplify.

48. $(m + 2)[(m^2 + 3m - 6) + (m^2 - 2m + 4)]$

49. $[(t^2 + 3t - 8) - (t^2 - 2t + 6)](t - 4)$

Find each product.

50. $(a - 2b)^2$

51. $(3c + 4d)^2$

52. $(x - 5y)^2$

53. $(2r - 3t)^3$

54. $(5g + 2h)^3$

55. $(4y + 3z)(4y - 3z)^2$

56. PRECISION Write each expression as a simplified polynomial.

    a. $(3c - 2)(4c^2 - c^3 + 3)$

    b. $(5x - y)(3x^2 - 2xy) + (2x + y)(y^2 - 4x^2)$

    c. $-2x(3 - x^2)(2x + 4)$

    d. $(z - 1)(2 - z)(z + 1)$

57. ART The museum where Julia works plans to have
a large wall mural painted in its lobby. First, Julia
wants to paint a large frame around where the
mural will be. She only has enough paint for the
frame to cover 100 square feet of wall surface. The
mural's length will be 5 feet longer than its width,
and the frame will be 2 feet wide on all sides.

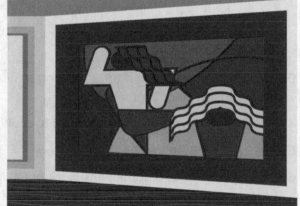

    a. Write an expression for the area of the mural.

    b. Write an expression for the area of the frame.

    c. Write and solve an equation to find how large
the mural can be.

58. STRUCTURE The dimensions of the composite figure shown are given
in terms of the triangle's height, h.

    a. Write and simplify a quadratic expression for the area of the figure.

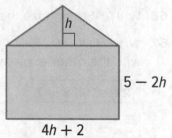

    b. If $h = 1.42$ units, what is the area of the figure? Round to the
nearest hundredth, if necessary.

59. STRUCTURE Consider the expression $x^{4p + 1}(x^{1 - 2p})^{2p + 3}$.

    a. Use the laws of exponents to simplify the expression.

    b. Find any integer values of p that make this expression equal to 1 for all values
of x.

60. USE A MODEL The relationship between monthly profit P, monthly sales n, and unit price p is $P = n(p - U) - F$, where U is the unit cost per sale and F is a fixed cost that does not depend on the number of sales. For an online business advice service, the unit cost is $30 per hour-long session and the monthly fixed cost is $3000.

    a. Given a model for monthly sales of $n = 5000 - 40p$ for a given price p per session, write and simplify a quadratic expression for P in terms of p.

    b. If the unit price, p, is $77.50, what is the monthly profit, P?

61. STRUCTURE Find and simplify an expression for the volume of the rectangular prism shown.

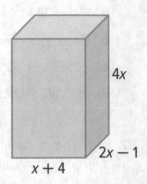

4x

2x − 1

x + 4

62. ANALYZE Determine if the following statement is *sometimes*, *always*, or *never* true. Justify your argument.

    The FOIL method can be used to multiply a binomial and a trinomial.

63. PERSEVERE Find $(x^m + x^p)(x^{m-1} - x^{1-p} + x^p)$.

64. CREATE Write a binomial and a trinomial involving a single variable. Then find their product.

65. WRITE Compare and contrast the procedure used to multiply a trinomial by a binomial using the vertical method with the procedure used to multiply a three-digit number by a two-digit number.

66. WRITE Summarize the methods that can be used to multiply polynomials.

67. WHICH ONE DOESN'T BELONG? Which polynomial expression does not belong with the other expressions? Explain your reasoning.

$(2x + 11)(3x - 7)$	$(6y + 1)(y - 4)$
$(3y - 4)(2y + 5)$	$(2x - 1)(x^2 + x - 1)$

68. FIND THE ERROR Jariah and Malia are multiplying the expression $(2x + 1)(x - 4)$. Is either of them correct? Explain your reasoning.

Jariah
$(2x + 1)(x - 4) = 2x^2 + 8x + 1x - 4$
$= 2x^2 + 9x - 4$

Malia
$(2x + 1)(x - 4) = 2x^2 - 8x + 1x - 4$
$= 2x^2 - 7x - 4$

Special Products

Today's Goals
- Multiply binomials by applying the pattern formed by squares of sums.
- Multiply binomials by applying the pattern formed by squares of differences.

Explore Using Algebra Tiles to Find the Squares of Sums

 Online Activity Use algebra tiles to complete the Explore.

> ⓠ **INQUIRY** How can you write the square of a sum?

Learn Square of a Sum

Key Concept • Square of a Sum	
Words	The square of $a + b$ is the square of a plus twice the product of a and b plus the square of b.
Symbols	$(a + b)^2 = (a + b)(a + b)$ $\quad\quad\quad = a^2 + 2ab + b^2$
Example	$(x + 3)^2 = (x + 3)(x + 3)$ $\quad\quad\quad\quad = x^2 + 6x + 9$

Example 1 Square of a Sum

Find each product.

a. $(x + 6)^2$

$(a + b)^2 = a^2 + 2ab + b^2$ Square of a sum

$(x + 6)^2 = (\underline{\quad})^2 + 2(\underline{\quad})(\underline{\quad}) + \underline{\quad}^2$ $a = x, b = 6$

$\quad\quad\quad\quad = x^2 + 12x + 36$ Simplify.

b. $(3g + 10h)^2$

$(a + b)^2 = a^2 + 2ab + b^2$ Square of a sum

$(3g + 10h)^2 = (\underline{\quad})^2 + 2(\underline{\quad})(\underline{\quad}) + (\underline{\quad})^2$ $a = 3g, b = 10h$

$\quad\quad\quad\quad = 9g^2 + 60gh + 100h^2$ Simplify.

💭 **Think About It!**

How can you check that using the square of a sum pattern gives the correct product?

Check

Find $(2x + 9)^2$. Select the correct product. ____

A. $2x^2 + 18x + 18$

B. $2x^2 + 36x + 81$

C. $4x^2 + 18x + 81$

D. $4x^2 + 36x + 81$

Find $(6m + 11n)^2$. Select the correct product. ____

A. $6m^2 + 132mn + 11n^2$

B. $12m^2 + 66mn + 121n^2$

C. $36m^2 + 66mn + 121n^2$

D. $36m^2 + 132mn + 121n^2$

🌐 Example 2 Use Squares of Sums

GENETICS **A Punnett square is used to predict the probability of offspring inheriting certain genetic characteristics. In Doberman Pinschers, black fur *B* is dominant over the recessive gene for brown fur *b*. If two parents have both a dominant and a recessive gene, use the square of a sum to determine the possible combinations of their offspring.**

	B	b
B	BB	Bb
b	bB	bb

Both parents can be represented by $(B + b)$, and the combinations of the offspring are the product of $(B + b)^2$.

$(a + b)^2 = a^2 + 2ab + b^2$ Square of a sum

$(B + b)^2$ _____ $a = B, b = b$

Check

SURFACE AREA The surface area of a cube is given by $A = 6s^2$, where s is the length of one side. Select the expression that represents the surface area of a cube with side length $3n + 6$. ____

A. $6(3n^2 + 18n + 36)$

B. $6(9n^2 + 36)$

C. $6(9n^2 + 36n + 36)$

D. $6(9n^2 + 81n + 36)$

🔾 **Go Online** You can complete an Extra Example online.

Online Activity Use algebra tiles to complete the Explore.

> **INQUIRY** How can you write the square of a difference?

Learn Square of a Difference

Key Concept • Square of a Difference	
Words	The square of $a - b$ is the square of a minus twice the product of a and b plus the square of b.
Symbols	$(a - b)^2 = (a - b)(a - b)$
	$ = a^2 - 2ab + b^2$
Example	$(x - 8)^2 = (x - 8)(x - 8)$
	$ = x^2 - 2 \cdot x \cdot 8 + 8^2$
	$ = x^2 - 16x + 64$

Example 3 Square of a Difference

Find $(7d - 2f)^2$.

$(a - b)^2 = a^2 - 2ab + b^2$ Square of a difference

$(7d - 2f)^2 = (\underline{})^2 - 2(\underline{})(\underline{}) + (\underline{})^2$ $a = 7d, b = 2f$

$ = \underline{}$ Simplify.

Check

Find $(4k - 1)^2$. ____

A. $4k^2 - 8k - 1$

B. $16k^2 - 1$

C. $16k^2 - 8k + 1$

D. $16k^2 - 16k + 1$

Go Online You can complete an Extra Example online.

Copyright © McGraw-Hill Education

Study Tip

Watching Signs
Because the square of a sum and the square of a difference vary only by the sign of the middle term, pay close attention to the sign being used within the square of the trinomial.

Think About It!
Demarco says that the Square of a Difference pattern is a special case of the Square of a Sum pattern. Is he correct? Explain.

Explore Using Algebra Tiles to Find Products of Sums and Differences

Online Activity Use algebra tiles to complete the Explore.

⊘ **INQUIRY** How can you write the product of a sum and a difference?

Learn Product of a Sum and a Difference

Key Concept • Product of a Sum and a Difference	
Words	The product of $a + b$ and $a - b$ is the square of a minus the square of b.
Symbols	$(a + b)(a - b) = a^2 - ab + ab - b^2$ $\qquad\qquad\quad = a^2 - b^2$
Example	$(x + 6)(x - 6) = x^2 - 6x + 6x - 6^2$ $\qquad\qquad\qquad = x^2 - 36$

Example 4 Product of a Sum and a Difference

Find each product.

a. $(z + 5)(z - 5)$

$(a + b)(a - b) = a^2 - b^2$ Product of a sum and a difference

$(z + 5)(z - 5) = $ _____ $a = z, b = 5$

$\qquad\qquad\quad = $ _____   Simplify.

b. $(3y^3 + 4)(3y^3 - 4)$

$(a + b)(a - b) = a^2 - b^2$ Product of a sum and a difference

$(3y^3 + 4)(3y^3 - 4) = ($___$)^2 - ($___$)^2$ $a = 3y^3, b = 4$

$\qquad\qquad\qquad = $ _____   Simplify.

Check

Find each product.

a. $(x^2 + 3y)(x^2 - 3y)$ _____

b. $(2x + 9y^2)(2x - 9y^2)$ _____

c. $(x + 9y)(x - 9y)$ _____

d. $(x + 9)(x - 9)$ _____

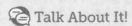 **Go Online** You can complete an Extra Example online.

Study Tip

Identical Values Note that the values of a and b must be identical within each quantity to use the pattern for the product of a sum and a difference. The only difference between the quantities is the operation between a and b.

Watch Out!

Power of a Power Remember to square all parts of a and b, including powers. When a power is raised to another power, multiply the exponents.

💬 Talk About It!

Explain how you would tell someone to find the product of a sum and a difference.

Name _____ Period _____ Date _____

Practice

🕐 **Go Online** You can complete your homework online.

Examples 1 and 3
Find each product.

1. $(a + 10)(a + 10)$

2. $(b - 6)(b - 6)$

3. $(h + 7)^2$

4. $(x + 6)^2$

5. $(8 - m)^2$

6. $(9 - 2y)^2$

7. $(2b + 3)^2$

8. $(5t - 2)^2$

9. $(8h - 4n)^2$

10. $(4m - 5n)^2$

Example 2

11. ROUNDABOUTS A city planner is proposing a roundabout to improve traffic flow at a busy intersection. Write a polynomial equation for the area A of the traffic circle if the radius of the outer circle is r and the width of the road is 18 feet.

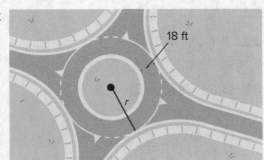

12. NUMBER CUBES Kivon has two number cubes. Each edge of number cube A is 3 millimeters less than each edge of number cube B. Each edge of number cube B is x millimeters. Write an equation that models the surface area of number cube A.

Number Cube A   Number Cube B

$(x - 3)$ mm
$(x - 3)$ mm
$(x - 3)$ mm
x mm
x mm
x mm

13. PROBABILITY The spinner has two equal sections, blue (B) and red (R). Use the square of a sum to determine the possible combinations of spinning the spinner two times.

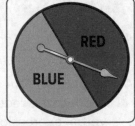

14. BUSINESS The Combo Lock Company finds that its profit data from 2015 to the present can be modeled by the function $y = (2n + 11)^2$, where y is the profit, in thousands of dollars, n years since 2015. Which special product does this polynomial demonstrate? Simplify the polynomial.

Example 4

Find each product.

15. $(u + 3)(u - 3)$

16. $(b + 7)(b - 7)$

17. $(2 + x)(2 - x)$

18. $(4 - x)(4 + x)$

19. $(2q + 5r)(2q - 5r)$

20. $(3a^2 + 7b)(3a^2 - 7b)$

Mixed Exercises

Find each product.

21. $(n + 3)^2$

22. $(x + 4)(x + 4)$

23. $(y - 7)^2$

24. $(t - 3)(t - 3)$

25. $(b + 1)(b - 1)$

26. $(a - 5)(a + 5)$

27. $(p - 4)^2$

28. $(z + 3)(z - 3)$

29. $(\ell + 2)(\ell + 2)$

30. $(r - 1)(r - 1)$

31. $(3g + 2)(3g - 2)$

32. $(2m - 3)(2m + 3)$

33. $(6 + u)^2$

34. $(r + t)^2$

35. $(3q + 1)(3q - 1)$

36. $(c - d)^2$

37. $(2k - 2)^2$

38. $(w + 3h)^2$

39. $(3p - 4)(3p + 4)$

40. $(t + 2u)^2$

41. $(x - 4y)^2$

42. $(3b + 7)(3b - 7)$

43. $(3y - 3g)(3y + 3g)$

44. $(n^2 + r^2)^2$

45. $(2k + m^2)^2$

46. $(3t^2 - n)^2$

47. GEOMETRY The length of a rectangle is the sum of two whole numbers. The width of the rectangle is the difference of the same two whole numbers. Write a verbal expression for the area of the rectangle.

48. Find the product of $(10 - 4t)$ and $(10 + 4t)$. What type of special product does this represent?

49. STORAGE A cylindrical tank is placed along a wall. A cylindrical PVC pipe will be hidden in the corner behind the tank. See the side-view diagram shown. The radius of the tank is r inches, and the radius of the PVC pipe is s inches.

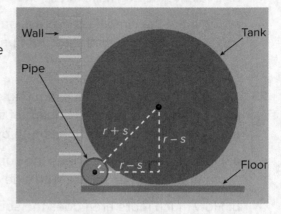

a. Use the Pythagorean Theorem to write an equation for the relationship between the two radii. Simplify your equation so that there is a zero on one side of the equal sign.

b. Write a polynomial equation you could solve to find the radius s of the PVC pipe if the radius of the tank is 20 inches.

Find each product.

50. $(2a - 3b)^2$

51. $(5y + 7)^2$

52. $(8 - 10a)^2$

53. $(10x - 2)(10x + 2)$

54. $(3t + 12)(3t - 12)$

55. $(a + 4b)^2$

56. $(3q - 5r)^2$

57. $(2c - 9d)^2$

58. $(g + 5h)^2$

59. $(6y - 13)(6y + 13)$

60. $(3a^4 - b)(3a^4 + b)$

61. $(5x^2 - y^2)^2$

62. $(8a^2 - 9b^3)(8a^2 + 9b^3)$

63. $\left(\frac{3}{4}k + 8\right)^2$

64. $\left(\frac{2}{5}y - 4\right)^2$

65. $(7z^2 + 5y^2)(7z^2 - 5y^2)$

66. $(2m + 3)(2m - 3)(m + 4)$

67. $(r + 2)(r - 5)(r - 2)(r + 5)$

Find each product.

68. $(c + d)(c + d)(c + d)$

69. $(2a - b)^3$

70. $(f + g)(f - g)(f + g)$

71. $(k - m)(k + m)(k - m)$

72. $(n - p)^2(n + p)$

73. $(q - r)^2(q - r)$

74. Consider the product $(a + b)(a - b)(a + b)(a - b)$.

   a. Show the steps required to determine the product.

   b. Evaluate the original expression for $a = 5$ and $b = 2$.

   c. Evaluate the simplified expression you wrote in **part a** for $a = 5$ and $b = 2$. Compare this to the result from **part b**.

75. STRUCTURE Tanisha is investigating growth patterns for the area of a square. She begins with a square of side length s and looks at the effects of enlarging the side length by one unit at a time.

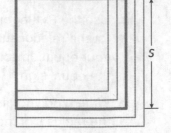

   a. How much more area does the square with side length $s + 1$ have compared to the square with side length s?

   b. How much more area does the square with side length $s + 2$ have compared to the square with side length $s + 1$?

   c. How much more area does the square with side length $s + 3$ have compared to the square with side length $s + 2$?

76. WHICH ONE DOESN'T BELONG? Which expression does not belong? Explain your reasoning.

| $(2c - d)(2c - d)$ | $(2c + d)(2c - d)$ | $(2c + d)(2c + d)$ | $(c + d)(c + d)$ |

77. PERSEVERE Does a pattern exist for the cube of a sum $(a + b)^3$?

   a. Investigate this question by finding the product $(a + b)(a + b)(a + b)$.

   b. Use the pattern you discovered in **part a** to find $(x + 2)^3$.

   c. Draw a diagram of a geometric model for $(a + b)^3$.

   d. What is the pattern for the cube of a difference $(a - b)^3$?

78. ANALYZE The square of a sum is called a *perfect square trinomial*. Find c that makes $25x^2 - 90x + c$ a perfect square trinomial.

79. CREATE Write two binomials with a product that is a binomial. Then write two binomials with a product that is not a binomial.

80. WRITE Describe how to square the sum of two quantities, how to square the difference of two quantities, and how to find the product of a sum of two quantities and a difference of two quantities.

Using the Distributive Property

- Factor polynomials by using the Distributive Property
- Factor polynomials by using the Distributive Property and grouping

Today's Vocabulary
factoring

factoring by grouping

Explore Using Algebra Tiles to Factor Polynomials

Online Activity Use algebra tiles to complete the Explore.

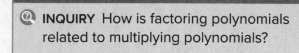

INQUIRY How is factoring polynomials related to multiplying polynomials?

Learn Factoring by Using the Distributive Property

You can use the Distributive Property to multiply a polynomial by a monomial.

$$3y(2y + 5) = 3y(2y) + 3y(5)$$
$$= 6y^2 + 15y$$

You can also use the Distributive Property to factor a polynomial. **Factoring** is the process of expressing a polynomial as the product of monomials and polynomials.

$$6y^2 + 15y = 3y(2y) + 3y(5)$$
$$= 3y(2y + 5)$$

So, $3y(2y + 5)$ is the factored form of $6y^2 + 15y$. When factoring a polynomial, it must be factored completely. If you are using the Distributive Property to factor, one factor will be the greatest common factor (GCF) for all terms of the polynomial.

Example 1 Use the Distributive Property

Use the Distributive Property to factor each polynomial.

a. $12a^2 + 16a$

Step 1 Factor each term.

$$12a^2 = 2 \cdot 2 \cdot 3 \cdot a \cdot a$$

$$16a = 2 \cdot 2 \cdot 2 \cdot 2 \cdot a$$

Step 2 Underline the common terms.

$$12a^2 = \underline{2} \cdot \underline{2} \cdot 3 \cdot \underline{a} \cdot a$$

$$16a = \underline{2} \cdot \underline{2} \cdot 2 \cdot 2 \cdot \underline{a}$$

Step 3 Find the GCF.

$$GCF = \underline{\hspace{2cm}} \text{ or } \underline{\hspace{1cm}}$$

(continued on the next page)

Study Tip

Greatest Common Factor To find the GCF of a polynomial, write all factors of each term as prime numbers or variables to the first degree.

Step 4 Write each term as the product of the GCF and the remaining factors. Use the Distributive Property.

$$12a^2 + 16a = 4a(\underline{\quad}) + 4a(\underline{\quad}) \quad \text{Rewrite each term using the GCF.}$$

$$= 4a(\underline{\quad\quad}) \quad \text{Distributive Property.}$$

b. $20x^2y^2 - 45x^2y - 35x^2$

$$20x^2y^2 = 2 \cdot 2 \cdot 5 \cdot x \cdot x \cdot y \cdot y$$

$$-45x^2y = -1 \cdot 3 \cdot 3 \cdot 5 \cdot x \cdot x \cdot y$$

$$-35x^2 = -1 \cdot 5 \cdot 7 \cdot x \cdot x$$

$$\text{GCF} = \underline{\quad}$$

Write each term as the product of the GCF and the remaining factors. Use the Distributive Property.

$$20x^2y^2 - 45x^2y - 35x^2 = 5x^2(\underline{\quad}) + 5x^2(\underline{\quad}) + 5x^2(\underline{\quad})$$

$$= 5x^2(\underline{\quad\quad\quad})$$

Check

Factor each polynomial.

a. $33n^3 - 121n^2$ **b.** $14a^2b^2c - 6ac^2 + 10ac$

🌐 **Example 2** Use Factoring

VOLCANOS **The 1980 eruption of Washington's Mt. St. Helens had an initial blast with a velocity of 440 feet per second. The expression $440t - 16t^2$ models the height of a rock erupted from the volcano after t seconds. Factor the expression.**

$$440t = \underline{\quad} \cdot \underline{\quad} \cdot \underline{\quad} \cdot 5 \cdot 11 \cdot \underline{\quad}$$

$$-16t^2 = \underline{\quad} \cdot 2 \cdot 2 \cdot 2 \cdot \underline{\quad} \cdot \underline{\quad} \cdot \underline{\quad}$$

$$\text{GCF} = \underline{\quad}$$

$$440t - 16t^2 = 8t(\underline{\quad}) + 8t(\underline{\quad})$$

$$= \underline{\quad\quad\quad}$$

Check

GOLF A golfer hits a golf ball with a velocity of 112 feet per second. The expression $112t - 16t^2$ represents the height of the golf ball after t seconds. Factor the expression.

🐻 **Go Online** You can complete an Extra Example online.

Learn Factor by Grouping

When a polynomial has four or more terms, you can sometimes use a method called **factoring by grouping**. Similar terms are grouped, and the Distributive Property is applied to a common binomial.

Key Concept • Factor by Grouping

Words	A polynomial can be factored by grouping only if all of the following conditions exist.
	• There are four or more terms.
	• Terms have common factors that can be grouped together.
	• There are at least two common factors that are identical or additive inverses of each other.
Symbols	$ax + bx + ay + by = (ax + bx) + (ay + by)$
	$= x(a + b) + y(a + b)$
	$= (x + y)(a + b)$

Example 3 Factor by Grouping

Factor $2uv + 6u + 5v + 15$.

$2uv + 6u + 5v + 15 = (2uv + 6u) + (\underline{\hspace{1.5cm}})$ Group terms with common factors

$= \underline{\hspace{0.8cm}} (v + 3) + \underline{\hspace{0.5cm}}(v + 3)$ Factor the GCF from each group.

Notice that $(v + 3)$ is common to both groups, so it becomes the GCF.

$= \underline{\hspace{2.5cm}}$ Distributive Property.

Check

Factor $tw + 10t - 2w - 20$.

$\underline{\hspace{3cm}}$

 Go Online You can complete an Extra Example online.

Talk About It!

Explain why you cannot use factoring by grouping on a polynomial with three terms.

Study Tip

Creating Groups If you are unable to get identical or additive inverse binomials after factoring out the GCF, try grouping the terms in a different way.

Think About It!

Which property is used to simplify $3m[(-1)(p - 7)] + 4(p - 7)$ to $-3m(p - 7) + 4(p - 7)$?

Example 4 Factor by Grouping with Additive Inverses

It is helpful to be able to recognize when binomials are additive inverses of each other. For example, $5 - x = -1(x - 5)$.

Factor $21m - 3mp + 4p - 28$.

$21m - 3mp + 4p - 28$	Original expression
$= (21m - 3mp) + (4p - 28)$	Group terms with common factors.
$= 3m(\underline{\quad}) + 4(\underline{\quad})$	Factor the GCF from each group.
$= 3m[(\underline{\quad})(p - 7)] + 4(p - 7)$	$7 - p = -1(p - 7)$
$= \underline{\quad}(p - 7) + 4(p - 7)$	Associative Property
$= \underline{\qquad\qquad}$	Distributive Property

Alternate Method

$21m - 3mp + 4p - 28$	Original expression
$= 4p - 3mp + 21m - 28$	Rearrange the expression.
$= (4p - 3mp) + (21m - 28)$	Group terms with common factors.
$= \underline{\quad}(4 - 3m) + \underline{\quad}(3m - 4)$	Factor the GCF from each group.
$= p(-3m + 4) + 7(3m - 4)$	$4 - 3m = -3m + 4$
$= p(-3m + 4) + 7[-1(-3m + 4)]$	$3m - 4 = -1(-3m + 4)$
$= \underline{\quad}(-3m + 4) - 7(-3m + 4)$	Associative Property
$= (p - 7)(-3m + 4)$	Distributive Property
$= (-3m + 4)(p - 7)$	Commutative Property

Check

Factor $-3x^3 + 33x^2 + 4x - 44$.

_____

Pause and Reflect

Did you struggle with anything in this lesson? If so, how did you deal with it?

Record your observations here.

 Go Online You can complete an Extra Example online.

 Go Online to learn how to prove the Elimination Method in Expand 10-5.

Practice

🅡 **Go Online** You can complete your homework online.

Example 1

Use the Distributive Property to factor each polynomial.

1. $16t - 40y$

2. $30v + 50x$

3. $2k^2 + 4k$

4. $5z^2 + 10z$

5. $4a^2b^2 + 2a^2b - 10ab^2$

6. $5c^2v - 15c^2v^2 + 5c^2v^3$

Example 2

7. PHYSICS The distance d an object falls after t seconds is given by $d = 16t^2$ (ignoring air resistance). To find the height of an object launched upward from ground level at a rate of 32 feet per second, use the expression $32t - 16t^2$, where t is the time in seconds. Factor the expression.

8. SWIMMING POOL The area of a rectangular swimming pool is given by the expression $12w - w^2$, where w is the width of one side. Factor the expression.

9. VERTICAL JUMP Your vertical jump height is measured by subtracting your standing reach height from the height of the highest point you can reach by jumping without taking a running start. Typically, NBA players have vertical jump heights of up to 34 inches. If an NBA player jumps this high, his height in inches above his standing reach height after t seconds can be modeled by the expression $162t - 192t^2$. Factor the expression.

10. PETS Conner is playing with his dog. He tosses a treat upward with an initial velocity of 13.7 meters per second. His hand starts at the same height as the dog's mouth, so the height of the treat above the dog's mouth in meters after t seconds is given by the expression $13.7t - 4.9t^2$. Factor the expression.

Examples 3 and 4

Factor each polynomial.

11. $fg - 5g + 4f - 20$

12. $a^2 - 4a - 24 + 6a$

13. $hj - 2h + 5j - 10$

14. $xy - 2x - 2 + y$

15. $45pq - 27q - 50p + 30$

16. $24ty - 18t + 4y - 3$

17. $3dt - 21d + 35 - 5t$

18. $8r^2 + 12r$

19. $21th - 3t - 35h + 5$

20. $vp + 12v + 8p + 96$

21. $5br - 25b + 2r - 10$

22. $2nu - 8u + 3n - 12$

23. $b^2 - 2b + 3b - 6$

24. $2j^2 + 2j + 3j + 3$

25. $2a^2 - 4a + a - 2$

Mixed Exercises

Factor each polynomial.

26. $7x + 49$

27. $8m - 6$

28. $5a^2 - 15$

29. $10q - 25q^2$

30. $a^2b^2 + a$

31. $x + x^2y + x^3y^2$

32. $3p^2r^2 + 6pr + p$

33. $4a^2b^2 + 16ab + 12a$

34. $10h^3n^3 - 2hn^2 + 14hn$

35. $48a^2b^2 - 12ab$

36. $6x^2y - 21y^2w + 24xw$

37. $x^2 + 3x + x + 3$

38. $2x^2 - 5x + 6x - 15$

39. $3n^2 + 6np - np - 2p^2$

40. $4x^2 - 1.2x + 0.5x - 0.15$

41. $9x^2 - 3xy + 6x - 2y$

42. $3x^2 + 24x - 1.5x - 12$

43. $2x^2 - 0.6x + 3x - 0.9$

44. ARCHERY The height, in feet, of an arrow can be modeled by the expression $80t - 16t^2$, where t is the time in seconds. Factor the expression.

🧁 Higher-Order Thinking Skills

45. REGULARITY You have factored some polynomials using the Distributive Property, and others were factored by grouping. What are the similarities between the two methods?

46. CREATE Write a four-term polynomial that can be factored by grouping. Then factor the polynomial.

47. FIND THE ERROR Given the polynomial expression $3x^2 + 7x - 18x - 42$, Theresa says you need to factor by grouping using the binomials $(3x^2 - 18x)$ and $(7x - 42)$. Akash says you need to use the binomials $(3x^2 + 7x)$ and $(-18x - 42)$. Is either of them correct? Justify your answer.

48. PRECISION Choose a value, q, so that $2x^2 + qx + 7x - 21$ can be factored by grouping. Then factor the expression.

49. WRITE Explain how to factor the polynomial $12a^2b^2 - 16a^2b^3$.

Factoring Quadratic Trinomials

Explore Using Algebra Tiles to Factor Trinomials

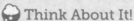 **Online Activity** Use algebra tiles to complete the Explore.

> ⓠ **INQUIRY** How can you use the constant and coefficients in a polynomial to find its factors?

Learn Factoring Trinomials with a Leading Coefficient of 1

Key Concept • Factoring Trinomials with a Leading Coefficient of 1	
Words	To factor trinomials in the form $x^2 + bx + c$, find two integers, m and p, with a sum of b and a product of c. Then write $x^2 + bx + c$ as $(x + m)(x + p)$.
Symbols	$x^2 + bx + c = (x + m)(x + p)$ when $m + p = b$ and $mp = c$.
Example	$x^2 + 8x + 15 = (x + 3)(x + 5)$ because $3 + 5 = 8$ and $3 \cdot 5 = 15$.

A polynomial that cannot be written as a product of two polynomials with integer coefficients is called a **prime polynomial**.

Example 1 *c* Is Positive

Factor $x^2 - 9x + 18$.

In this trinomial, $b =$ _____ and $c =$ _____. Because c is positive and b is negative, you need to find two negative factors with a sum of -9 and a product of 18.

Complete the table to make an organized list of the factors of 18 and look for the pair of factors with a sum of -9.

Factors of 18	Sum of Factors
$-1, -18$	_____
$-2, -9$	_____
$-3, -6$	_____

The correct factors are -3 and -6.

$$x^2 - 9x + 18 = (x + m)(x + p) \qquad \text{Write the pattern.}$$
$$= [x + \underline{\quad}][x + \underline{\quad}] \qquad m = -3 \text{ and } p = -6$$
$$= \underline{\qquad\qquad} \qquad \text{Simplify.}$$

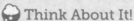 **Go Online** You can complete an Extra Example online.

Today's Goals
• Determine the factors of trinomials with a leading coefficient of 1.
• Determine the factors of trinomials with a leading coefficient not equal to 1.

Today's Vocabulary
prime polynomial

Go Online
You may want to complete the Concept Check to check your understanding.

💭 **Think About It!**
If c is negative, then what is the relationship between the signs of the factors? Explain your reasoning.

_____

_____

_____

_____

_____

_____

Problem-Solving Tip

Guess and Check
When factoring a trinomial, make an educated guess, check for reasonableness, and then adjust the guess until you find the correct answer.

Copyright © McGraw-Hill Education

Example 2 c Is Negative and b Is Positive

Factor $x^2 + 5x - 14$.

In this trinomial, $b = 5$ and $c = -14$. Since c is negative, the factors m and p have opposite signs. So, either m or p is negative, but not both. Since b is positive, the factor with the greater absolute value is also positive.

Complete the table to make a list of the factors of -14, where one factor of each pair is negative and the factor with the greater absolute value is positive. Look for the pair of factors with a sum of 5.

Factors of -14	Sum of Factors
$-1, 14$	___
$-2, 7$	___

The correct factors are -2 and 7.

$x^2 + 5x - 14 = (x + $___$)(x + p)$

$= (x$ ___$)(x + 7)$

Example 3 c Is Negative and b Is Negative

Factor $x^2 - 3x - 4$.

In this trinomial, $b = $___ and $c = $___. Either m or p is negative, but not both. Since b is negative, the factor with the greater absolute value is also negative.

Complete the table to make a list of the factors of -4, where one factor of each pair is negative and the factor with the greater absolute value is negative. Look for the pair of factors with a sum of -3.

Factors of -4	Sum of Factors
$1, -4$	___
$2, -2$	___

The correct factors are 1 and -4.

$x^2 - 3x - 4 = (x + m)(x + p)$

$= (x + $___$)(x$ ___$)$

Example 4 Factor a Polynomial

Factor $x^2 - 4x + 8$, if possible. If the polynomial cannot be factored using integers, write _prime_.

In this trinomial, $b = $___ and $c = $___. Since b is negative, $m + p$ is _____. Since c is positive, mp is _____. So, m and p are both _____.

Next, list the factors of 8. Look for the pair with a sum of -4.

Factors of 8	Sum of Factors
$-1, -8$	___
$-2, -4$	___

There are _____ with a sum of -4. So, the trinomial cannot be factored using integers. Therefore, $x^2 - 4x + 8$ is _____.

Check

Write the factored form of each polynomial. If the polynomial cannot be factored using integers, write _prime_.

a. $x^2 + 7x + 6$　　　　**b.** $x^2 - 8x + 12$　　　　**c.** $x^2 + 3x - 40$

_____　　　　　_____　　　　　_____

Think About It!

Is $(x + 1)(x - 4)$ equal to $(x - 4)(x + 1)$? to $(x - 1)(x + 4)$? Explain your reasoning.

Talk About It

How does the process of factoring $ax^2 + bx + c$ compare to the process of factoring $x^2 + bx + c$?

🌐 Example 5 Solve a Problem by Factoring

FLAG DESIGN Switzerland's flag has a very unique shape; it is a square. However, the flag used by the country's naval vessels is rectangular, as shown. If the area of the square flag is $x^2 - 6x + 9$ square feet, and the length is increased by 4 feet, then what is the area of the naval flag in terms of x?

Step 1 Factor $x^2 - 6x + 9$. In this trinomial, $b = -6$ and $c = 9$. Because c is positive and b is negative, you need to find two _____ factors with a sum of _____ and a product of _____.

Factors of 9	Sum of Factors
−1, −9	_____
−3, −3	_____

The correct factors are −3 and −3.

$$x^2 - 6x + 9 = (x + m)(x + p) \qquad \text{Write the pattern.}$$
$$= (x - 3)(x - 3) \qquad m = -3 \text{ and } p = -3$$

Step 2 Increase length and multiply. The length is increased by 4 feet, so the factor representing the length must be increased by 4.

$$(x - 3 + 4)(x - 3) = (x + 1)(x - 3) \qquad \text{Add 4 to the length.}$$
$$= x^2 - 3x + x - 3 \qquad \text{FOIL}$$
$$= x^2 - 2x - 3 \qquad \text{Simplify.}$$

The new area is $x^2 - 2x - 3$ square feet.

Learn Factoring Trinomials

Key Concept • Factoring Trinomials with a Leading Coefficient ≠ 1

To factor trinomials in the form $ax^2 + bx + c$, find two integers, m and p, with a sum of b and a product of ac. Then write $ax^2 + bx + c$ as $ax^2 + mx + px + c$, and factor by grouping.

Example 6 c Is Negative

Factor $4x^2 + 18x - 10$.

In this trinomial, a GCF of 2 can be factored out.

$$4x^2 + 18x - 10 = \_\_\_\_(2x^2 + 9x - 5)$$

Then in the trinomial $2x^2 + 9x - 5$, $a = \_\_\_\_$, $b = \_\_\_\_$, and $c = \_\_\_\_$. You need to find two numbers with a sum of 9 and a product of $2(-5)$ or −10.

(continued on the next page)

Study Tip

Assumptions Because the polynomial represents the area of a flag and has integer coefficients, you can assume that it can be factored. The area of a rectangle can always be written as the product of two sides.

🤓 Think About It!

Which property is applied to go from $2[2x(x + 5) - (x + 5)]$ to $2(2x - 1)(x + 5)$?

Math History Minute

Abraham bar Hiyya (1070–1136) was a Spanish mathematician, astronomer, and philosopher. His most famous work is *Treatise on Measurement and Calculation,* which contains the first complete solution of the quadratic equation $x^2 - ax + b = 0$ known in Europe. It is said that his work influenced the later work of Leonardo Fibonacci.

Go Online An alternate method is available for this example.

Complete the table to make a list of the factors of −10.

Factors of −10	Sum of Factors
1, −10	_____
2, −5	_____
5, −2	_____
10, −1	_____

Look for a pair of factors with a sum of 9. The correct factors are _____ and _____.

$4x^2 + 18x - 10 = 2(2x^2 + mx + px - 5)$ Write the pattern.

$\qquad = 2(2x^2 + \underline{\quad}x + (\underline{\quad})x - 5)$ $m = 10$ and $p = -1$

$\qquad = 2[(2x^2 + 10x) + (\underline{\qquad})]$ Group terms with common factors.

$\qquad = 2[\underline{\quad}(x + 5) - (x + 5)]$ Factor the GCFs.

$\qquad = \underline{\qquad\qquad}$ $x + 5$ is the common factor.

Example 7 *c* Is Positive

Factor $2x^2 - 17x + 21$.

In this trinomial, $a = \underline{\quad}$, $b = \underline{\quad}$, and $c = \underline{\quad}$. Since b is negative, $m + p$ will be negative. Since c is positive, mp will be positive. To determine m and p, list the negative factors of ac. The sum of m and p should be equal to b.

Complete the table to make a list of the negative factors of 42, and look for a pair of factors with a sum of −17.

Factors of 42	Sum of Factors
−1, −42	_____
−2, −21	_____
−3, −14	_____
−6, −7	_____

The correct factors are _____ and _____.

$2x^2 - 17x + 21 = 2x^2 + mx + px + 21$ Write the pattern.

$\qquad = 2x^2 + (\underline{\quad})x + (\underline{\quad})x + 21$ $m = -3$ and $p = -14$

$\qquad = (2x^2 - 14x) + (-3x + 21)$ Group terms with common factors.

$\qquad = \underline{\quad}(x - 7) + (\underline{\quad})(x - 7)$ Factor the GCFs.

$\qquad = \underline{\qquad\qquad}$ $x - 7$ is the common factor.

Go Online You can complete an Extra Example online.

Practice

Go Online You can complete your homework online.

Examples 1–4

Factor each polynomial, if possible. If the polynomial cannot be factored using integers, write *prime*.

1. $x^2 + 17x + 42$

2. $y^2 - 17y + 72$

3. $a^2 + 8a - 48$

4. $n^2 - 2n - 35$

5. $44 + 15h + h^2$

6. $40 - 22x + x^2$

7. $-24 - 5x + x^2$

8. $-42 - m + m^2$

9. $t^2 + 8t + 12$

10. $d^2 + 5d - 13$

11. $y^2 - 6y + 17$

12. $n^2 + 7n + 12$

13. $b^2 - 12b - 101$

14. $p^2 + 9p + 20$

15. $h^2 + 9h + 18$

16. $c^2 + c + 21$

Example 5

17. **COSMETICS CASE** The top of a cosmetics case is a rectangle in which the width is 2 centimeters greater than the length. The expression $x^2 + 26x + 168$ represents the area of the top of the case. Factor the expression.

18. **CARPENTRY** Miko wants to build a crate to hold record albums. The expression $2x^2 - 6x - 80$ represents the volume of the crate. Factor the expression.

19. **BRIDGE ENGINEERING** A suspension bridge is a bridge in which the deck is supported by cables with towers spaced throughout the span of the bridge. The height of a cable n inches above the deck measured at distance d in yards from the first tower is given by $d^2 - 36d + 324$. Factor the expression.

20. **FINANCE** The break-even point for a business occurs when the revenues equal the cost. A local children's museum studied their costs and revenues from paid admission. They found that their break-even point is given by the expression $2h^2 - 2h - 24$, where h is the number of hours the museum is open per day. Factor the expression.

Factor each polynomial, if possible. If the polynomial cannot be factored using integers, write *prime*.

21. $5x^2 + 34x + 24$

22. $2x^2 + 19x + 24$

23. $4x^2 + 22x + 10$

24. $4x^2 + 38x + 70$

25. $2x^2 - 3x - 9$

26. $4x^2 - 13x + 10$

27. $2x^2 + 3x + 6$

28. $5x^2 + 3x + 4$

29. $12x^2 + 69x + 45$

30. $4x^2 - 5x + 7$

31. $3x^2 - 8x + 15$

32. $5x^2 + 23x + 24$

33. $2x^2 + 3x - 6$

34. $2t^2 + 9t - 5$

35. $2y^2 + y - 1$

36. $4h^2 + 8h - 5$

Mixed Exercises

Factor each polynomial, if possible. If the polynomial cannot be factored using integers, write *prime*.

37. $n^2 + 3n - 18$

38. $x^2 + 2x - 8$

39. $r^2 + 4r - 12$

40. $x^2 - x - 12$

41. $w^2 - w - 6$

42. $y^2 - 6y + 8$

43. $t^2 - 15t + 56$

44. $-4 - 3m + m^2$

45. $2x^2 + 5x + 2$

46. $3n^2 + 5n + 2$

47. $3g^2 - 7g + 2$

48. $2t^2 - 11t + 15$

49. $4x^2 - 3x - 3$

50. $4b^2 + 15b - 4$

Factor each polynomial, if possible. If the polynomial cannot be factored using integers, write *prime*.

51. $9p^2 + 6p - 8$ **52.** $6q^2 - 13q + 6$

53. $a^2 - 10a + 21$ **54.** $x^2 + 2x - 15$

55. $2x^2 + 7x + 3$ **56.** $6x^2 + x + 2$

57. $x^2 + x - 20$ **58.** $x^2 - 6x - 7$

59. $p^2 - 10p + 21$ **60.** $5x^2 - 6x + 1$

61. $q^2 + 11qr + 18r^2$ **62.** $x^2 - 14xy - 51y^2$

63. $x^2 - 6xy + 5y^2$ **64.** $a^2 + 10ab - 39b^2$

65. $-6x^2 - 23x - 20$ **66.** $-4x^2 - 15x - 14$

67. $-5x^2 + 18x + 8$ **68.** $-6x^2 + 31x - 35$

69. $-4x^2 + 5x - 12$ **70.** $-12x^2 + x + 20$

71. MONUMENTS Susan is designing a pyramidal stone monument for a local park. The design specifications tell her that the width of the base must be 5 feet less than the length and there must be 2 feet of space on each side of the pyramid. The expression $x^2 + 3x - 4$ represents the area that the pyramidal stone monument will require.

a. Factor the expression that represents the area that the pyramidal stone monument will require.

b. What does x represent?

72. PROJECTILES The height of a projectile in feet is given by $-16t^2 + vt + h_0$, where t is the time in seconds, v is the initial upward velocity in feet per second, and h_0 is the initial height in feet. A T-shirt is propelled from 32 feet above ground level into the air at an initial velocity of 16 feet per second.

a. Write an expression to represent how much time the T-shirt is in the air.

b. Factor the expression that represents the amount of time the T-shirt is in the air.

Higher-Order Thinking Skills

73. WHICH ONE DOESN'T BELONG? Which expression does not belong with the others? Explain your reasoning.

| $x^2 + 2x - 24$ | $x^2 + 11x + 24$ | $x^2 - 10x - 24$ | $x^2 + 12x + 24$ |

74. FIND THE ERROR Jamaall and Charles have factored $x^2 + 6x - 16$. Is either of them correct? Explain your reasoning.

Jamaall	Charles
$x^2 + 6x - 16 = (x + 2)(x - 8)$	$x^2 + 6x - 16 = (x - 2)(x + 8)$

ANALYZE Find all values of k so that each polynomial can be factored using integers.

75. $x^2 + kx - 19$

76. $x^2 + kx + 14$

77. $x^2 - 8x + k, k > 0$

78. $x^2 - 5x + k, k > 0$

79. ANALYZE For any factorable trinomial, $x^2 + bx + c$, will the absolute value of b *sometimes*, *always*, or *never* be less than the absolute value of c? Justify your argument.

80. CREATE Give an example of a trinomial that can be factored using the factoring techniques presented in this lesson. Then factor the trinomial.

81. PERSEVERE Factor $(4y - 5)^2 + 3(4y - 5) - 70$.

82. WRITE Explain how to factor trinomials of the form $x^2 + bx + c$ and how to determine the signs of the factors of c.

83. ANALYZE A square has an area of $9x^2 + 30xy + 25y^2$ square inches. The dimensions are binomials with positive integers coefficients. What is the perimeter of the square? Explain.

84. PERSEVERE Find all values of k so that $2x^2 + kx + 12$ can be factored as two binomials using integers.

85. WRITE Explain how to determine which values should be chosen for m and p when factoring a polynomial of the form $ax^2 + bx + c$.

Factoring Special Products

Today's Goals
- Factor binomials that are differences of squares.
- Factor trinomials that are perfect squares.

Today's Vocabulary
difference of two squares
perfect square trinomials

Explore Using Algebra Tiles to Factor Differences of Squares

 Online Activity Use algebra tiles to complete the Explore.

> **INQUIRY** How is factoring a difference of squares related to the product of a sum and a difference?

Learn Factoring Differences of Squares

The product of the sum and difference of two quantities results in a **difference of two squares.** So, the factored form of a difference of squares is the product of the sum and difference of two quantities.

Key Concept • Factoring Differences of Squares	
Symbols	$a^2 - b^2 = (a + b)(a - b)$ or $(a - b)(a + b)$
Examples	$x^2 - 9 = (x + 3)(x - 3)$ or $(x - 3)(x + 3)$
	$4u^2 - 1 = (2u + 1)(2u - 1)$ or $(2u - 1)(2u + 1)$

Example 1 Factor Differences of Squares

Factor each polynomial.

a. $81v^2 - 64w^2$

$81v^2 - 64w^2 = (\underline{\quad})^2 - (\underline{\quad})^2$ Write in the form $a^2 - b^2$.

$= \underline{\hspace{3cm}}$ Factor the difference of squares.

b. $1 - 144q^2$

$1 - 144q^2 = (\underline{\quad})^2 - (\underline{\quad})^2$ Write in the form $a^2 - b^2$.

$= \underline{\hspace{3cm}}$ Factor the difference of squares.

Check

Factor each polynomial.

$25x^2 - 64y^2$ $\underline{\hspace{3cm}}$

$196 - g^4$ $\underline{\hspace{3cm}}$

 **Go Online** You can complete an Extra Example online.

Copyright © McGraw-Hill Education

Copyright © McGraw-Hill Education

Watch Out!

Sum of Squares The sum of squares cannot be factored; that is, $a^2 + b^2 \neq (a + b)(a + b)$. The sum of squares is a prime polynomial.

Example 2 Factor More Than Once

Factor $x^4 - 256$.

$x^4 - 256 = (\underline{\quad})^2 - (\underline{\quad})^2$ Write in the form $a^2 - b^2$.

$= (x^2 + 16)(\underline{\quad\quad})$ Factor the difference of squares.

$= (x^2 + 16)(\underline{\quad\quad})$ $x^2 - 16$ is also a difference of squares.

$= (x^2 + 16)\underline{\quad\quad\quad}$ Factor the difference of squares.

Check

Factor each polynomial.

$81n^4 - 1$

_____

$16x^8 - y^4$

_____

_____

Think About It!

What would the factors be if Rosita's family knew that they needed to tile a 1-meter square area near the door, but did not know the dimensions of their square living room?

🌐 Example 3 Use Factors to Find Area

AREA Rosita's family is buying carpet for their living room, which is 6 meters wide and 6 meters long. They plan to carpet the whole area except for a square area near the door, which will be tiled. Find the factors representing the area of the living room that will be carpeted.

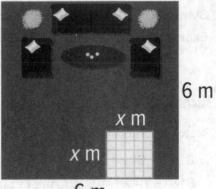

The area of the living room is $6 \cdot 6$ or 36 square meters, and the tiled area is $x \cdot x$ or x^2 square meters. So, the carpet will cover an area of $36 - x^2$.

$36 - x^2 = (\underline{\quad})^2 - (\underline{\quad})^2$ Write in the form $a^2 - b^2$.

$= \underline{\quad\quad\quad}$ Factor the difference of squares.

Check

VOLUME A rectangular solid has a volume of $x^4 - 16$. Factor the polynomial to determine the dimensions of the rectangular solid. _____

A. $x^2(x + 4)(x - 4)$

B. $(x^2 + 4)(x + 2)(x - 2)$

C. $(x^2 - 4)(x + 2)(x + 2)$

D. $(x^3 + 4)(x - 2)(x - 2)$

▶ Go Online You can complete an Extra Example online.

Learn Factoring Perfect Squares

Squares of binomials, such as $(a + b)^2$ and $(a - b)^2$, have special products called **perfect square trinomials.**

$$(a + b)^2 = (a + b)(a + b) \qquad\qquad (a - b)^2 = (a - b)(a - b)$$
$$= a^2 + ab + ab + b^2 \qquad\qquad = a^2 - ab - ab + b^2$$
$$= a^2 + 2ab + b^2 \qquad\qquad = a^2 - 2ab + b^2$$

For a trinomial to be factorable as a perfect square, the following must be true:

- The first term is a perfect square.

- The last term is a perfect square.

- The middle term is two times the product of the square roots of the first and last terms.

Key Concept · Factoring Perfect Square Trinomials	
Symbols	$a^2 + 2ab + b^2 = (a + b)(a + b) = (a + b)^2$
	$a^2 - 2ab + b^2 = (a - b)(a - b) = (a - b)^2$
Examples	$x^2 + 10x + 25 = (x + 5)(x + 5) = (x + 5)^2$
	$4x^2 - 28x + 49 = (2x - 7)(2x - 7) = (2x - 7)^2$

Example 4 Identify a Perfect Square Trinomial

Determine whether $4j^2 + 8j + 16$ is a perfect square trinomial. If so, factor it.

Is the first term a perfect square? _____

Is the last term a perfect square? _____

Is the second term equal to $2(2j)(4)$? _____

Because this trinomial does not satisfy all conditions, it is not a perfect square trinomial.

Go Online You can complete an Extra Example online.

Think About It!

To make $?j^2 + 8j + 16$ a perfect square trinomial, what would the coefficient of the first term need to be? _____

To make $4j^2 + ?j + 16$ a perfect square trinomial, what would the coefficient of the middle term need to be? _____

To make $4j^2 + 8j + ?$ a perfect square trinomial, what would the constant term need to be? _____

Copyright © McGraw-Hill Education

 **Go Online** to practice what you've learned about factoring quadratic expressions in the Put It All Together over Lessons 10-6 and 10-7.

Example 5 Recognize and Factor a Perfect Square Trinomial

Determine whether $36h^2 - 12h + 1$ is a perfect square trinomial. If so, factor it.

Is $36h^2$ a perfect square? _____

Is 1 a perfect square? _____

Is $-12h$ equal to $-2(6h)(1)$? _____

This trinomial satisfies all the conditions for a perfect square trinomial.

$36h^2 - 12h + 1 = (6h)^2 - 2(6h)(1) + (1)^2$ Write as $a^2 - 2ab + b^2$.

               = _____   Factor using the pattern.

Check

If the trinomial is a perfect square trinomial, factor it. If not, write *not a perfect square trinomial*.

$36x^2 - 60x + 25$

_____

$36x^2 - 30x + 25$

_____

$36x^2 + 60x + 25$

_____

Pause and Reflect

Did you struggle with anything in this lesson? If so, how did you deal with it?

Record your observations here.

Go Online You can complete an Extra Example online.

Name _____ Period _____ Date _____

Practice

® **Go Online** You can complete your homework online.

Examples 1 and 2
Factor each polynomial.

1. $q^2 - 121$

2. $r^4 - k^4$

3. $w^4 - 625$

4. $r^2 - 9t^2$

5. $h^4 - 256$

6. $2x^3 - x^2 - 162x + 81$

7. $x^2 - 4y^2$

8. $3c^3 + 2c^2 - 147c - 98$

9. $f^3 + 2f^2 - 64f - 128$

10. $r^3 - 5r^2 - 100r + 500$

11. $3t^3 - 7t^2 - 3t + 7$

12. $a^2 - 49$

13. $4m^3 + 9m^2 - 36m - 81$

14. $3x^3 + x^2 - 75x - 25$

Example 3

15. TICKETING A ticketing company for sporting events analyzes the ticket purchasing patterns. The expression $9a^2 - 4b^2$ is developed to help officials calculate the likely number of people who will buy tickets for a certain sporting event. Factor the expression.

16. BINGO A bingo card contains 25 square spaces arranged into a larger square. 24 of the squares are labeled with numbers and the center square is labeled as a free space. The expression $16 - x^2$ represents total area of the squares labeled with numbers. Factor the expression.

17. DECORATING Marvin saw a rug in a store that he would like to purchase. It has an area represented by the expression shown on the rug. He cannot remember the length and width, but he remembers that the length and the width were the same.

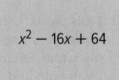

$x^2 - 16x + 64$

  a. Factor the expression that represents the area of the rug.

  b. What do the factors in the factored expression represent?

Examples 4 and 5
Determine whether each trinomial is a perfect square trinomial. Write *yes* or *no*. If so, factor it.

18. $4x^2 - 42x + 110$

19. $16x^2 - 56x + 49$

20. $81x^2 - 90x + 25$

21. $x^2 + 26x + 168$

Copyright © McGraw-Hill Education

Lesson 10-7 • Factoring Special Products   **595**

Mixed Exercises

Factor each polynomial, if possible. If the polynomial cannot be factored using integers, write *prime*.

22. $36t^2 - 24t + 4$

23. $4h^2 - 56$

24. $17a^2 - 24ab$

25. $q^2 - 14q + 36$

26. $y^2 + 24y + 144$

27. $6d^2 - 96$

28. $1 - 49d^2$

29. $-16 + p^2$

30. $k^2 + 25$

31. $36 - 100w^2$

32. $64m^2 - 9y^2$

33. $4h^2 - 25g^2$

34. $x^3 + 3x^2 - 4x - 12$

35. $8x^2 - 72p^2$

36. $20q^2 - 5r^2$

37. $32a^2 - 50b^2$

38. $16b^2 - 100$

39. $49x^2 - 64y^2$

40. $3n^4 - 42n^3 + 147n^2$

41. $8m^3 - 24m^2 + 18m$

42. **GARDEN DESIGN** Marren is planning to build a raised garden bed. The area of the rectangular plot can be represented by $x^2 - 49$. Factor the expression to determine the possible length and width of the garden bed.

43. **PARKING LOT** The area of a rectangular parking lot is represented by the expression $a^2 - 25$, where the length is longer than the width. Factor the expression to determine the possible dimensions of the length and width of the parking lot. If the length of the parking lot is 105 yards, what is the width of the parking lot?

44. **USE A SOURCE** Research the dimensions of the outside diameter and inside diameter of metal washers. Write an expression for the surface area of the top of a metal washer with outside diameter D and inside diameter d. Factor your expression. Then use your expression and the dimensions you researched to find the surface area of the top of a metal washer.

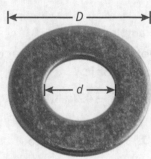

Determine whether each trinomial is a perfect square trinomial. Write *yes* or *no*. If so, factor it.

45. $m^2 - 6m + 9$

46. $r^2 + 4r + 4$

47. $g^2 - 14g + 49$

48. $2w^2 - 4w + 9$

49. $4d^2 - 4d + 1$

50. $9n^2 + 30n + 25$

51. $9z^2 - 6z + 1$

52. $36x^2 - 60x + 25$

53. $49r^2 + 14r + 4$

54. $a^2 + 14a + 49$

55. $t^2 - 18t + 81$

56. $4c^2 + 2cd + d^2$

57. ARCHITECTURE The drawing shows a triangular roof truss with a base measuring the same as its height. The expression $\frac{1}{2}x^2 - 98$ represents the area of the triangular roof truss, where x is the length of the base and the height. Factor the expression.

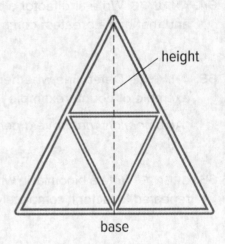

height

base

58. MANUFACTURING A company that manufactures cardboard boxes sells three sizes of boxes. The volume of the small box is represented by $n^3 + 4n^2 - 16n - 64$ in³. The volume of the medium box is represented by $n^3 + 6n^2 - 36n - 216$ in³. The volume of the large box is represented by $n^3 + 8n^2 - 64n - 512$ in³. The company wants to start selling an extra-large box. Predict the dimensions and the volume of the extra-large box.

a. Find the dimensions of the small, medium, and large boxes. Explain the method you used.

b. Look for a pattern in the areas. What pattern is represented by the areas of the small, medium, and large boxes? Use this pattern to predict the dimensions of the extra-large box.

c. Use your prediction in **part b** to find the volume of the extra-large box.

59. ANALYZE Paul suggests that to factor $x^4 - 81$, you can use difference of squares twice. He says that the result will have the same factors as $x^2 - 9$. Prove or disprove his statement.

60. REASONING Debony claims the expression $9x^2 + 50x + 25$ is a perfect square trinomial. Is she correct? If she is incorrect, show how the expression can be changed so that it is a perfect square.

61. STRUCTURE Find a value for m that will make the expression $4x^4 - 44x^2 + m$ a perfect square. Then use the value to factor the expression completely.

62. FIND THE ERROR Elizabeth and Lorenzo are factoring an expression. Is either of them correct? Explain your reasoning.

Elizabeth	Lorenzo
$16x^4 - 25y^2 =$	$16x^4 - 25y^2 =$
$(4x - 5y)(4x + 5y)$	$(4x^2 - 5y)(4x^2 + 5y)$

63. PERSEVERE Factor and simplify $9 - (k + 3)^2$, a difference of squares.

64. ANALYZE Write and factor a binomial that is the difference of two perfect squares and that has a greatest common factor of $5mk$.

65. ANALYZE Determine whether the following statement is *true* or *false*. Give an example or counterexample to justify your answer.

   All binomials that have a perfect square in each of the two terms can be factored.

66. CREATE Write a binomial in which the difference of squares pattern must be repeated to factor it completely. Then factor the binomial.

67. WRITE Describe why the difference of squares has no middle term.

68. WHICH ONE DOESN'T BELONG? Identify the trinomial that does not belong. Explain.

$4x^2 - 36x + 81$	$25x^2 + 10x + 1$	$4x^2 + 10x + 4$	$9x^2 - 24x + 16$

69. WRITE Explain how to determine whether a trinomial is a perfect square trinomial.

70. PERSEVERE Use the difference of squares to factor and simplify the expression $121x^2y^6z^4 - 16y^2z^2$.

ⓔ Essential Question

How can you perform operations on polynomials and use them to represent real-world situations?

Module Summary

Lesson 10-1

Adding and Subtracting Polynomials

- The degree of a monomial is the sum of the exponents of all its variables.
- The degree of a polynomial is the greatest degree of any term in the polynomial.
- When adding polynomials, you can group like terms by using a horizontal or vertical format.
- You can subtract a polynomial by adding its additive inverse.
- Adding or subtracting polynomials results in a polynomial, so the set of polynomials is closed under addition and subtraction.

Lessons 10-2 through 10-4

Multiplying Polynomials

- To find the product of a monomial and a binomial, use the Distributive Property.
- The FOIL method helps to multiply binomials: To multiply two binomials, find the sum of the products of *F* the *First Terms*, *O* the *Outer terms*, *I* the *Inner terms*, and *L* the *Last terms*.
- Square of a sum: $(a + b)^2 = (a + b)(a + b)$
  $= a^2 + 2ab + b^2$
- Square of a difference: $(a - b)^2 =$
  $(a - b)(a - b) = a^2 - 2ab + b^2$
- The product of $a + b$ and $a - b$ is the square of a minus the square of b: $(a + b)(a - b) =$
  $a^2 - ab + ab - b^2 = a^2 - b^2$

Lessons 10-5 through 10-7

Factoring Polynomials

- By using the Distributive Property in reverse, you can factor a polynomial as the product of a monomial and a polynomial.
- To factor trinomials in the form $x^2 + bx + c$, find two integers, *m* and *p*, with a sum of *b* and a product of *c*. Then write $x^2 + bx + c$ as $(x + m)(x + p)$.
- To factor trinomials in the form $ax^2 + bx + c$, find two integers, *m* and *p*, with a sum of *b* and a product of *ac*. Then write $ax^2 + bx + c$ as $ax^2 + mx + px + c$, and factor by grouping.
- The factored form of a difference of squares is the product of the sum and difference of two quantities.
  $$a^2 - b^2 = (a + b)(a - b) \text{ or } (a - b)(a + b)$$
- Squares of binomials, such as $(a + b)^2$ and $(a - b)^2$, have special products called perfect square trinomials.
  $$a^2 + 2ab + b^2 = (a + b)(a + b) = (a + b)^2$$
  $$a^2 - 2ab + b^2 = (a - b)(a - b) = (a - b)^2$$

Study Organizer

📖 **Foldables**

Use your Foldable to review this module. Working with a partner can be helpful. Ask for clarification of concepts as needed.

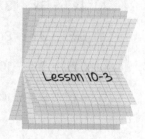

Lesson 10-3

Test Practice

1. MULTI-SELECT Select all of the expressions that are polynomials. (Lesson 10-1)

(A) $4xy - 3x$

(B) $4ab + \dfrac{5}{x^2}$

(C) $45 + 3d^2 + d - w^3$

(D) 12

(E) $\dfrac{x^{-3}y^2}{5}$

2. OPEN RESPONSE The perimeter of this triangular plot of land can be represented by $3x^2 + 10x + 20$. (Lesson 10-1)

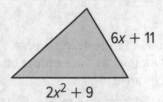

Write a polynomial that represents the measure of the third side of the plot of land.

┌──────────────────────────────────────┐
└──────────────────────────────────────┘

3. MULTIPLE CHOICE Find the difference. $(5d - 8) - (-2d + 3)$ (Lesson 10-1)

(A) $3d - 5$

(B) $7d - 11$

(C) $7d - 5$

(D) $7d + 11$

4. MULTIPLE CHOICE Find the sum. $(x^2 + 3x - 9) + (4x^2 - 6x + 1)$ (Lesson 10-1)

(A) $4x^2 - 3x - 8$

(B) $5x^2 - 3x - 8$

(C) $5x^2 + 9x - 10$

(D) $4x^2 + 9x - 10$

5. OPEN RESPONSE Mr. Soto is installing a new trapezoid shaped window. He plans to cover the window with a protective coating that prevents UV rays from entering the house. The height h of the window is 24 inches.

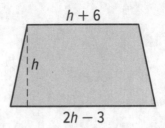

The formula for the area of a trapezoid is $A = \dfrac{1}{2}h(b_1 + b_2)$ where h is the height and b_1 and b_2 are the bases of the trapezoid. How many square inches of protective coating will Mr. Soto need? (Lesson 10-2)

┌──────────────────────────────────────┐
└──────────────────────────────────────┘

6. OPEN RESPONSE Use the Distributive Property to simplify $-3(g^2 - 4g + 1)$. (Lesson 10-2)

┌──────────────────────────────────────┐
└──────────────────────────────────────┘

7. MULTIPLE CHOICE Find the product of $(2x + 7)$ and $(x - 5)$. (Lesson 10-3)

(A) $3x + 2$

(B) $2x^2 - 3x - 35$

(C) $2x^2 - 10x - 35$

(D) $2x^2 - 35$

8. OPEN RESPONSE Saurabh is designing a rectangular vegetable garden with a stone path around it. The total area of the path is 296 square feet. Use the diagram to determine the value of x. (Lesson 10-3)

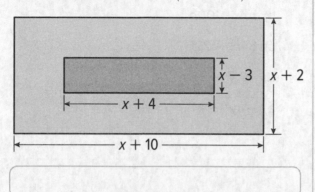

9. OPEN RESPONSE Find the product. $(2r - 3)(r^2 - 5r + 1)$ (Lesson 10-3)

10. OPEN RESPONSE *True* or *false*: $(8k^2 + 3k - 1)(k^2 - k + 1)$ is equivalent to $8k^4 - 5k^3 + 4k^2 + 4k - 1$. (Lesson 10-3)

11. MULTIPLE CHOICE Find $(8a + 3b)^2$. (Lesson 10-4)

Ⓐ $73a^2b^2 + 48ab$

Ⓑ $64a^2 + 48ab + 9b^2$

Ⓒ $64a^2 + 9b^2$

Ⓓ $32a^2 + 22ab + 12b^2$

12. OPEN RESPONSE Find the product. $(6h + 7)(6h - 7)$ (Lesson 10-4)

13. OPEN RESPONSE Describe a general rule for finding the square of a sum. (Lesson 10-4)

14. MULTIPLE CHOICE Kala is making a tile design for her kitchen floor. Each tile has sides that are 3 inches less than twice the side length of the smaller square inside the design. (Lesson 10-4)

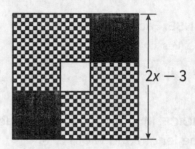

Select the polynomial that represents the area of the tile.

Ⓐ $2x^2 - 3x$

Ⓑ $4x^2 - 12x + 9$

Ⓒ $4x^2 + 12x + 9$

Ⓓ $4x^2 - 9$

15. OPEN RESPONSE Factor $-10xy - 15x + 12y + 18$. (Lesson 10-5)

16. OPEN RESPONSE Factor $36x^3 - 24x^2$. (Lesson 10-5)

17. MULTI-SELECT A golf ball is hit with a velocity of 128 feet per second. The expression $128t - 16t^2$ represents the height of the ball after t seconds. Select all that are equivalent to the expression. (Lesson 10-5)

Ⓐ $16t(8 - t)$

Ⓑ $-16t(t - 8)$

Ⓒ $t(t - 8)(8 - t)$

Ⓓ $8t(7 - 2t)$

Ⓔ $8(- 2t + 16)$

18. OPEN RESPONSE Factor $x^2 + 8x + 15$. (Lesson 10-6)

19. MULTIPLE CHOICE Mrs. Torres wants to add x feet onto the length and width of an existing rectangular patio. The new patio would have an area of $x^2 + 14x + 48$ square feet. (Lesson 10-6)

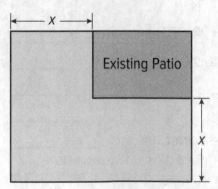

What are the dimensions of the existing patio?

Ⓐ 14 ft by 48 ft

Ⓑ 6 ft by 8 ft

Ⓒ 7 ft by 24 ft

Ⓓ 4 ft by 12 ft

20. OPEN RESPONSE Factor $9x^2 + 9x + 2$.
(Lesson 10-6)

21. OPEN RESPONSE Identify if each polynomial is a difference of squares. Write *yes* or *no*. (Lesson 10-7)

a) $16y^2 - 25b^2$

b) $m^2 - 100n^2$

c) $64r^2 - 225$

d) $49z^2 - 36a^3$

e) $-4y^2 - k^2$

22. MULTI-SELECT Select all of the perfect square trinomials. (Lesson 10-7)

Ⓐ $49x^2 + 112x + 64$

Ⓑ $16x^2 - 24x + 9$

Ⓒ $49x^2 + 30x + 64$

Ⓓ $9x^2 - 6x + 16$

Ⓔ $x^2y^2 - 10xy^2 + 25y^2$

23. MULTIPLE CHOICE A square piece of cloth has an area of $4y^2 - 28y + 49$ square meters. Find the length of each side.
(Lesson 10-7)

Ⓐ $(2y + 7)$ m

Ⓑ $(2y - 7)$ m

Ⓒ $(4y - 49)$ m

Ⓓ $(2y - 14)$ m

Quadratic Functions

e Essential Question

Why is it helpful to have different methods to analyze quadratic functions and solve quadratic equations?

What Will You Learn?

Place a check mark (✓) in each row that corresponds with how much you already know about each topic **before** starting this module.

KEY

👎 — I don't know.   👈 — I've heard of it.   👍 — I know it!

	Before			After		
	👎	👈	👍	👎	👈	👍
graph quadratic functions using a table						
graph quadratic functions using key features						
transform quadratic functions						
solve systems of linear and quadratic equations						
solve quadratic equations by factoring						
solve quadratic equations by completing the square						
find maximum and minimum values						
solve quadratic equations using the Quadratic Formula						
fit a quadratic function to data						
combine functions						

📖 **Foldables** Make this Foldable to help you organize your notes about quadratic functions. Begin with a sheet of notebook paper.

1. **Fold** the sheet of paper along the length so that the edge of the paper aligns with the margin rule of the paper.

2. **Fold** the sheet twice widthwise to form four sections.

3. **Unfold** the sheet and cut along the folds on the front flap only.

4. **Label** each section as shown.

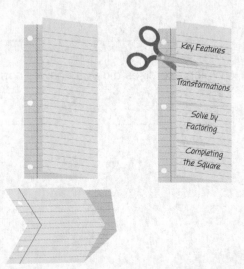

What Vocabulary Will You Learn?

Check the box next to each vocabulary term that you may already know.

- ☐ axis of symmetry
- ☐ coefficient of determination
- ☐ completing the square
- ☐ curve fitting
- ☐ discriminant

- ☐ double root
- ☐ maximum
- ☐ minimum
- ☐ parabola
- ☐ quadratic equation

- ☐ quadratic function
- ☐ standard form of a quadratic function
- ☐ vertex form

Are You Ready?

Complete the Quick Review to see if you are ready to start this module.
Then complete the Quick Check.

Quick Review

Example 1

Describe the transformation in $g(x) = 5^x + 3$ as it relates to the graph of the parent function $f(x) = 5^x$.

A constant term is being added to the function, so it is a translation of that many units up.

The graph of $g(x) = 5^x + 3$ is the translation of the graph of the parent function $f(x) = 5^x$ 3 units up.

Example 2

Determine whether $x^2 - 10x + 25$ is a perfect square trinomial. Write *yes* or *no*. If so, factor it.

Is the first term, x^2, a perfect square? yes

Is the last term, 25, a perfect square? yes

Is the middle term equal to the opposite of twice the product of the first and last terms?

yes; $2(5)(x) = 10x$ and the opposite of $10x$ is $-10x$.

$x^2 - 10x + 25 = (x - 5)^2$

Quick Check

Describe the transformation in $g(x)$ as it relates to the graph of the parent function $f(x) = 2^x$.

1. $g(x) = 2^{x-4}$ **2.** $g(x) = -2^x$

3. $g(x) = 2^x - 1$ **4.** $g(x) = 0.25 \cdot 2^x$

Determine whether each trinomial is a perfect square trinomial. Write *yes* or *no*. If so, factor it.

5. $a^2 + 12a + 36$

6. $w^2 + 5w + 25$

7. $m^2 - 22m + 121$

8. $5t^2 + 12t + 100$

How Did You Do?

Which exercises did you answer correctly in the Quick Check? Shade those exercise numbers below.

       ⑧

Graphing Quadratic Functions

Learn Analyzing Graphs of Quadratic Functions

A second-degree polynomial, for example $x^2 + 2x - 8$, is a quadratic polynomial, which has a related quadratic function $f(x) = x^2 + 2x - 8$. A **quadratic function** has an equation of the form $y = ax^2 + bx + c$, where $a \neq 0$. The graph of a quadratic function is called a **parabola**.

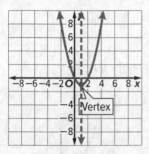

Parabolas are symmetric about a central line called the **axis of symmetry**. Every point on the parabola to the left of the axis of symmetry has a corresponding point on the right half.

The axis of symmetry intersects a parabola at the vertex. The vertex is either the lowest point or the highest point on a parabola.

When $a > 0$, the graph of $y = ax^2 + bx + c$ opens up. In this case, the graph has a **minimum** at the lowest point. When $a < 0$, the graph opens down. In this case, the graph has a **maximum** at the highest point.

The leading coefficient determines the end behavior of a quadratic function.

Example 1 Identify Characteristics: Graph with x-intercept

Identify the axis of symmetry, the vertex, and the y-intercept of the graph. Then describe the end behavior.

The equation of the axis of symmetry is $x =$ _____.

The vertex is located at the maximum point, (_____).

The parabola crosses the y-axis at (0, 4), so the y-intercept is _____.

As x increases or decreases, y _____.

Today's Goals
- Analyze graphs of quadratic functions.
- Graph quadratic functions by using key features and tables.
- Use graphing calculators to analyze key features of quadratic functions.

Today's Vocabulary
quadratic function

parabola

axis of symmetry

minimum

maximum

end behavior

standard form of a quadratic function

🗪 Think About It!

What do you notice about the equation of the axis of symmetry and the x-coordinate of the vertex?

Check

Identify the axis of symmetry, the vertex, and the *y*-intercept of the graph. Then describe the end behavior.

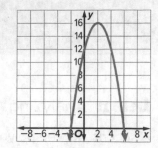

axis of symmetry: *x* = _____.

vertex: (_____, _____).

y-intercept: _____

end behavior:

As *x* increases, *y* _____.

As *x* decreases, *y* _____.

Example 2 Identify Characteristics: Graph with No *x*-intercept

Identify the axis of symmetry, the vertex, and the *y*-intercept of the graph. Then describe the end behavior of the function.

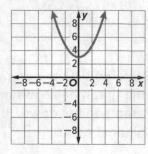

The equation of the axis of symmetry is *x* = _____.

The vertex is located at the minimum point, (_____, _____).

The parabola crosses the *y*-axis at (0, 3), so the *y*-intercept is _____.

The parabola opens _____.

As *x* increases, *y* _____.

As *x* decreases, *y* _____.

Check

Identify the axis of symmetry, the vertex, and the *y*-intercept of the graph. Then describe the end behavior of the function.

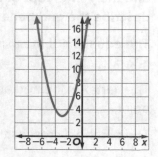

axis of symmetry: *x* = _____.

vertex: (_____, _____).

y-intercept: _____

end behavior:

As *x* increases, *y* _____.

As *x* decreases, *y* _____.

Go Online You can complete an Extra Example online.

Explore Graphing Parabolas

Online Activity Use graphing technology to complete the Explore.

> **INQUIRY** How can you use the equation of a quadratic function to visualize its graph? ×

Learn Graphing Quadratic Functions

Key Concept • Standard Form of a Quadratic Function

Words	The **standard form of a quadratic function** is $f(x) = ax^2 + bx + c$, where $a \neq 0$.
Examples	In $f(x) = 3x^2 - 2x + 9$, $a = 3$, $b = -2$, and $c = 9$. In $f(x) = -16x^2$, $a = -16$, $b = 0$, and $c = 0$.

Key Concept • Graphing Quadratic Functions

Step 1 Find the equation of the axis of symmetry, $x = -\dfrac{b}{2a}$.

Step 2 Find the vertex, and determine whether it is a maximum or minimum.

Step 3 Find the y-intercept.

Step 4 Use symmetry to find additional points on the graph, if necessary.

Step 5 Connect the points with a smooth curve.

Example 3 Graph a Quadratic Function by Using Key Features

Graph $f(x) = x^2 + 2x - 6$.

Step 1 Find the axis of symmetry.

Use the formula to find the equation of the axis of symmetry.

$x = -\dfrac{b}{2a}$ Equation of the axis of symmetry

$x = -\dfrac{2}{2(1)} =$ _____ $a = 1$ and $b = 2$

Step 2 Find the vertex.

Because the axis of symmetry passes through the vertex, use −1 as the x-coordinate of the vertex. Find the y-coordinate using the original equation.

$f(x) = x^2 + 2x - 6$ Original equation

$= ($_____$)^2 + 2($_____$) - 6$ $x = -1$

$=$ _____   Simplify.

The vertex lies at $(-1, -7)$. Because a is _____, the graph opens _____. So the vertex is a _____.

(continued on the next page)

Study Tip

Open and Closed Intervals The interval at which a function increases or decreases is always an open interval along the x-axis and is expressed using parentheses. A closed interval is expressed using brackets.

Talk About It!

Compare and contrast the graphs of exponential and quadratic functions.

Watch Out!

Minimum and Maximum Values Don't forget to find both coordinates of the vertex (x, y). The minimum or maximum value is the y-coordinate.

Copyright © McGraw-Hill Education

Study Tip

Symmetry and Points
When locating points
that are on opposite
sides of the axis of
symmetry, not only are
the points equidistant
from the axis of
symmetry, they are
also equidistant from
the vertex.

Go Online

You can watch a video
to see how to graph
quadratic functions by
using a table.

Think About It!

Why do you think that
the range is {y | y ≥ −6}?

Step 3 Find the y-intercept.

$f(x) = x^2 + 2x - 6$ Original equation

$= (\underline{\quad})^2 + 2(\underline{\quad}) - 6$ $x = 0$

$= \underline{\quad}$ Simplify.

The y-intercept is _____.

Step 4 Find additional points.

Use the axis of symmetry to find the corresponding points.

Let $x = 2$.

$f(x) = x^2 + 2x - 6$ Original equation

$= (\underline{\quad})^2 + 2(\underline{\quad}) - 6$ $x = 2$

$= \underline{\quad}$ Simplify.

Step 5 Connect the points.

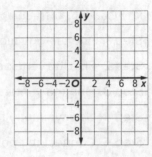

Example 4 Graph a Quadratic Function by Using a Table

Use a table of values to graph $y = 2x^2 - 8x + 2$.

First find the x-coordinate of the vertex by using the equation for the axis of symmetry.

$x = -\dfrac{b}{2a}$ Equation of the axis of symmetry

$x = -\dfrac{-8}{2(2)} = \underline{\quad}$ $a = 2, b = -8$

Complete the table.

x	0	1	2	3	4
y					

y-intercept (0, _____)

The vertex of this function is a _____.

Graph the function.

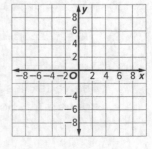

The parabola extends to infinity, so the domain
is _____. The range is {y | y ≥ ____}.

Go Online You can complete an Extra Example online.

🌐 Example 5 Use the Graph of a Quadratic Function

CATAPULT In 1304, Edward I built Warwolf, one of the largest catapults ever used in battle. It had the capacity to launch a 300-pound boulder over 900 feet. The height of a projectile launched from Warwolf can be modeled by the function $h(x) = -16x^2 + 96x + 40$, where $h(x)$ represents the height in feet of the projectile after x seconds. Graph the function. Interpret the key features of the graph in terms of the quantities.

Step 1 Find the axis of symmetry and vertex.

$$x = -\frac{b}{2a}$$ Equation of the axis of symmetry

$$x = -\frac{96}{2(-16)} = \underline{\qquad}$$ $a = -16$ and $b = 96$

$$h(x) = -16x^2 + 96x + 40$$ Original equation

$$= -16(\underline{\quad})^2 + 96(\underline{\quad}) + 40$$ $x = 3$

$$= \underline{\qquad}$$ Simplify.

The vertex is at $(\underline{\quad}, \underline{\quad})$.

Step 2 Complete the table by substituting each x-value.

x	0	1	2	3	4	5
$h(x)$						

Step 3 Graph the function.

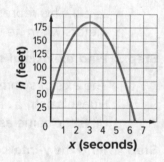

Step 4 Interpret the key features.

intercepts: $x =$ _____, so the projectile _____ about ____ seconds after it was launched.

vertex: (____ , ____), so the projectile reached its maximum _____ of ____ feet ____ seconds after launch.

increasing: The function is increasing for $x <$ _____, so it is _____ height up to ____ seconds after launch.

decreasing: The function is decreasing for $x >$ _____, so it is _____ seconds after launch.

positive: intervals where the function represents the _____

end behavior: represents the projectile _____ and _____ to the ground

domain: all nonnegative numbers _____

range: $\{h(x) \mid$ _____ $\leq h(x) \leq$ _____ $\}$

Common Error A common error when determining the axis of symmetry of a parabola is to forget the negative sign in the equation $x = -\frac{b}{2a}$. Remember that if your parabola opens down, a is negative and, therefore, $2a$ will also be negative.

Study Tip

Labeling Axes When modeling a real-world quadratic function on the coordinate plane, it is important to appropriately label the axes using the given information. In this case, the independent variable is height in feet h; therefore, the y-axis is labeled h (feet).

Go Online You can watch a video to see how to graph and analyze a quadratic function on a graphing calculator.

Go Online to see how to use a graphing calculator with this example.

Math History Minute

During the 19th century, several designers of the Brooklyn Bridge were unable to complete it. But **Emily Warren Roebling** (1843–1903) supervised the bridge's construction until its completion. The bridge was the first major suspension bridge constructed in the U.S. Its cables form a catenary, which approximates a parabola.

Learn Analyzing Key Features of Quadratic Functions

You can use a graphing calculator to analyze key features of quadratic functions by graphing or by using the table feature.

🌐 Example 6 Interpret the Graph of a Quadratic Function

HORSES In 1949, Huaso, ridden by Alberto Larraguibel Morales, set the world record for the highest jump by a horse when he cleared a 2.47-meter-high obstacle. His path can be approximately modeled by $f(x) = -0.418x^2 + 2.033x$, where $f(x)$ is the height above the ground in meters and x is the distance from the point of take-off in meters. Find and interpret the key features of the path of Huaso's jump.

Step 1 Graph the function.

The graph only exists in Quadrant I in the context of the situation.

Step 2 Find the vertex.

Use the maximum feature from the CALC menu to find the vertex.

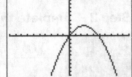

The vertex is located at about (____, ____). This represents Huaso's maximum height of ____ meters when he was ____ meters from the point of take-off.

Step 3 Find the axis of symmetry.

The axis of symmetry is located at $x = $ ____. This means that Huaso's path ____ he is ____ meters from the point of take-off is the same as ____ he reaches ____ meters.

Step 4 Find the y-intercept.

Use the value feature from the CALC menu to find the y-intercept. Enter 0 for x. The y-intercept is located at (____, ____), so the y-intercept is ____. This means that when Huaso took off, he was ____ meters high and ____ meters from the point of take-off.

Step 5 Find the zeros.

Use the zero feature from the CALC menu to find the zeros, or x-intercepts.

The zeros are located at x equals ____ and x equals ____. These represent the points of ____ and ____, when Huaso's height above the ground was 0 meters.

Step 6 Examine the end behavior.

As x increases or decreases from the maximum, the value of y ____ until it reaches 0. This represents Huaso ____ and ____ to the ground.

Name _____ Period _____ Date _____

Practice

🅡 **Go Online** You can complete your homework online.

Examples 1 and 2

Identify the axis of symmetry, the vertex, and the *y*-intercept of each graph. Then describe the end behavior.

1.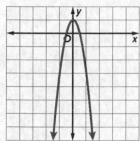

2.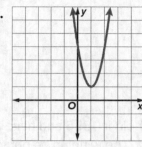

3.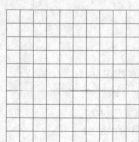

4.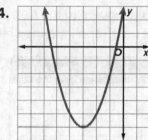

Example 3

Graph each function.

5. $y = -3x^2 + 6x - 4$

6. $y = -2x^2 - 4x - 3$

7. $y = -2x^2 - 8x + 2$

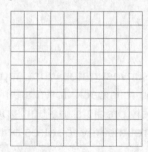

8. $y = x^2 + 6x - 6$

9. $y = x^2 - 2x + 2$

10. $y = 3x^2 - 12x + 5$

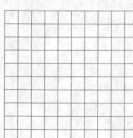

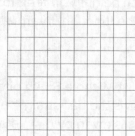

Copyright © McGraw-Hill Education

Copyright © McGraw-Hill Education

Copyright © McGraw-Hill Education

Copyright © McGraw-Hill Education

Copyright © McGraw-Hill Education

Copyright © McGraw-Hill Education

Copyright © McGraw-Hill Education

Example 4

Use a table of values to graph each function. State the domain and range.

11. $y = x^2 + 4x + 6$

12. $y = 2x^2 + 4x + 7$

13. $y = 2x^2 - 8x - 5$

14. $y = 3x^2 + 12x + 5$

15. $y = 3x^2 - 6x - 2$

16. $y = x^2 - 2x - 1$

Examples 5 and 6

17. OLYMPICS Olympics were held in 1896 and have been held every four years except 1916, 1940, and 1944. The winning height y in men's pole vault at any number Olympiad x can be approximated by the equation $y = 0.37x^2 + 4.3x + 126$. Complete the table to estimate the winning pole vault heights in each of the Olympic Games. Round your answers to the nearest tenth. Graph the function. Interpret the key features of the graph in terms of the quantities.

Year	Olympiad (x)	Height (y inches)
1896	1	
1900	2	
1924	7	
1936	10	
1964	15	
2008	26	
2012	27	
2016	28	

18. PHYSICS Mrs. Capwell's physics class investigates what happens when a ball is given an initial push, rolls up, and then rolls back down an inclined plane. The class finds that $y = -x^2 + 6x$ accurately predicts the ball's position y after rolling x seconds. Graph the function. Interpret the key features of the graph in terms of the quantities.

19. ARCHITECTURE A hotel's main entrance is in the shape of a parabolic arch. The equation $y = -x^2 + 10x$ models the arch height y, in feet, for any distance x, in feet, from one side of the arch. Graph the function. Interpret the key features of the graph in terms of the quantities.

20. SOFTBALL Olympic softball gold medalist Michele Smith pitches a curveball with a speed of 64 feet per second. If she throws the ball straight upward at this speed, the ball's height h in feet after t seconds is given by $h = -16t^2 + 64t$. Graph the function. Interpret the key features of the graph in terms of the quantities.

Mixed Exercises

21. CONSTRUCTION Teddy is building the rectangular deck shown.

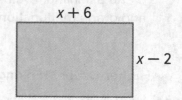

$x + 6$

$x - 2$

   a. Write an equation representing the area of the deck y.

   b. What is the equation of the axis of symmetry?

   c. Graph the equation and label its vertex.

22. USE TOOLS Write a quadratic function whose graph opens up. Write an exponential function. Graph both functions on the same coordinate plane. Compare and contrast the domains and ranges of the two functions and any symmetry of the two graphs.

23. STRUCTURE Consider the quadratic function $y = -x^2 - 2x + 2$.

   a. Find the equation for the axis of symmetry.

   b. Find the coordinates of the vertex and determine if it is a maximum or minimum.

   c. Graph the function.

Identify the axis of symmetry, the vertex, and the y-intercept of each graph. Then describe the end behavior.

24. $y = 2x^2 - 8x + 6$ **25.** $y = x^2 + 4x + 6$ **26.** $y = -3x^2 - 12x + 3$

27. STRUCTURE DeMarcus is sitting in a lifeguard's chair at the beach. He tosses a beanbag into the air and lets it land on the beach below him. The function $h(t) = -16t^2 + 8t + 12$ models the beanbag's height in feet, t seconds after DeMarcus tosses it into the air.

   a. Graph the function on a coordinate plane. Be sure to label the x- and y-axes and provide a scale for each axis.

   b. Find the intercepts of the graph. Describe what they represent in the context of the situation.

   c. What is the maximum value of the function, and what does it represent? At what time is the maximum reached?

   d. What is the value of $h(0.5)$? Describe its meaning in the context of the situation.

   e. State a reasonable domain for this function. What does it represent in the context of the situation?

28. REGULARITY Write the equation for a quadratic function that has a *y*-intercept of 0 and a minimum value at $x = 2$. Explain the steps you used to write the equation, and graph the function on a coordinate plane.

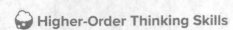

Higher-Order Thinking Skills

29. CREATE Write a quadratic function for which the graph has an axis of symmetry of $x = -\frac{3}{8}$. Summarize your steps.

30. FIND THE ERROR Noelia thinks that the parabolas represented by the graph and the description have the same axis of symmetry. Chase disagrees. Is either correct? Explain your reasoning.

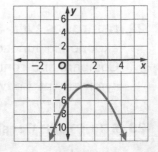

> a parabola that opens downward, passing through (0, 6) and having a vertex at (2, 2)

31. PERSEVERE Using the axis of symmetry, the *y*-intercept, and one *x*-intercept, write an equation for the graph shown.

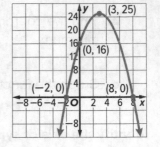

32. ANALYZE The graph of a quadratic function has a vertex (2, 0). One point on the graph is (5, 9). Find another point on the graph. Explain how you found it.

33. CREATE Describe a real-world situation that involves a quadratic equation. Explain what the vertex represents.

34. ANALYZE Provide a counterexample that shows that the following statement is false. Justify your argument.

The vertex of a parabola is always the minimum of the graph.

35. WRITE Use tables and graphs to compare and contrast an exponential function $f(x) = ab^x + c$, where $a \neq 0$, $b > 0$, and $b \neq 1$, a quadratic function $g(x) = ax^2 + c$, and a linear function $h(x) = ax + c$. Include intercepts, portions of the graph where the functions are increasing, decreasing, positive, negative, relative maxima, relative minima, symmetries, and end behavior. Which function eventually exceeds the others?

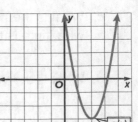

36. ANALYZE Consider $g(x)$ shown in the graph and $f(x) = -x^2 + 3x - 2$. Determine which function has the lesser minimum.

Transformations of Quadratic Functions

Explore Transforming Quadratic Functions

 Online Activity Use graphing technology to complete the Explore.

> ⑦ **INQUIRY** How does performing an operation on a quadratic function change its graph? ✕

Learn Translations of Quadratic Functions

Key Concept • Vertical Translations of Quadratic Functions

The graph of $g(x) = x^2 + k$ is the graph of $f(x) = x^2$ translated vertically.

If $k > 0$, the graph of $f(x)$ is translated k units up.

If $k < 0$, the graph of $f(x)$ is translated $|k|$ units down.

Key Concept • Horizontal Translations of Quadratic Functions

The graph of $g(x) = (x - h)^2$ is the graph of $f(x) = x^2$ translated horizontally.

If $h > 0$, the graph of $f(x)$ is translated h units right.

If $h < 0$, the graph of $f(x)$ is translated $|h|$ units left.

Example 1 Vertical Translations of Quadratic Functions

Describe the translation in $g(x) = x^2 - 4$ as it relates to the graph of the parent function.

Graph the parent function, $f(x) = x^2$, for quadratic functions.

Because $f(x) = x^2$, $g(x) = f(x) + k$ where $k = $ _____.

The constant k is added to the function after it has been evaluated, so k affects the output, or y-values. The value of k is less than 0, so the graph of $f(x) = x^2$ is translated

_____ units.

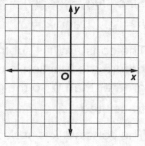

$g(x) = x^2 - 4$ is the translation of the graph of

the parent function _____.

Today's Goals
- Apply translations to quadratic functions.
- Apply dilations to quadratic functions.
- Use transformations to identify quadratic functions from graphs and write equations of quadratic functions.

Today's Vocabulary
vertex form

 Go Online
You can watch a video to see how to translate functions.

😊 **Think About It!**
Is the following statement *sometimes*, *always*, or *never* true? Explain.
The graph of $g(x) = x^2 + k$ has its vertex at the origin.

Check

Describe the transformation of $g(x) = x^2 - 8$ as it relates to the graph of the parent function.

The graph of $g(x) = x^2 - 8$ is the translation of the graph of the parent function ____ units _____.

Example 2 Horizontal Translations of Quadratic Functions

Describe the translation in $g(x) = (x + 2)^2$ as it relates to the graph of the parent function.

Graph the parent function, $f(x) = x^2$, for quadratic functions.

Because $f(x) = x^2$, $g(x) = f(x - h)$ where $h = $ ____.

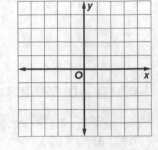

The constant h is subtracted from x before the function is performed, so h affects the input, or x-values. The value of h is less than 0, so the graph of $f(x) = x^2$ is translated $|-2|$ units left.

$g(x) = (x + 2)^2$ is the translation of the graph of the parent function _____.

Check

Describe the transformation of $g(x) = (x - 6)^2$ as it relates to the graph of the parent function.

The graph of $g(x) = (x - 6)^2$ is the translation of the graph of the parent function ____ units _____.

Example 3 Multiple Translations of Quadratic Functions

Describe the translation in $g(x) = (x - 3)^2 + 1$ as it relates to the graph of the parent function.

Graph the parent function, $f(x) = x^2$, for quadratic functions.

Since $f(x) = x^2$, $g(x) = f(x - h) + k$ where $h = $ _____ and $k = $ _____.

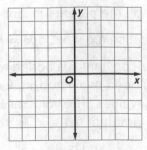

The constant h is subtracted from x before the function is performed and is greater than 0, so the graph of $f(x) = x^2$ is translated _____.

The constant k is added to the function after it has been evaluated and is greater than 0, so the graph of $f(x) = x^2$ is also translated _____.

🄝 **Go Online** You can complete an Extra Example online.

Check

Describe the translation in $g(x) = (x + 3)^2 - 5$ as it relates to the graph of the parent function.

The graph of $g(x) = (x + 3)^2 - 5$ is the translation of the graph of the parent function 3 units _____ and 5 units _____.

Graph $g(x) = (x + 3)^2 - 5$.

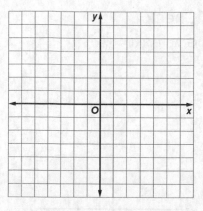

Learn Dilations of Quadratic Functions

Key Concept • Vertical Dilations of Quadratic Functions

The graph of $g(x) = ax^2$ is the graph of $f(x) = x^2$ stretched or compressed vertically by a factor of $|a|$.

If $|a| > 1$, the graph of $f(x)$ is stretched vertically away from the x-axis.

If $0 < |a| < 1$, the graph of $f(x)$ is compressed vertically toward the x-axis.

Key Concept • Horizontal Dilations of Quadratic Functions

The graph of $g(x) = (ax)^2$ is the graph of $f(x) = x^2$ stretched or compressed horizontally by a factor of $\frac{1}{|a|}$.

If $|a| > 1$, the graph of $f(x)$ is compressed horizontally toward the y-axis.

If $0 < |a| < 1$, the graph of $f(x)$ is stretched horizontally away from the y-axis.

Go Online
You can watch a video to see how to describe dilations of functions.

Example 4 Vertical Dilations of Quadratic Functions

Describe the dilation in $g(x) = 3x^2$ as it relates to the graph of the parent function.

Graph the parent function, $f(x) = x^2$, for quadratic functions.

Since $f(x) = x^2$, $g(x) = a \cdot f(x)$ where $a = $ _____.

The function is multiplied by the positive constant a after it has been evaluated and $|a|$ is greater than 1, so the graph of $f(x) = x^2$ is stretched vertically by a factor of $|a|$, or 3.

$g(x) = 3x^2$ is a _____ of the graph of the parent function.

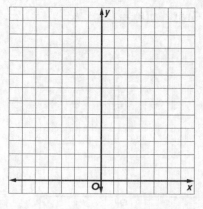

Study Tip

Vertical Dilations For a vertical dilation, if you multiply each y-coordinate of the function $f(x)$ by a, you'll get the corresponding y-coordinate of the function $g(x)$. For example, for the function above, the point (2, 4) on $f(x)$ corresponds to the point (2, 12) on $g(x)$. The y-coordinate of $f(x)$, 4, is multiplied by a, which is 3.

Check

Graph $g(x) = \frac{1}{4}x^2$.

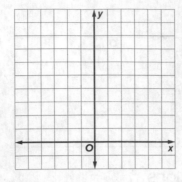

Example 5 Horizontal Dilations of Quadratic Functions

Describe the dilation in $g(x) = \left(\frac{1}{2}x\right)^2$ as it relates to the graph of the parent function.

Graph the parent function, $f(x) = x^2$, for quadratic functions.

Since $f(x) = x^2$, $g(x) = f(a \cdot x)$ where

$a = \underline{\quad}$.

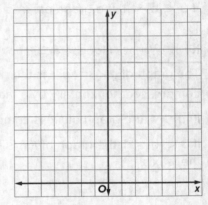

x is multiplied by the positive constant a before the function is performed and $|a|$ is between 0 and 1, so the graph of $f(x) = x^2$ is stretched horizontally by a factor of $\frac{1}{|a|}$, or 2.

$g(x) = \left(\frac{1}{2}x\right)^2$ is a _____ of the graph of the parent function.

Check

Graph $g(x) = (2x)^2$

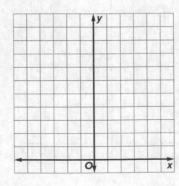

Talk About It!

Could the graph of $g(x) = \left(\frac{1}{2}x\right)^2$ also be described by a vertical dilation? Justify your argument.

Study Tip

Horizontal Dilations For a horizontal dilation, if you multiply each x-coordinate of the function $f(x)$ by $\frac{1}{|a|}$, you'll get the corresponding x-coordinate of the function $g(x)$. For example, for the function above, the point $(-1, 1)$ on $f(x)$ corresponds to the point $(-2, 1)$ on $g(x)$. The x-coordinate of $f(x)$, -1, is multiplied by $\frac{1}{|a|}$.

Learn Reflections of Quadratic Functions

> **Key Concept • Reflections of Quadratic Functions Across the x-axis**
>
> The graph of $-f(x)$ is the reflection of the graph of $f(x) = x^2$ across the x-axis.

> **Key Concept • Reflections of Quadratic Functions Across the y-axis**
>
> The graph of $f(-x)$ is the reflection of the graph of $f(x) = x^2$ across the y-axis.

Example 6 Vertical Reflections of Quadratic Functions

Describe how the graph of $g(x) = -\frac{2}{3}x^2$ is related to the graph of the parent function.

Graph the parent function, $f(x) = x^2$, for quadratic functions.

Because $f(x) = x^2$, $g(x) = -1 \cdot a \cdot f(x)$ where

$a = $ _____.

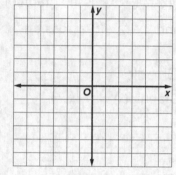

The function is multiplied by -1 and the constant a after it has been evaluated and $|a|$ is between 0 and 1, so the graph of $f(x)$ is compressed vertically and reflected across the x-axis.

$g(x) = -\frac{2}{3}x^2$ is the graph of the parent function _____ and _____ across the _____.

Check

Graph $g(x) = -\frac{3}{5}x^2$.

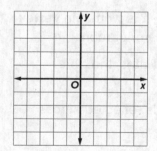

Think About It!

How does a reflection of the parent function $f(x) = x^2$ across the x-axis affect the end behavior of the graph?

Go Online You can complete an Extra Example online.

Example 7 Horizontal Reflections of Quadratic Functions

Describe how the graph of $g(x) = (-3x)^2$ is related to the graph of the parent function.

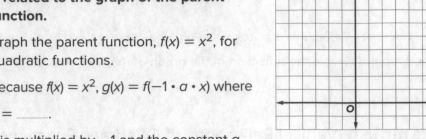

Graph the parent function, $f(x) = x^2$, for quadratic functions.

Because $f(x) = x^2$, $g(x) = f(-1 \cdot a \cdot x)$ where

$a = $ _____.

x is multiplied by -1 and the constant a before the function is performed and $|a|$ is greater than 1, so the graph of $f(x)$ is compressed horizontally and reflected across the y-axis.

$g(x) = (-3x)^2$ is the graph of the parent function _____

and _____ across the _____.

Check

Graph $g(x) = \left(-\frac{7}{2}x\right)^2$.

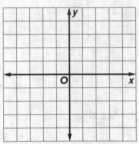

Learn Transformations of Quadratic Functions

A quadratic function in the form $f(x) = a(x - h)^2 + k$ is written in **vertex form**. Each constant in the equation affects the parent graph.

- The value of $|a|$ stretches or compresses (dilates) the parent graph.

- When the value of a is negative, the graph is reflected across the x-axis.

- The value of k shifts (translates) the parent graph up or down.

- The value of h shifts (translates) the parent graph left or right.

 Go Online
You can watch a video to see how to use a graphing calculator to graph transformations of quadratic functions.

Example 8 Multiple Transformations of Quadratic Functions

Describe how the graph of $g(x) = -3(x-1)^2 - 2$ is related to the graph of the parent function.

Graph the parent function, $f(x) = x^2$.

Since $f(x) = x^2$, $g(x) = a \cdot f(x-h) + k$ where $a = -3$.

$a < 0$ and $|a| > 1$, so the graph of $f(x) = x^2$ is compressed vertically and reflected across the x-axis and stretched vertically by a factor of $|a|$, or 3.

$h > 0$, so the graph is then translated h units right, or 1 unit right.

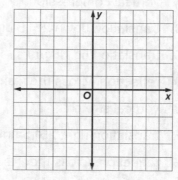

$k < 0$, so the graph is then translated $|k|$ units down, or 2 units down.

$g(x) = -3(x-1)^2 - 2$ is the graph of the parent function _____, reflected across the _____, and translated _____ and _____.

Check

Graph $g(x) = -\frac{1}{3}(x+4)^2 - 1$.

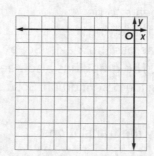

🔵 **Think About It!**

Write a quadratic function that opens down and is translated 6 units left and 1 unit down.

🌐 Example 9 Apply Transformations of Quadratic Functions

FOOTBALL Although they may appear flat, properly designed football fields arc to allow water to drain. Fields rise from each sideline to the center of the field, known as the crown, which should be between 1 and 1.5 feet in height. A cross section of a football field that is 160 feet wide and has a 1.5 foot crown can be modeled by $g(x) = -0.000234(x-80)^2 + 1.5$, where $g(x)$ is the height of the field and x is the distance from the sideline in feet. Describe how $g(x)$ is related to the graph of $f(x) = x^2$.

$a < 0$, so the graph of $f(x) = x^2$ is a _____ across the x-axis.

$0 < |-0.000234| < 1$, so the graph of $f(x) = x^2$ is _____ vertically.

$80 > 0$, so the graph of $f(x) = x^2$ is translated 80 units _____.

$1.5 > 0$, so the graph of $f(x) = x^2$ is translated 1.5 units _____.

🔵 **Go Online** You can complete an Extra Example online.

🔵 **Think About It!**

What do the values of h and k represent in the context of the situation?

Check

BRIDGES The lower arch of the Sydney Harbor Bridge can be modeled by $g(x) = -0.0018(x - 251.5)^2 + 118$. Select all of the transformations that occur in $g(x)$ as it relates to the graph of $f(x) = x^2$. _____

A. vertical compression

B. translation down 251.5 units

C. translation up 118 units

D. reflection across the x-axis

E. vertical stretch

F. translation right 251.5 units

G. reflection across the y-axis

Watch Out!

Choosing a Point
When substituting for x and y in the equation, use a point other than the vertex.

 Think About It!

How does the equation you found compare to the prediction you made in Step 1?

Example 10 Identify a Quadratic Equation from a Graph

Use the graph of the function to write its equation.

Step 1 Analyze the graph.

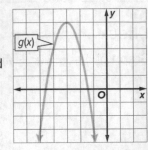

The graph appears to be _____ than the parent function, implying a _____, and is _____ across the x-axis. So, a ___ 0 and |a| ___ 1. The graph has also been shifted left and up from the parent graph. So, h ___ 0 and k ___ 0.

Step 2 Identify the translations.

The vertex is shifted 3 units left, so h = ___.

It is also shifted 5 units up, so k = ___.

Step 3 Identify the dilation and/or reflection.

The point $(-2, 3)$ lies on the graph. Substitute the coordinates in for x and y to solve for a.

$$y = a(x + 3)^2 + 5 \qquad \text{Vertex form of the graph}$$

$$\underline{} = a(\underline{} + 3)^2 + 5 \qquad (-2, 3) = (x, y)$$

$$3 = a(\underline{})^2 + 5 \qquad \text{Add.}$$

$$3 = a + 5 \qquad \text{Evaluate } 1^2.$$

$$\underline{} = a \qquad \text{Subtract 5 from each side.}$$

So, the equation is $g(x) = \underline{}(x + \underline{})^2 + \underline{}$.

 Go Online You can complete an Extra Example online.

Practice

🔘 **Go Online** You can complete your homework online.

Examples 1–3

Describe the translation in each function as it relates to the graph of the parent function.

1. $g(x) = -10 + x^2$

2. $g(x) = x^2 + 2$

3. $g(x) = (x - 1)^2$

4. $g(x) = x^2 - 8$

5. $g(x) = (x + 3)^2$

6. $g(x) = x^2 + 7$

7. $g(x) = (x + 2.5)^2$

8. $g(x) = (x + 5)^2 - 2$

9. $g(x) = 6 + x^2$

10. $g(x) = -4 + (x - 3)^2$

11. $g(x) = (x - 9.5)^2$

12. $g(x) = (x - 1.5)^2 + 3.5$

Examples 4 and 5

Describe the dilation in each function as it relates to the graph of the parent function.

13. $g(x) = 7x^2$

14. $g(x) = \frac{1}{5}x^2$

15. $g(x) = (4x)^2$

16. $g(x) = \left(\frac{1}{2}x\right)^2$

17. $g(x) = \left(\frac{5}{3}x\right)^2$

18. $g(x) = 5x^2$

19. $g(x) = \frac{3}{4}x^2$

20. $g(x) = \left(\frac{7}{8}x\right)^2$

Examples 6 and 7

Describe how the graph of each function is related to the graph of the parent function.

21. $g(x) = -6x^2$

22. $g(x) = (-9x)^2$

23. $g(x) = -\frac{1}{3}x^2$

24. $g(x) = \left(-\frac{2}{3}x\right)^2$

25. $g(x) = -2x^2$

26. $g(x) = \left(-\frac{6}{5}x\right)^2$

Example 8

Describe how the graph of each function is related to the graph of the parent function.

27. $h(x) = -7 - x^2$

28. $g(x) = 2(x - 3)^2 + 8$

29. $h(x) = 6 + \frac{2}{3}x^2$

30. $g(x) = -5 - \frac{4}{3}x^2$

31. $h(x) = 3 + \frac{5}{2}x^2$

32. $g(x) = -x^2 + 3$

Example 9

33. SPRINGS The potential energy stored in a spring is given by $U_s = \frac{1}{2}kx^2$, where k is a constant known as the spring constant, and x is the distance the spring is stretched or compressed from its initial position. How is the graph of the function for a spring where $k = 10$ newtons/meter related to the graph of the function for a spring where $k = 2$ newtons/meter?

34. PHYSICS A ball is dropped from a height of 20 feet. The function $h = -16t^2 + 20$ models the height of the ball in feet after t seconds. Compare this graph to the graph of its parent function.

35. ACCELERATION The distance d in feet a car accelerating at 6 ft/s^2 travels from the starting line of a race after t seconds is modeled by the function $d = 3t^2$. Suppose a second car that is 100 feet ahead of the first car and accelerates at 4 ft/s^2 begins the race at the same time. The distance of the second car from the starting line after t seconds is modeled by the function $d = 2t^2 + 100$.

    a. Explain how each graph is related to the graph of $d = t^2$.

    b. After how many seconds will the first car pass the second car?

Example 10

Use the graph of each function to write its equation.

36.

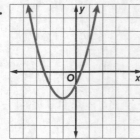

37.

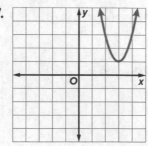

38.

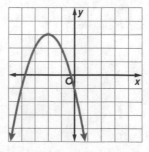

Mixed Exercises

Describe how the graph of each function is related to the graph of the parent function.

39. $g(x) = 5 - \frac{1}{5}x^2$

40. $g(x) = 4(x - 1)^2$

41. $g(x) = 0.25x^2 - 1.1$

42. $h(x) = 1.35(x + 1)^2 + 2.6$

43. $g(x) = \frac{3}{4}x^2 + \frac{5}{6}$

44. $h(x) = 1.01x^2 - 6.5$

STRUCTURE **Use transformations to graph each quadratic function. Then describe the transformation.**

45. $h(x) = -2(x + 2)^2 + 2$ 　　**46.** $g(x) = \left(\frac{1}{2}x\right)^2 - 3$ 　　**47.** $h(x) = -\frac{1}{4}(x - 1)^2 + 4$

Match each function to its graph.

A.

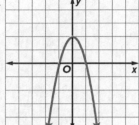

B.

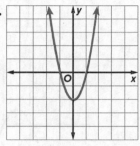

C.

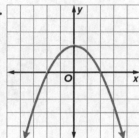

D.

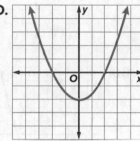

48. $f(x) = 2x^2 - 2$ 　　　　　　　**49.** $g(x) = \frac{1}{2}x^2 - 2$

50. $h(x) = -\frac{1}{2}x^2 + 2$ 　　　　　**51.** $j(x) = -2x^2 + 2$

52. **PRECISION** Write a set of instructions that a classmate could use to transform the graph of $f(x) = x^2$ and end up with the graph of $g(x) = -10(x - 17)^2$.

53. **CONSTRUCT ARGUMENTS** A function is an even function if $f(-x) = f(x)$ for all x in the domain of the function. A function is an odd function if $f(-x) = -f(x)$ for all x in the domain of the function. Is the parent function of a quadratic function an even function or an odd function? Justify your answer.

54. **USE A MODEL** An animator is using a coordinate plane to design a scene in a movie. In the scene, a comet enters the screen in Quadrant II, moves around the screen in a parabolic path, and leaves the screen in Quadrant I.

　a. The figure shows the path of the comet. What equation represents the path?

　b. The animator decides to translate the path of the comet 8 units up and 7 units left. What equation represents the new path?

Use the graph of each function to write its equation.

55.
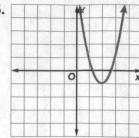

56.
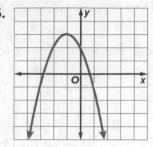

57.
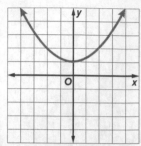

58.
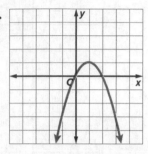

59.
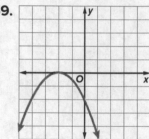

60.
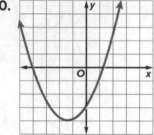

Higher-Order Thinking Skills

61. ANALYZE Are the following statements *sometimes*, *always*, or *never* true? Justify your argument.

    a. The graph of $y = x^2 + k$ has its vertex at the origin.

    b. The graphs of $y = ax^2$ and its reflection across the x-axis are the same width.

    c. The graph of $f(x) = x^2 + k$, where $k \geq 0$, and the graph of a quadratic function $g(x)$ with vertex at $(0, -3)$ have the same maximum or minimum.

62. PERSEVERE Write a function of the form $y = ax^2 + k$ with a graph that passes through the points $(-2, 3)$ and $(4, 15)$.

63. ANALYZE Determine whether all quadratic functions that are reflected across the y-axis produce the same graph. Justify your argument.

64. CREATE Write a quadratic function with a graph that opens down and is wider than the parent graph.

65. WRITE Describe how the values of a and k affect the graphical and tabular representations of the functions $y = ax^2$, $y = x^2 + k$, and $y = ax^2 + k$.

Solving Quadratic Equations by Graphing

Today's Goal
• Solve quadratic equations by graphing.

Today's Vocabulary
quadratic equation
double root

Explore Roots and Zeros of Quadratics

Online Activity Use graphing technology to complete the Explore.

INQUIRY How can you use the *x*-intercepts of a quadratic function to identify the solutions of its related equation?

Learn Solving Quadratic Equations by Graphing

A **quadratic equation** can be written in the standard form $ax^2 + bx + c = 0$, where $a \neq 0$. Because the graphs of quadratic functions have zero, one, or two zeros, the corresponding quadratic equations also have zero, one, or two solutions.

Key Concept • Solutions of Quadratic Equations

two unique real solutions	*one* unique real solution	*no* real solutions
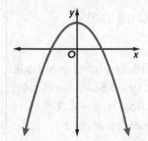	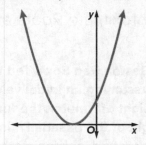	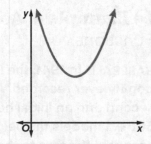

When the vertex of the related quadratic function lies on the *x*-axis, the solution is a **double root**, which means that there are two roots of a quadratic equation that are the same number.

Example 1 Solve a Quadratic Equation with Two Roots

Solve $-x^2 + 4x + 5 = 0$ by graphing.

Graph the related function $f(x) = -x^2 + 4x + 5$. The *x*-intercepts of the graph appear to be at _____ and _____, so the solutions are _____ and _____.

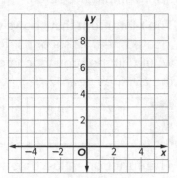

Check

Use the graph of the related function to solve $x^2 + 5x + 6 = 0$ by graphing. The solutions are _____ and _____.

Think About It!
How can the solutions of a quadratic equation help you graph the related function?

Think About It!
How do you know that the solutions are correct?

The function $f(x) = x^2 - 8x + 16$ is a perfect square trinomial. Explain why there is a double root for this type of function.

Go Online An alternate method is available for this example.

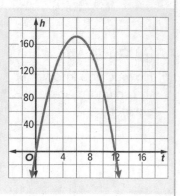

Example 2 Solve a Quadratic Equation with a Double Root

Solve $x^2 - 8x = -16$ by graphing.

Rewrite the equation in standard form and then graph the related function $f(x) = x^2 - 8x + 16$.

Notice that the vertex of the parabola is the only x-intercept. Therefore, there is only one solution, _____.

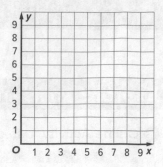

Example 3 Solve a Quadratic Equation with No Real Roots

Solve $-2x^2 - x - 2 = 0$ by graphing.

Graph the related function $f(x) = -2x^2 - x - 2$.

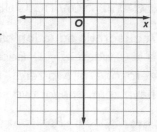

This graph has _____ x-intercepts. Therefore, this equation has no real number solutions.

🌐 Example 4 Approximate Roots of Quadratic Functions

BASEBALL In 1941, the Boston Red Sox's Ted Williams hit the highest pop fly ever recorded. Assuming an initial velocity of 58 meters per second and an initial height of 1 meter, the function $h = -4.9t^2 + 58t + 1$ models the height of the baseball h in meters after t seconds. If it falls to the ground, approximately how long would the ball be in the air?

You need to find the roots of the equation $-4.9t^2 + 58t + 1 = 0$. Graph the related function $h = -4.9t^2 + 58t + 1$, and use estimation to approximate the zeros.

The t-intercept that shows when the ball would hit the ground is located near 12 seconds. To find a more exact time when the ball would hit the ground, complete the table using an increment of 0.1 for the t-values near 12.

t	11.7	11.8	11.9	12	12.1	12.2	12.3
h	8.84	_____	_____	_____	_____	-20.72	-26.92

The function value that is closest to zero when the sign changes is _____. Thus, the ball would be in the air for approximately _____ seconds before hitting the ground.

Go Online You can complete an Extra Example online.

Practice

Go Online You can complete your homework online.

Examples 1–3

Solve each equation by graphing.

1. $x^2 + 7x + 14 = 0$

2. $x^2 + 2x - 24 = 0$

3. $x^2 + 16x + 64 = 0$

4. $x^2 - 5x + 12 = 0$

5. $x^2 + 14x = -49$

6. $x^2 = 2x - 1$

7. $x^2 - 10x = -16$

8. $-2x^2 - 8x = 13$

9. $2x^2 - 16x = -30$

10. $2x^2 = -24x - 72$

11. $-3x^2 + 2x = 15$

12. $x^2 = -2x + 80$

Example 4

13. **SOCCER** Claudia kicked a soccer ball off of a platform. The equation $y = -x^2 + 3x + 12$ models the height of the ball y in feet after x seconds. Approximately how long is the ball in the air?

14. **TRAMPOLINE** A gymnast jumped on a trampoline. The equation $y = -16x^2 + 58x$ models the height of the gymnast y in feet after x seconds for one of the jumps. Approximately how long is the gymnast in the air?

Mixed Exercises

Estimate the solution(s) to each equation by graphing the related function. Round to the nearest tenth.

15. $p^2 + 4p + 2 = 0$

16. $x^2 + x - 3 = 0$

17. $d^2 + 6d = -3$

18. $h^2 + 1 = 4h$

19. $3x^2 - 5x = -1$

20. $x^2 + 1 = 5x$

21. **FARMING** In order for Mr. Moore to decide how much fertilizer to apply to his corn crop this year, he reviews records from previous years. His crop yield y depends on the amount of fertilizer he applies to his fields x according to the equation $y = -x^2 + 4x + 12$. Graph the function, and find the point at which Mr. Moore gets the highest yield possible. Then find and interpret the zero(s) of the function.

22. **FRAMING** A rectangular photograph is 7 inches long and 6 inches wide. The photograph is framed using a material that is x inches wide. If the area of the frame and photograph combined is 156 square inches, what is the width of the framing material?

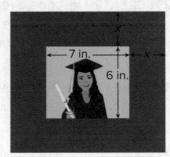

23. **WRAPPING PAPER** Can a rectangular piece of wrapping paper with an area of 81 square inches have a perimeter of 60 inches? (*Hint*: Let length = 30 − width.) Explain.

24. **ENGINEERING** The shape of a satellite dish is often parabolic because of the reflective qualities of parabolas. Suppose a particular satellite dish is modeled by the function $0.5x^2 = 2 + y$.
    a. Approximate the zeros of this function by graphing.
    b. On the coordinate plane, translate the parabola so that there is only one zero. Label this curve A.
    c. Translate the parabola so that there are no real zeros. Label this curve B.

25. REASONING The three equations below are shown on the graph.

$y = x^2 + 12x + m; y = 2x^2 - nx + 72; y = -x^2 + 3x - 10$

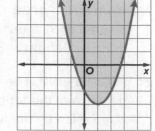

a. Describe the solutions for each function. Then, match each graph to its related equation. Explain your reasoning.

b. How could you choose an appropriate value for m?

c. How could you choose an appropriate value for n?

26. ROCKETS The height h of a model rocket launched from ground level into the air after t seconds can be modeled by the equation $h = -16t^2 + 160t$. The equation is graphed on the coordinate grid.

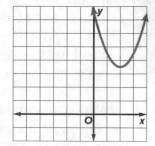

a. Where does the graph intersect the x-axis? What do these points represent?

b. How long does it take the rocket to reach its maximum height? Explain your reasoning.

27. USE TOOLS Through market research, a company finds that its profit in dollars can be modeled by the function $P(x) = -50,000x^2 + 300,000x - 250,000$, where x is the price in dollars at which they sell their product.

a. Describe how to use graphing technology to graph the function, and then provide a sketch of the graph.

b. Find the x-intercepts. What do these represent in the context of the situation?

c. Write an equation for the price at which the company should sell their product to make a profit of $150,000. Sketch a graph of the function relating to this equation and use it to solve the equation graphically.

d. Is there a price at which the company can sell their product to make a profit of $300,000? Explain your reasoning, and use a graph to support your answer.

28. FIND THE ERROR Iku and Zachary are finding the number of real zeros of the function graphed at the right. Iku says that the function has no real zeros because there are no x-intercepts. Zachary says that the function has one real zero because the graph has a y-intercept. Is either of them correct? Explain your reasoning.

29. CREATE Describe a real-world situation in which a thrown object travels in the air. Write an equation that models the height of the object with respect to time, and determine how long the object travels in the air.

30. ANALYZE The graph shown is a *quadratic inequality*. Analyze the graph, and find 3 solutions with y-values greater than 2.

31. PERSEVERE Write a quadratic equation that has the roots described.

a. one double root

b. no real solutions

c. two unique real solutions

32. WRITE Explain how to approximate the roots of a quadratic equation when the roots are not integers.

Solving Quadratic Equations by Factoring

Learn Solving Quadratic Equations by Using the Square Root Property

Key Concept • Square Root Property

Words: To solve a quadratic equation in the form $x^2 = n$, take the square root of each side.

Symbols: For any number $n \geq 0$, if $x^2 = n$, then $x = \pm\sqrt{n}$.

Example:

$x^2 = 49$

$x = \pm\sqrt{49}$ or $x = \pm 7$

In the equation $x^2 = n$, if n is a perfect square, you will have an integer answer. If n is not a perfect square, you will need to keep your answer in radical form or approximate the square root. You can use a calculator or estimation to find an approximation.

Example 1 Use the Square Root Property

Solve $(y + 5)^2 = 21$.

$(y + 5)^2 = 21$	Original equation
$\rule{2cm}{0.4pt} = \pm\sqrt{21}$	Square Root Property
$y = \rule{1cm}{0.4pt} \pm \sqrt{21}$	Subtract 5 from each side.
$y = -5 \rule{1cm}{0.4pt} \sqrt{21}$ or $y = -5 \rule{1cm}{0.4pt} \sqrt{21}$	Separate into two equations.

The solutions are $-5 + \sqrt{21}$ or $-5 - \sqrt{21}$.
Using a calculator $-5 + \sqrt{21} \approx \rule{1cm}{0.4pt}$ and $-5 - \sqrt{21} \approx \rule{1cm}{0.4pt}$.

Check

a. Select all the solutions of $(x - 2)^2 = 16$.

$\rule{2cm}{0.4pt}$

A. −6	B. −4
C. −2	D. 2
E. 4	F. 6

b. Select all the solutions of $(y + 18)^2 = 77$.

$\rule{2cm}{0.4pt}$

A. ≈ −26.77	B. ≈ −9.23
C. ≈ −7.68	D. ≈ 7.68
E. ≈ 9.23	F. ≈ 26.77

 Go Online You can complete an Extra Example online.

Today's Goals
- Solve quadratic equations by using the Square Root Property.
- Solve quadratic equations by factoring and sketch graphs of quadratic functions by using the zeros.
- Write equations of quadratic functions given their graphs or roots.

💭 Think About It!

Without using a calculator, describe how you can approximate $\sqrt{30}$.

Study Tip

Reading Math $\pm\sqrt{30}$ is read as "plus or minus the square root of 30."

Watch Out!

Square Root Be careful not to confuse the *roots* of an equation with taking the *square root* of an expression. The *roots* of an equation are any values of the equation that make it true. The *square root* of an expression is one of two equal factors.

🌐 **Example 2** Solve an Equation by Using the Square Root Property

EGG DROP During the Egg Drop Competition at Anika's school, students must create a container for an egg that prevents the egg from breaking when dropped from a height of 30 feet. The formula $h = -16t^2 + h_0$ can be used to approximate the number of seconds t it takes for the container to reach height h from an initial height of h_0 in feet. Find the time it takes the container to reach the ground.

At ground level, $h = 0$ and the initial height is 30, so $h_0 = 30$.

$h = -16t^2 + h_0$	Original equation
$0 = -16t^2 + 30$	$h = 0$ and $h_0 = 30$
_____ $= -16t^2$	Subtract 30 from each side.
_____ $= t^2$	Divide each side by -16.
_____ $\approx t$	Use the Square Root Property.

It takes $\approx$ _____ seconds for the container to reach the ground.

Study Tip

Modeling The function $h = -16t^2 + h_0$ does not model the path of the container after it is dropped. This function is used to model the height of the container as a function of time.

Explore Using Factors to Solve Quadratic Equations

🧭 **Online Activity** Use graphing technology to complete the Explore.

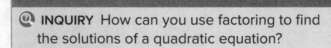

@ **INQUIRY** How can you use factoring to find the solutions of a quadratic equation?

Learn Solving Quadratic Equations by Factoring

Key Concept • Zero Product Property

Words: If the product of two factors is 0, then at least one of the factors must be 0.

Symbols: For any real numbers a and b, if $ab = 0$, then $a = 0$, $b = 0$, or both a and b equal zero.

Factor Using the Distributive Property

$ax + bx + zy + by$
$= x(a + b) + y(a + b)$
$= (a + b)(x + y)$

Factor Quadratic Trinomials

$ax^2 + bx + c = ax^2 + mx + px + c$ when $m + p = b$ and $mp = ac$

Factor Differences of Squares

$a^2 - b^2 = (a + b)(a - b)$

Factor Perfect Squares

$a^2 + 2ab + b^2 = (a + b)^2$

🧠 **Think About It!**

Why do you think that there are so many different methods for factoring an equation?

🧭 **Go Online** You can complete an Extra Example online.

Example 3 Solve a Quadratic Equation by Using the Distributive Property

Solve $4x^2 - 12x = 0$. Check your solution.

$4x^2 - 12x = 0$	Original equation
$4x(x \underline{\quad}) = 0$	Factor by using the Distributive Property.
$4x = 0$ and $x - 3 = 0$	Zero Product Property
$x = \underline{\quad}$ $x = \underline{\quad}$	Simplify.

Check the roots by substituting them into the original equation.

$4(0)^2 - 12(0) = 0$ ✓ $4(3)^2 - 12(3) = 0$ ✓

Example 4 Solve a Quadratic Equation by Factoring a Trinomial

Part A Solve $x^2 - 2x - 8 = 0$. Check your solution.

$x^2 - 2x - 8 = 0$	Original equation
$(x \underline{\quad})(x \underline{\quad}) = 0$	Factor the trinomial.
$x + 2 = 0$ and $x - 4 = 0$	Zero Product Property
$x = \underline{\quad}$ $x = \underline{\quad}$	Simplify.

Check the roots by substituting them into the original equation.

Part B Use the roots of $x^2 - 2x - 8 = 0$ to sketch the graph of the related function.

Graph ($\underline{\quad}$, 0) and ($\underline{\quad}$, 0). The vertex is
($\underline{\quad}$, $\underline{\quad}$). Because a is positive, the graph
opens $\underline{\quad}$.

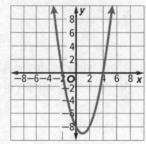

Check

Solve $x^2 - 10x + 24 = 0$. $\underline{\qquad}$

Problem-Solving Tip

Make a Chart Making a chart to organize the factors of c can help you identify the pair of factors that have a sum of b.

Example 5 Solve a Quadratic Equation by Factoring a Difference of Squares

Solve $49 - x^2 = 0$. Check your solution.

$49 - x^2 = 0$	Original equation
$7^2 - x^2 = 0$	Write in the form $a^2 - b^2$.
$(7 + x)(7 - x) = 0$	Difference of squares
$7 + x = \underline{\quad}$ and $7 - x = \underline{\quad}$	Zero Product Property
$x = \underline{\quad}$ and $x = \underline{\quad}$	Simplify.

Check the roots by substituting them into the original equation.

Copyright © McGraw-Hill Education

Example 6 Solve a Quadratic Equation by Factoring a Perfect Square Trinomial

Solve $25y^2 - 60y + 81 = 45$. Check your solution.

$25y^2 - 60y + 81 = 45$	Original equation
$25y^2 - 60y + \underline{\quad} = 0$	Subtract 45 from each side.
$(5y)^2 - 2(5y)(6) + 6^2 = 0$	Factor the perfect square trinomial.
$(\underline{\quad} - \underline{\quad})^2 = 0$	Factor the trinomial.
$\sqrt{(5y - 6)^2} = \sqrt{0}$	Square Root Property
$y = \underline{\quad}$	Simplify.

Check the root by substituting it into the original equation.

🌐 Apply Example 7 Factor a Trinomial to Solve a Problem

The area of an isosceles triangle is 108 square feet. Find the perimeter of the triangle if the length of each leg is 15 feet, the length of the base is $4x + 2$ feet, and the height is $3x$ feet.

1. **What is the task?**

Describe the task in your own words. Then list any questions that you may have. How can you find answers to your questions?

_____

_____

_____

2. **How will you approach the task? What have you learned that you can use to help you complete the task?**

_____

_____

3. **What is your solution?**

Use your strategy to solve the problem. Label the drawing and write an equation for the area of the triangle.

_____

Solve your equation by factoring.

_____

Find the perimeter of the triangle. _____

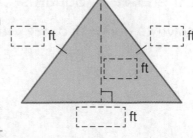

4. **How can you know that your solution is reasonable?**

✏️ **Write About It!** Write an argument that can be used to defend your solution.

_____

_____

Sidebar (left column)

🧁 **Think About It!**

Why does $25y^2 - 60y + 81 = 45$ have one solution instead of two?

🧁 **Think About It!**

Can $x = -\frac{9}{2}$ ever be a solution of $108 = \frac{1}{2}(4x + 2)(3x)$?

⭐ **Go Online**

You can complete an Extra Example online.

Check

The surface area of a prism is 264 square centimeters. Find the volume of the prism if its height is 15 centimeters, its length is $x + 4$ centimeters, and its width is x centimeters. (Hint: $S = Ph + 2B$ and $V = Bh$) _____

A. 2 cm^3

B. 26 cm^3

C. 180 cm^3

D. 264 cm^3

Learn Writing Quadratic Functions Given the Zeros

You can write the equation of a quadratic function if you know its zeros.

Key Concept • Writing Equations of Quadratic Functions

Step 1 Find the factors of a related expression.

Step 2 Determine whether the value of a is positive or negative.

Step 3 Use another point on the graph to determine the value of a.

Talk About It!

How many points on the graph of a quadratic function must be identified in order to write its equation? Explain.

Example 8 Write a Quadratic Function Given a Graph

Write a quadratic function for the given graph.

Step 1 Find the factors of a related expression.

The two zeros on the graph are -2 and _____, so $(x$ _____$)$ and $(x - 6)$ are factors of the related expression.
The function $f(x) =$ _____$(x + 2)(x - 6)$ represents the graph.

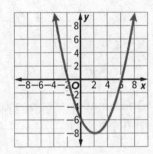

Step 2 Determine whether the value of a is positive or negative.
The graph opens upward, so a must be _____.

Step 3 Use another point on the graph to determine the value of a.
The point $(2, -8)$ is on the graph.

$f(x) = a(x + 2)(x - 6)$ Quadratic function with zeros of -2 and 6

$-8 = a ($_____$+ 2)($_____$- 6)$ $[x, f(x)] = (2, -8)$

$-8 =$ _____a Simplify.

_____ $= a$ Divide each side by -16.

One equation for the function is $f(x) = \frac{1}{2}(x + 2)(x - 6)$, or

$f(x) =$ _____.

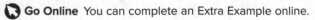 **Go Online** You can complete an Extra Example online.

Think About It!

How would this process have been different if the graph opened downward?

Check

Which quadratic function represents the given graph? _____

A. $f(x) = -\frac{1}{2}x^2 - \frac{3}{4}x + \frac{1}{2}$

B. $f(x) = \frac{1}{2}x^2 - \frac{3}{4}x + \frac{1}{2}$

C. $f(x) = -\frac{1}{2}x^2 + \frac{3}{4}x + \frac{1}{2}$

D. $f(x) = \frac{1}{2}x^2 + \frac{3}{4}x + \frac{1}{2}$

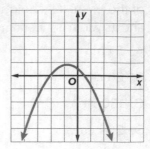

Example 9 Write a Quadratic Function Given Points

Write a quadratic function for the graph that contains (−7, 0), (−3, −32), and (1, 0).

We know that the zeros of a function occur at the *x*-intercepts, or when *y* = 0. Therefore, −7 and 1 are the zeros of this function, and *x* + 7 and *x* − 1 are factors of the related function. To find the value of *a* in the related function $f(x) = a(x + 7)(x − 1)$, use the third point (−3, −32).

$f(x) =$ _____(_____ + 7)(_____ − 1)   Quadratic function with zeros of −7 and 1

$-32 = a ($ _____ + 7)(_____ − 1) $[x, f(x)] = (−3, −32)$

$-32 =$ _____a Simplify.

_____ $= a$ Divide each side by −16.

One function that passes through the given points is $f(x) = 2(x + 7)(x − 1)$ or $f(x) =$ _____.

Check

Which quadratic function has a graph that contains $\left(-\frac{5}{2}, 0\right)$, (1, 35), and $\left(\frac{7}{2}, 0\right)$? _____

A. $f(x) = x^2 + x - \frac{35}{4}$

B. $f(x) = -4x^2 + 4x + 35$

C. $f(x) = 4x^2 + 4x - 35$

D. No quadratic function exists that passes through these points.

🔵 **Go Online** You can complete an Extra Example online.

Practice

Go Online You can complete your homework online.

Examples 1, 3, 5, 6

Solve each equation. Check your solutions.

1. $x^2 = 36$

2. $x^2 = 81$

3. $(k + 1)^2 = 9$

4. $81 - 4b^2 = 0$

5. $3b(9b - 27) = 0$

6. $(7x + 3)(2x - 6) = 0$

7. $b^2 = -3b$

8. $a^2 = 4a$

9. $x^2 - 6x = 27$

10. $a^2 + 11a = -18$

11. $n^2 - 120 = 7n$

12. $d^2 + 56 = -18d$

13. $y^2 - 90 = 13y$

14. $h^2 + 48 = 16h$

15. $x^2 + 9x + 18 = 0$

16. $4x^2 + 28x + 49 = 0$

17. $-9x^2 = 30x + 25$

18. $-x^2 - 12 = -8x$

19. $36w^2 = 121$

20. $100 = 25w^2$

21. $64x^2 - 1 = 0$

22. $4y^2 - \frac{9}{16} = 0$

23. $\frac{1}{4}b^2 = 16$

24. $81 - \frac{1}{25}x^2 = 0$

Example 4

25. Consider the equation $c^2 + 10c + 9 = 0$.
  a. Solve the equation by factoring.
  b. Use the roots to sketch the related function.

26. Consider the equation $x^2 - 18x = -32$.
  a. Solve the equation by factoring.
  b. Use the roots to sketch the related function.

27. NUMBER THEORY The product of the two consecutive positive integers is 11 more than their sum. What are the numbers?

28. LADDERS A ladder is resting against a wall. The top of the ladder touches the wall at a height of 15 feet, and the length of the ladder is one foot more than twice its distance from the wall. Find the distance from the wall to the bottom of the ladder. (*Hint*: Use the Pythagorean Theorem to solve the problem.)

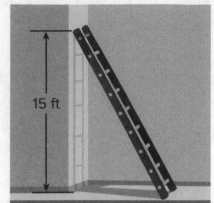

15 ft

29. FREE FALL The function $f(t) = -16t^2 + 576$ represents the height of a freely falling ballast bag that was accidentally dropped from a hot-air balloon 576 feet above the ground. After how many seconds t does the ballast bag hit the ground?

30. VOLUME Catalina can make an open-topped box out of a square piece of cardboard by cutting 3-inch squares from the corners and folding up the sides to meet. The volume of the resulting box is $V = 3x^2 - 36x + 108$, where x is the original length and width of the cardboard.

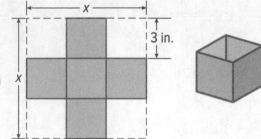

a. Factor the polynomial expression from the volume equation.

b. What is the volume of the box if the original length of each side of the cardboard was 9 inches?

Example 8

Write a quadratic function for the given graph.

31.

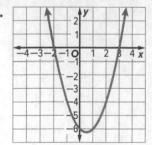

32.

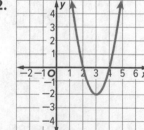

33.

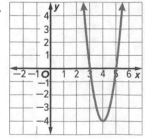

34.

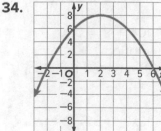

Example 9

Write a quadratic function that contains the given points.

35. (−9, 0), (−3, −30), (2, 0)

36. (−5, 0), (−4, 0), (3, 56)

37. (−4, 0), (2, −9), (8, 0)

38. (1, 0), (3, −18), (6, 0)

Mixed Exercises

39. CONSTRUCT ARGUMENTS Kayla says the zeros of the equation $2m^2 − 12m = 0$ are 0 and −12. Is she correct? Explain your reasoning.

40. STRUCTURE Write an expression for the perimeter of a rectangle that has an area $A = x^2 + 20x + 96$. Explain how you solved the problem.

Solve each equation. Check the solutions.

41. $25p^2 − 16 = 0$

42. $4p^2 + 4p + 1 = 0$

43. $n^2 − \frac{9}{25} = 0$

44. $9y^2 + 18y − 12 = 6y$

45. $6h^2 + 8h + 2 = 0$

46. $16d^2 = 4$

47. $\frac{1}{16} y^2 = 81$

48. $8p^2 − 16p = 10$

49. $x^2 + 30x + 150 = −75$

50. $10b^2 − 15b = 8b − 12$

51. $y^2 − 16y + 64 = 81$

52. $(m − 4)^2 = 9$

53. $4x^2 = 80x − 400$

54. $9 − 54x = −81x^2$

55. $4c^2 + 4c + 1 = 9$

56. $x^2 − 16x + 64 = 4$

57. STRUCTURE A triangle's height is 10 feet more than its base. If the area of the triangle is 100 square feet, use factoring to find its dimensions.

58. REGULARITY Explain how to use the Zero Product Property to solve a quadratic equation.

59. REGULARITY Consider a quadratic equation written in standard form, $ax^2 + bx + c = 0$, where $a \neq 1$. Explain how solving an equation by factoring a trinomial where $a \neq 1$ differs from solving a quadratic equation by factoring a trinomial where $a = 1$.

60. STRUCTURE A square has an area of $4x^2 + 16xy + 16y^2$ square inches. The dimensions are binomials with positive integer coefficients. Find the perimeter of the square. Is the perimeter a multiple of $(x + 2y)$? Explain.

61. STRUCTURE The equation $x^2 + qx - 12 = 6x$ can be solved by factoring. How can you use number relationships to predict values for q?

62. USE A MODEL The rectangle at the right has a perimeter of $14x + 30$ centimeters and an area of 225 square centimeters. Find the dimensions of the rectangle. Explain how you solved the problem.

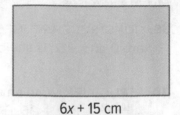

6x + 15 cm

63. AREA A triangle has an area of 64 square feet. If the height of the triangle is 8 feet more than its base, x, what are its height and base?

64. CONSTRUCT ARGUMENTS Jarrod claims that the quadratic equation $x^2 - 7x + 5 = 0$ has no solution because the left side does not factor. Do you agree or disagree? Use a graph to justify your argument.

Higher-Order Thinking Skills

65. PERSEVERE Given the equation $(ax + b)(ax - b) = 0$, solve for x. What do you know about the values of a and b?

66. WRITE Explain how to solve the quadratic equation $x^2 + 16x = -64$.

67. FIND THE ERROR Ignatio and Samantha are solving $6x^2 - x = 12$. Is either of them correct? Explain your reasoning.

Ignatio	Samantha
$6x^2 - x = 12$	$6x^2 - x = 12$
$x(6x - 1) = 12$	$6x^2 - x - 12 = 0$
$x = 12$ or $6x - 1 = 12$	$(2x - 3)(3x + 4) = 0$
$6x = 13$	$2x - 3 = 0$ or $3x + 4 = 0$
$x = \frac{13}{6}$	$x = \frac{3}{2} \qquad x = -\frac{4}{3}$

68. ANALYZE What should you consider when solving a quadratic equation that models a real-world situation?

69. CREATE Write a perfect square trinomial equation in which the coefficient of the middle term is negative and the last term is a fraction. Solve the equation.

70. PERSEVERE Factor the polynomial $x^3 + x^2 - 6x$. Make a table. Then sketch the graph of the related function.

Solving Quadratic Equations by Completing the Square

Explore Using Algebra Tiles to Complete the Square

Online Activity Use algebra tiles to complete the Explore.

> **INQUIRY** How does forming a square to create a perfect square trinomial help you solve quadratic equations?   ×

Learn Solving Quadratic Equations by Completing the Square

Key Concept • Completing the Square

To **complete the square** for any quadratic expression of the form $x^2 + bx$, follow the steps below.

Step 1 Find one-half of b, the coefficient of x.

Step 2 Square the result from Step 1.

Step 3 Add the result of Step 2 to $x^2 + bx$.

The pattern to complete the square is represented by $x^2 + bx + \left(\frac{b}{2}\right)^2 = \left(x + \frac{b}{2}\right)^2$.

Example 1 Complete the Square

Find the value of c that makes $x^2 + 8x + c$ a perfect square trinomial. Use the completing-the-square algorithm.

Step 1 Find half of 8. $\frac{8}{2} =$ ____

Step 2 Square the result of Step 1. $4^2 =$ ____

Step 3 Add the result of Step 2 to $x^2 + 8x$. $x^2 + 8x +$ ____

Thus, $c = 16$. Notice that $x^2 + 8x + 16 = ($____$)^2$.

Check

Find the value that makes the expression a perfect square trinomial.

$x^2 + 40x +$ _____

 Go Online You can complete an Extra Example online.

Today's Goals
- Solve quadratic equations by completing the square.
- Identify key features of quadratic functions by writing quadratic equations in vertex form.

Today's Vocabulary
completing the square

Think About It!
To complete the square, what constant would you add to $x^2 + 5x$?

Think About It!
How is the process of completing the square different when b is even and when b is odd?

Example 2 Solve an Equation by Completing the Square

Solve $x^2 - 10x + 14 = 5$ by completing the square.

In order to solve a quadratic equation by completing the square, first isolate $x^2 - bx$ on one side.

$x^2 - 10x + 14 = 5$	Original equation
$x^2 - 10x = -9$	_____
$x^2 - 10x + 25 = -9 + 25$	Because $\left(\frac{-10}{2}\right)^2 = 25$, add 25 to each side.
$(x - 5)^2 = 16$	_____
$x - 5 = \pm 4$	_____
$x = 5 \pm 4$	Add 5 to each side.
$x = 9$ or 1	_____

Example 3 Solve an Equation with a Not Equal to 1

Solve $-4x^2 + 32x - 72 = 0$ by completing the square.

To solve a quadratic equation when the leading coefficient is not 1, divide or multiply each term to eliminate the coefficient. Then, isolate $x^2 - bx$ and complete the square.

$-4x^2 + 32x - 72 = 0$	Original equation
$x^2 - \underline{\;\;}x + \underline{\;\;} = 0$	Divide each side by -4.
$x^2 - 8x = -18$	Subtract 18 from each side.
$x^2 - 8x + \underline{\;\;} = -18 + \underline{\;\;}$	Add $\left(\frac{8}{2}\right)^2$ to each side.
$(x - \underline{\;\;})^2 = \underline{\;\;\;\;}$	Factor $x^2 - 8x + 16$

No real number has a negative square. So, this equation has _____ solutions.

Learn Finding the Maximum or Minimum Value

Key Concept • Use Vertex Form to Graph

Step 1 Complete the square to write the function in vertex form.

Step 2 Identify the axis of symmetry and extrema based on the function in vertex form. When the leading coefficient is positive, the parabola will open up and the vertex will be a minimum. When the leading coefficient is negative, the parabola will open down and the vertex will be a maximum.

Step 3 Solve for x to find the zeros. The zeros are the x-intercepts of the graph.

Step 4 Use the key features to graph the function.

Example 4 Find a Minimum

Write $y = x^2 + 2x - 5$ in vertex form. Identify the axis of symmetry, extrema, and zeros. Then, use the key features to graph the function.

Step 1 Complete the square to write the function in vertex form.

$y = x^2 + 2x - 5$	Original function
$y + 5 = x^2 + 2x$	Add 5 to each side.
$y + 5 + 1 = x^2 + 2x + 1$	Add $\left(\frac{b}{2}\right)^2$ to each side.
$y + 6 = (x + 1)^2$	Factor.
$y = (x + 1)^2 - 6$	Subtract 6 from each side to write in vertex form.

Step 2 Identify the axis of symmetry and extrema.

In vertex form the vertex of the parabola (h, k) or _____. Since the x^2-term is positive, the vertex is a _____. The axis of symmetry is $x = h$ or _____.

Step 3 Solve for x to find the zeros.

$$(x + 1)^2 - 6 = 0$$
$$(x + 1)^2 = 6$$
$$x + 1 = \pm\sqrt{6}$$
$$x \approx -3.45 \text{ or } 1.45$$

Step 4 Use the key features to graph the function.

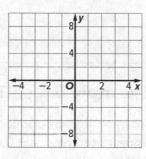

Talk About It!

Explain how you can determine whether a parabola will open up or open down.

Example 5 Find a Maximum

Write $y = -2x^2 + 20x - 42$ in vertex form. Identify the axis of symmetry, extrema, and zeros. Then, use the key features to graph the function.

Step 1 Complete the square to write the function in vertex form.

$y = -2x^2 + 20x - 42$	Original function
$y + 42 = -2x^2 + 20x$	Add 42 to each side.
$y + 42 = -2(x^2 - 10x)$	Factor out -2.
$y + 42 \underline{\hspace{1cm}} = -2(x^2 - 10x + 25)$	Since $-2\left[\left(\frac{10}{2}\right)^2\right] = -50$, add -50 to each side.
$y - 8 = -2(x - 5)^2$	Factor $x^2 - 10x + 25$.
$y = \underline{\hspace{3cm}}$	Add 8 to each side to write in vertex form.

Watch Out!

Leading Coefficients
When completing the square of a quadratic equation where $a \neq 1$, be sure to multiply the constant added by the coefficient before adding it to both sides.

(continued on the next page)

Step 2 Identify the axis of symmetry and extrema. In vertex form the vertex of the parabola is at (h, k) or (____, ____). Because the x^2-term is negative, the vertex is a _____. The axis of symmetry is $x = h$ or $x =$ ____.

Step 3 Solve for x to find the zeros.

$$0 = -2(x - 5)^2 + 8$$
$$2(x - 5)^2 = \underline{\quad}$$
$$(x - 5)^2 = \underline{\quad}$$
$$x - 5 = \underline{\quad}$$
$$x = \underline{\quad} \text{ or } \underline{\quad}$$

Step 4 Use the key features to graph the function.

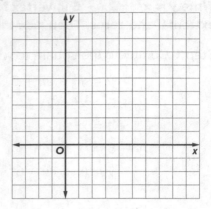

🌐 **Example 6** Use Extrema and Key Features

LUNAR LANDING Suppose the path of a golf ball hit on the moon can be represented by $y = -0.001x^2 + 0.248x$, where y is the height of the ball in meters and x is its horizontal distance in meters. Determine the maximum height of the golf ball and the horizontal distance it traveled.

Complete the square to write the equation of the function in vertex form.

$$y = -0.001x^2 + 0.248x \qquad \text{Original equation}$$
$$y = \underline{\qquad}(x^2 - \underline{\quad}x) \qquad \text{Factor out } -0.001.$$
$$y - \underline{\quad} = -0.001(x^2 - 248x + \underline{\quad}) \qquad \text{Add } -15.376 \text{ to each side.}$$
$$y - 15.376 = -0.001(x - \underline{\quad})^2 \qquad \text{Factor } x^2 - 248x + 15,376.$$
$$y = -0.001(x - 124)^2 + \underline{\qquad} \qquad \text{Add } 15.376 \text{ to each side.}$$

The vertex of the parabola is at (____, ____). Because the vertex represents the maximum, the ball reached a height of ____ meters.

To find the horizontal distance the ball traveled, find the zeros.

$$0 = -0.001(x - 124)^2 + 15.376$$
$$0.001(x - 124)^2 = \underline{\qquad}$$
$$(x - 124)^2 = \underline{\qquad}$$
$$x - 124 = \underline{\qquad}$$
$$x = \underline{\quad} \text{ or } \underline{\quad}$$

Because $x = 0$ represents the golf ball's initial point, the golf ball traveled ____ meters before hitting the surface of the Moon.

▶ **Go Online** You can complete an Extra Example online.

🍩 **Think About It!**

Suppose the golf ball has been hit from a platform 3 meters above the lunar surface. How would this affect the vertex form of the equation?

Practice

Go Online You can complete your homework online.

Example 1

Find the value of c that makes each trinomial a perfect square.

1. $x^2 + 26x + c$

2. $x^2 - 24x + c$

3. $x^2 - 19x + c$

4. $x^2 + 17x + c$

5. $x^2 + 5x + c$

6. $x^2 - 13x + c$

7. $x^2 - 22x + c$

8. $x^2 - 15x + c$

9. $x^2 + 24x + c$

Examples 2 and 3

Solve each equation by completing the square. Round to the nearest tenth, if necessary.

10. $x^2 + 6x - 16 = 0$

11. $x^2 - 2x - 14 = 0$

12. $x^2 - 8x - 1 = 8$

13. $x^2 + 3x + 21 = 22$

14. $x^2 - 11x + 3 = 5$

15. $5x^2 - 10x = 23$

16. $2x^2 - 2x + 7 = 5$

17. $3x^2 + 12x + 81 = 15$

18. $4x^2 + 6x = 12$

19. $4x^2 + 6 = 10x$

20. $-2x^2 + 10x = -14$

21. $-3x^2 - 12 = 14x$

Examples 4 and 5

Write each equation in vertex form. Identify the axis of symmetry, extrema, and zeros. Then, use the key features to graph the function.

22. $y = x^2 + 8x + 7$

23. $y = x^2 - 12x + 16$

24. $y = -x^2 - 4x + 5$

25. $y = x^2 - 8x + 10$

Example 6

26. **MARS** On Mars, the gravity acting on an object is less than that on Earth. On Earth, a golf ball hit with an initial upward velocity of 26 meters per second will hit the ground in about 5.4 seconds. The height h of an object on Mars that leaves the ground with an initial velocity of 26 meters per second is given by the equation $h = -1.9t^2 + 26t$. How much longer will it take for the golf ball hit on Mars to reach the ground? What is the maximum height of the golf ball on Mars? Round your answer to the nearest tenth.

27. **FROGS** A frog sitting on a stump 3 feet high hops off and lands on the ground. During its leap, its height h in feet is given by $h = -0.5d^2 + 2d + 3$, where d is the distance from the base of the stump. How far is the frog from the base of the stump when it lands on the ground? What is the maximum height of the frog during its leap?

28. **FALLING OBJECTS** Keisha throws a rock down an abandoned well. The distance d in feet the rock falls after t seconds can be represented by $d = 16t^2 + 64t$. If the water in the well is 80 feet below ground, how many seconds will it take for the rock to hit the water?

Mixed Exercises

29. GARDENING Peggy is planning a rectangular vegetable garden using 200 feet of fencing material. She only needs to fence three sides of the garden since one side borders an existing fence. Let x = the width of the rectangle. For what widths would the area of Peggy's garden equal 4800 square feet if she uses all the fencing material?

30. REASONING Find the value of q that makes $0.5x^2 + 0.5qx + 72$ a perfect square trinomial. Show that the same value of q makes $4x^2 + 24x + 1.5q$ a perfect square trinomial.

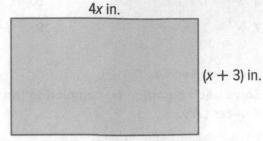

4x in.

(x + 3) in.

31. AREA The area of the rectangle shown is 352 square inches. What are the dimensions of the rectangle?

32. CONSTRUCT ARGUMENTS Use completing the square to show that no two consecutive positive even integers can have a product of 27.

33. STRUCTURE Lawrence writes a trinomial that can be solved by completing the square and cannot be solved by factoring. One of the solutions of his equation is between 4 and 5.

a. Write a possible equation.

b. What are the solutions to the equation?

c. Explain why your equation meets Lawrence's criteria.

34. STRUCTURE The quadratic function $f(t) = -5000t^2 + 70,000t + 5000$ is used to model the number of products a company sells in years t since the product was released. Complete the square to find the zeros of $f(t)$ and interpret each in the context of the situation. Then find the maximum sales for the product.

35. PERSEVERE Given $y = ax^2 + bx + c$ with $a \neq 0$, derive the equation for the axis of symmetry by completing the square, and rewrite the equation in the form $y = a(x - h)^2 + k$.

36. ANALYZE Determine the number of solutions $x^2 + bx = c$ has if $c < -\left(\frac{b}{2}\right)^2$. Justify your argument.

37. WHICH ONE DOESN'T BELONG? Identify the expression that does not belong with the other three. Explain your reasoning.

| $n^2 - n + \frac{1}{4}$ | $n^2 + n + \frac{1}{4}$ | $n^2 - \frac{2}{3}n + \frac{1}{9}$ | $n^2 + \frac{1}{3}n + \frac{1}{9}$ |

38. CREATE Write a quadratic equation for which the only solution is 4.

39. WRITE Compare and contrast the following strategies for solving $x^2 - 5x - 7 = 0$: completing the square, graphing, and factoring.

Solving Quadratic Equations by Using the Quadratic Formula

Learn Solving Quadratic Equations by Using the Quadratic Formula

Completing the square of the quadratic equation produces a formula that allows you to find the solutions of *any* quadratic equation. This formula is called the Quadratic Formula, $x = \dfrac{-b \pm \sqrt{b^2 - 4ac}}{2a}$.

Example 1 Use the Quadratic Formula

Solve $x^2 - 12x = -11$ by using the Quadratic Formula.

Step 1 Rewrite the equation in standard form.

$x^2 - 12x = -11$	Original equation
$x^2 - 12x + \underline{} = \underline{}$	Add 11 to each side.

Step 2 Apply the Quadratic Formula.

$x = \dfrac{-b \pm \sqrt{b^2 - 4ac}}{2a}$	Quadratic Formula
$= \dfrac{-(-12) \pm \sqrt{(-12)^2 - 4(1)(11)}}{2(1)}$	$a = 1, b = -12, c = 11$
$= \dfrac{12 \pm \sqrt{144 - 44}}{2}$	Multiply.
$= \dfrac{12 \pm \sqrt{100}}{2}$	Subtract.
$= \dfrac{12 \pm 10}{2}$	Take the square root.
$x = \dfrac{12 + 10}{2}$ or $x = \dfrac{12 - 10}{2}$	Separate the solutions.
$= 11 \qquad = 1$	Simplify.

The solutions are _____ and _____.

Check

Solve the equation by using the Quadratic Formula.

$x^2 - 19x = -70$

$x = \underline{}, \underline{}$

Today's Goals
- Solve quadratic equations by using the Quadratic Formula.
- Use the discriminant to determine the number of solutions of a quadratic equation.

Today's Vocabulary
discriminant

 Talk About It!

What does the $\pm$ symbol indicate? How does that affect the number of factors determined by using the Quadratic Formula? Explain.

 Think About It!

If the expression under the square root simplified to 0, then what would be the relation between the two factors?

Example 2 Use the Quadratic Formula When a Is Not Equal to 1

Solve $2x^2 + 5x = 12$ by using the Quadratic Formula.

Step 1 Rewrite the equation in standard form.

$$\underline{\quad}x^2 + \underline{\quad}x - \underline{\quad} = 0$$

Step 2 Apply the Quadratic Formula.

$$x = \frac{-b \pm \sqrt{b^2 - 4ac}}{2a} \qquad \text{Quadratic Formula}$$

$$= \frac{-(5) \pm \sqrt{(5)^2 - 4(2)(-12)}}{2(2)} \qquad a = 2, b = 5, c = -12$$

$$= \frac{-5 \pm \sqrt{25 + 96}}{4} \qquad \text{Multiply.}$$

$$= \frac{-5 \pm \sqrt{121}}{4} \qquad \text{Add.}$$

$$= \frac{-5 \pm 11}{4} \qquad \text{Take the square root.}$$

$$x = \frac{-5 - 11}{4} \text{ or } x = \frac{-5 + 11}{4} \qquad \text{Separate the solutions.}$$

$$= -4 \qquad\quad = \frac{3}{2} \qquad \text{Simplify.}$$

The solutions are _____ and _____.

🌐 **Example 3** Solve a Quadratic Equation with Irrational Roots

HEALTH The normal blood pressure of an adult woman P in millimeters of mercury given her age t in years can be modeled by the equation $P = 0.01t^2 + 0.5t + 107$. If Seiko's blood pressure is 120, how old might she be?

Step 1 Rewrite the equation in standard form.

$$0.01t^2 + 0.5t + 107 = 120 \qquad \text{Original equation}$$

$$0.01t^2 + 0.5t - \underline{\quad} = \underline{\quad} \qquad \text{Subtract 120 from each side.}$$

Step 2 Apply the Quadratic Formula.

$$x = \frac{-b \pm \sqrt{b^2 - 4ac}}{2a} \qquad \text{Quadratic Formula}$$

$$= \frac{-0.5 \pm \sqrt{(0.5)^2 - 4(0.01)(-13)}}{2(0.01)} \qquad a = 0.01, b = 0.5, c = -13$$

$$= \frac{-0.5 \pm \sqrt{0.25 + 0.52}}{0.02} \qquad \text{Multiply.}$$

$$= \frac{-0.5 \pm \sqrt{0.77}}{0.02} \qquad \text{Simplify.}$$

$$x = \frac{-0.5 + \sqrt{0.77}}{0.02} \text{ or } x = \frac{-0.5 - \sqrt{0.77}}{0.02} \qquad \text{Separate the solutions.}$$

$$\approx 18.9 \qquad\qquad \approx -68.9 \qquad \text{Simplify.}$$

The solutions are _____ and _____.

🅡 **Go Online** You can complete an Extra Example online.

Step 3 Eliminate unreasonable solutions.

Because t measures age in years, it is _____ to assume that Seiko is 18.9 years old.

Because t measures age in years, which cannot be negative, -68.9 is an _____ solution.

Check

FOOTBALL The quarterbacks on a high school football team want to see who can get the most hang-time when throwing the ball. Connor knows that using his initial velocity and height, he can model the trajectory of his throw with the equation $h = -16t^2 + 95t + 5.2$, where h is the height of the ball and t is the time after throwing. Solve the equation by using the Quadratic Formula, eliminating any extraneous solutions. Round your answer(s) to the nearest tenth.

$t =$ _____

Explore Deriving the Quadratic Formula Algebraically

Online Activity Use an interactive tool to complete the Explore.

> **⊘**
>
> **INQUIRY** How does having a formula make it possible to solve quadratic equations where the other methods are not easy to apply?

Explore Deriving the Quadratic Formula Visually

Online Activity Use an interactive tool to complete the Explore.

> **⊘**
>
> **INQUIRY** Why can you use the Quadratic Formula to solve any quadratic equation?

Think About It!

If $P = 0$, then there are no real solutions to the equation. Does this make sense in the context of the situation?

Go Online You can complete an Extra Example online.

Learn The Discriminant

In the Quadratic Formula, $x = \dfrac{-b \pm \sqrt{b^2 - 4ac}}{2a}$, the expression under the radical sign, $b^2 - 4ac$, is called the **discriminant**. You can use the discriminant to determine the number of real solutions of a quadratic equation.

Key Concept • Using the Discriminant

$b^2 - 4ac < 0$	$b^2 - 4ac = 0$	$b^2 - 4ac > 0$
Graph		
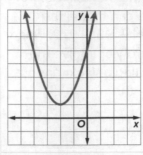	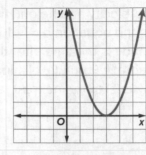	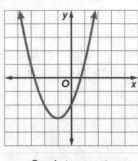
0 x-intercepts	1 x-intercept	2 x-intercepts
Real Solutions		
0	1	2

> **Study Tip**
>
> **Real Solutions** The real solutions are the cases where $y = 0$, which means that they are values of x where the graph of the function crosses the x-axis.

Example 4 Use the Discriminant

State the value of the discriminant of $4x^2 - 3x = -1$. Then determine the number of real solutions of the equation.

Step 1 Rewrite the equation in standard form.

$4x^2 - 3x = -1$ Original equation

$4x^2 - 3x + \underline{} = \underline{}$ Add 1 to each side.

Step 2 Find the discriminant.

$b^2 - 4ac = (-3)^2 - 4(4)(1)$ $a = 4, b = -3, c = 1$

$= \underline{}$ Simplify.

Because the discriminant is _____, the equation has _____ _____.

Check

State the value of the discriminant of $8x^2 - 15x = -9$.

The discriminant is _____.

Determine the number of real solutions of the equation.

The equation has _____ solution(s).

> **Go Online** to practice what you've learned about solving quadratic equations in the Put It All Together over Lessons 11-3 through 11-6.

Go Online You can complete an Extra Example online.

Practice

Go Online You can complete your homework online.

Examples 1 and 2

Solve each equation by using the Quadratic Formula. Round to the nearest tenth, if necessary.

1. $x^2 - 49 = 0$

2. $x^2 - x - 20 = 0$

3. $x^2 - 5x - 36 = 0$

4. $4x^2 + 5x - 6 = 0$

5. $x^2 + 16 = 0$

6. $6x^2 - 12x + 1 = 0$

7. $5x^2 - 8x = 6$

8. $2x^2 - 5x = -7$

9. $5x^2 + 21x = -18$

10. $81x^2 = 9$

11. $8x^2 + 12x = 8$

12. $4x^2 = -16x - 16$

13. $10x^2 = -7x + 6$

14. $-3x^2 = 8x - 12$

15. $2x^2 = 12x - 18$

Example 3

16. BUSINESS Tanya runs a catering business. Based on her records, her weekly profit can be approximated by the function $f(x) = x^2 + 2x - 37$, where x is the number of meals she caters. If $f(x)$ is negative, it means that the business has lost money. What is the least number of meals that Tanya needs to cater in order to make a profit?

17. FREE FALL Josh drops his phone from a height of 50 feet. The situation is best modeled by $h = -16t^2 + 50$, where h is the height in feet and t is the time in seconds. How long will it take for the phone to hit the ground? Round to the nearest tenth.

18. ARCHITECTURE The Golden Ratio appears in the design of the Greek Parthenon because the width and height of the façade are related by the equation $\frac{W+H}{W} = \frac{W}{H}$. If the height of a model of the Parthenon is 16 inches, what is its width? Round your answer to the nearest tenth.

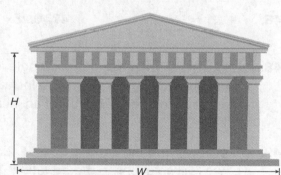

19. CRAFTS Ariadna cut a 60-inch chenille stem into two unequal pieces, and then she used each piece to make a square. The sum of the areas of the squares was 117 square inches. Let x be the length of one piece. Write and solve an equation to represent the situation and find the lengths of the two original pieces.

Example 4

State the value of the discriminant for each equation. Then determine the number of real solutions of the equation.

20. $0.2x^2 - 1.5x + 2.9 = 0$

21. $2x^2 - 5x + 20 = 0$

22. $x^2 - \frac{4}{5}x = 3$

23. $0.5x^2 - 2x = -2$

24. $2.25x^2 - 3x = -1$

25. $2x^2 = \frac{5}{2}x + \frac{3}{2}$

26. $x^2 + 2x + 1 = 0$

27. $x^2 - 4x + 10 = 0$

28. $x^2 - 6x + 7 = 0$

29. $x^2 - 2x - 7 = 0$

30. $x^2 - 10x + 25 = 0$

31. $2x^2 + 5x - 8 = 0$

32. $2x^2 + 6x + 12 = 0$

33. $2x^2 - 4x + 10 = 0$

34. $3x^2 + 7x + 3 = 0$

Mixed Exercises

Solve each equation by using the Quadratic Formula. Round to the nearest tenth, if necessary.

35. $x^2 + 4x = -1$

36. $x^2 - 9x + 22 = 0$

37. $x^2 + 6x + 3 = 0$

38. $2x^2 + 5x - 7 = 0$

39. $2x^2 - 3x = -1$

40. $2x^2 + 5x + 4 = 0$

41. $2x^2 + 7x = 9$

42. $3x^2 + 2x - 3 = 0$

43. $3x^2 - 7x - 6 = 0$

Without graphing, determine the number of x-intercepts of the graph of the related function for each equation.

44. $x^2 + 4x + 3 = 0$

45. $4.25x + 3 = -3x^2$

46. $x^2 + \frac{2}{25} = \frac{3}{5}x$

47. $0.25x^2 + x = -1$

48. RECTANGLES The base of a rectangle is 4 inches greater than the height. The area of the rectangle is 15 square inches. What are the dimensions of the rectangle to the nearest tenth of an inch?

49. REASONING For the equation $3x^2 + 2x + q = 0$, find all values of q so that there are two real solutions to the equation. Then find all values of q so that there are two complex solutions for the equation. Explain.

50. SITE DESIGN The town of Smallport plans to build a new water treatment plant on a rectangular piece of land 75 yards wide and 200 yards long. The buildings and facilities need to cover an area of 10,000 square yards. The town's zoning board wants the site designer to allow as much room as possible between each edge of the site and the buildings and facilities. Let x represent the width of the border.

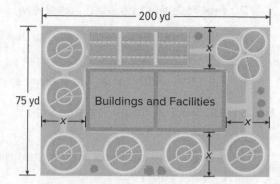

a. Write an equation to represent the area covered by the building and facilities.

b. Write the standard form of the quadratic equation.

c. Find the width of the border. Round your answer to the nearest tenth.

51. USE A MODEL New City's public works is putting together a fireworks presentation to celebrate the Fourth of July. They will be launching fireworks at different heights. The path of one of the fireworks is defined by the equation $h_1 = -16t^2 + 90t + 120$, with t in seconds and h_1 in feet.

a. The equation provides the height the firework is above the ground at any time between when it is launched and when it hits the ground. What does 120 represent in the context of the situation?

b. Write the equation that models when the firework will hit the ground. Solve the equation. Round to the nearest hundredth.

c. How high is the firework when it is at its highest point? How long did it take to get to this point? Round to the nearest hundredth. Explain.

d. Without graphing the equation, use the information from parts **a, b,** and **c** to describe what the graph would look like. Explain.

e. Write and solve an equation for the times at which the firework reaches a height of 200 feet. Round to the nearest tenth.

52. STRUCTURE A cylinder is filled completely with water. Its base has a radius of x cm, and its height is 8 cm. After Ella removes $40x$ cm^3 of water, the volume of the remaining water is 250 cm^3. What is the radius of the cylinder? Round to the nearest whole number, if necessary.

53. REGULARITY Solve the equation $x^2 + bx + c = 0$ for x in terms of b and c. Then determine whether each statement is *sometimes, always,* or *never* true.

a. If both b and c are negative, there will be at least one real solution.

b. If both b and c are positive, there will be complex nonreal solutions.

54. REASONING Braden wrote three quadratic equations: $-2x^2 - 3x - 8 = 0$, $-x^2 + 6x - 9 = 0$, and $-2x^2 + 4x + 3 = 0$. He graphs one of the equations as shown. He shows Ava the graph and the three equations and challenges her to match the correct equation with the graph. Does Ava need to solve all the equations to match one with the graph? Explain your reasoning.

55. STRUCTURE A rectangular area is to be enclosed with its length 20 feet less than its width, as shown.

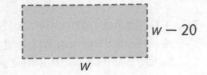

a. What should the width be in order for the enclosure to have an area of 100 ft²?

b. Find the width for which the area enclosed is R ft² for $R > 0$.

56. STRUCTURE A frame is made with width w inches and length $15 - w$ inches, as shown in the figure.

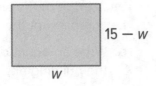

a. Is there any width for which the frame will have an area of 150 in²? Explain your reasoning.

b. Find the maximum area of the frame and the dimensions that produce the maximum area.

57. USE A SOURCE The St. Louis Arch can be approximated by a parabola. Research its dimensions and use the dimensions to write a quadratic equation representing the St. Louis Arch. Then set the equation equal to 0 and use the Quadratic Formula to solve for x. Interpret the solutions.

Determine whether there are *two*, *one*, or *no* real solutions of each equation.

58. The graph of the related quadratic function does not have an x-intercept.

59. The graph of the related quadratic function touches, but does not intersect the x-axis.

60. The graph of the related quadratic function intersects the x-axis twice.

61. Both a and b are greater than 0 and c is less than 0 in a quadratic equation.

62. WRITE Why can the discriminant be used to confirm the number of real solutions of a quadratic equation?

63. ANALYZE Use factoring techniques to determine the number of real zeros of $f(x) = x^2 - 8x + 16$. Compare this method to using the discriminant.

64. PERSEVERE Find all values of k such that $2x^2 - 3x + 5k = 0$ has two solutions.

65. WRITE Describe the advantages and disadvantages of each method of solving quadratic equations. Why are the methods equivalent? Which method do you prefer, and why?

66. CREATE Write a quadratic equation that has no real roots, and find its discriminant. Explain how the discriminant shows that a quadratic equation has no real roots.

67. CREATE Write a quadratic equation that has one real root, and find its discriminant. Explain why the Quadratic Formula yields only one solution when the discriminant of a quadratic equation is equal to zero.

68. CREATE Write a quadratic equation that has two real roots. Determine whether completing the square, graphing, or factoring would be the best method to use to solve your quadratic equation.

Solving Systems of Linear and Quadratic Equations

Today's Goals
- Solve systems of linear and quadratic equations by graphing.
- Solve systems of linear and quadratic equations by using algebraic methods.

Explore Using Algebra Tiles to Solve Systems of Linear and Quadratic Equations

Online Activity Use algebra tiles to complete the Explore.

> ☓
> @ **INQUIRY** How can you use algebra tiles to solve systems of linear and quadratic equations?

Think About It!
Can you put the steps in any other order? Explain your reasoning.

Learn Solving Systems of Linear and Quadratic Equations by Graphing

To solve a system of linear and quadratic equations by graphing, you can follow a set of steps.

Step 1 Graph the quadratic function.

Step 2 Graph the linear function.

Step 3 Find the point(s) of intersection.

Check Substitute the values in the original equation.

Talk About It!
Cordell says that the graphs of a linear function and a quadratic function always have 2 points of intersection. Do you agree or disagree? Justify your reasoning or provide a counterexample.

Example 1 Solve a System of Linear and Quadratic Equations Graphically

Solve the system of equations by graphing.

$y = x^2 + 4x - 1$

$y = 2x + 2$

Step 1 Graph $y = x^2 + 4x - 1$.

Step 2 Graph $y = 2x + 2$.

Step 3 Find the points of intersection.

The graphs appear to intersect at (___, ___) and (___, ___)

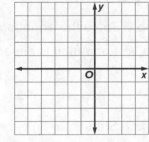

Go Online You can complete an Extra Example online.

Check

The graphs of $f(x) = -2x + 3$ and $g(x) = x^2 - 4x - 5$ are shown.

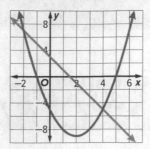

Complete the ordered pair(s) to represent each situation.

$f(x) = 0$ (___, ___)

$g(x) = 0$ (___, ___) (___, ___)

$f(x) = g(x)$ (___, ___) (___, ___)

☁ **Think About It!**

What does it mean if there is no solution when you substitute one expression for y in the other equation and solve?

Learn Solving Systems of Linear and Quadratic Equations Algebraically

To solve a system of linear and quadratic equations algebraically, you can follow a set of steps.

Step 1 Solve the equations for y.

Step 2 Substitute one expression for y in the other equation.

Step 3 Solve for x.

Step 4 Substitute the x-value(s) in either equation.

Step 5 Solve for y.

Check Graph the equations.

Example 2 Solve a System of Linear and Quadratic Equations Algebraically

Solve the system of equations algebraically.

$y = x^2 - 2x - 3$

$x + y = 3$

Step 1 Solve the equations for y.

The first equation is already solved for y.

$x + y = 3$ Original second equation

$y = $ ___ $ + 3$ Subtract x from each side.

Step 2 Substitute an expression for y.

$y = -x + 3$ Second equation solved for y

_____ $ = -x + 3$ Substitution

🅝 **Go Online** You can complete an Extra Example online.

Copyright © McGraw-Hill Education

Step 3 Solve for x.

$x^2 - 2x - 3 = -x + 3$	Original equation
$x^2 - x - 3 = +3$	Add x to each side.
$x^2 - x - 6 = 0$	Subtract 3 from each side.
$(x \underline{\quad})(x \underline{\quad}) = 0$	Factor.
$x - 3 = 0 \quad x + 2 = 0$	Zero Product Property
$x = \underline{\quad} \qquad x = \underline{\quad}$	Simplify.

Step 4 Substitute the x-value(s) in either equation.

$$y = -x + 3 \qquad\qquad y = -x + 3$$
$$y = -\underline{\quad} + 3 \qquad\qquad y = -(\underline{\quad}) + 3$$

Step 5 Solve for y.

$$y = -3 + 3 \qquad\qquad y = -(-2) + 3$$
$$y = \underline{\quad} \qquad\qquad\quad y = \underline{\quad}$$

Check Graph the equations.

The graphs of the functions intersect at
$(3, 0)$ and $(-2, 5)$, so $(\underline{\quad}, \underline{\quad})$ and $(\underline{\quad}, \underline{\quad})$
are solutions of this system.

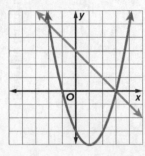

Check

Solve the system of equations algebraically.

$$y = x^2 - 8x + 19 \qquad x - 0.5y = 3$$

$(\underline{\quad}, \underline{\quad})$

🌐 Example 3 Use a System of Linear and Quadratic Equations

SALES Bathing suit sales at a clothing store can be modeled by the function $y = -x^2 + 12x + 25$, and gift card sales can be modeled by the function $y = 5x + 7$, where x represents the number of months past January and y represents the total revenue in thousands of dollars. Solve a system of equations algebraically to find the month when the revenue from bathing suit sales is equal to the revenue from gift card sales.

(continued on the next page)

> ## 💭 Think About It!
> Are there any other methods that can be used to solve the system? Explain your reasoning.

🔵 **Go Online** You can complete an Extra Example online.

Set the expressions equal to each other and solve for x.

$-x^2 + 12x + 25 = \underline{\quad} + \underline{\quad}$	Substitute.
$12x + 25 = \underline{\quad} + 5x + 7$	Add $-x^2$ to each side.
$25 = x^2 - \underline{\quad}x + 7$	Subtract $12x$ from each side.
$0 = x^2 - 7x - \underline{\quad}$	Subtract 25 from each side.
$0 = (x - \underline{\quad})(x + \underline{\quad})$	Factor.
$x - 9 = 0 \qquad x + 2 = 0$	Zero Product Property
$x = \underline{\quad} \qquad\quad x = \underline{\quad}$	Simplify.

Because the expressions are equivalent when $x = 9$, the revenue of bathing suit sales and gift card sales is equal $\underline{\quad}$ months past January, or in October.

Graph the equations to check your solution.

Sales

Check

SALES Revenue from single-day ticket sales to a local amusement park can be modeled by the function $y = -x^2 + 11x + 42$, and revenue from season pass sales can be modeled by the function $y = -5x + 81$, where x represents the number of months past January and y represents the total revenue in tens of thousands of dollars. Solve a system of equations algebraically to find the first month when the revenue from single-day tickets is equal to the revenue from season pass sales. $\underline{\quad}$

A. March

B. April

C. May

D. June

🔵 **Go Online** You can complete an Extra Example online.

Practice

Go Online You can complete your homework online.

Example 1

Solve each system of equations by graphing.

1. $y = x^2 - 4$
$y = -3$

2. $y = x^2 + x - 2$
$y = -x + 1$

3. $y = 2x^2 + 1$
$y = 1$

4. $y = x^2 + 3x + 1$
$y = x + 1$

Example 2

Solve each system of equations algebraically.

5. $y = x^2 - 2x - 5$
$y = 3$

6. $y = x^2 + 4x - 1$
$y = 3x + 1$

7. $y = x^2 - 6x + 5$
$x + y = -1$

8. $y = x^2 + x + 1$
$y - 1 = x$

9. $y + 3x = x^2 - 3$
$y = -2x + 3$

10. $y - 1 = 2x^2 - x$
$-2x + y = 3$

Example 3

11. GYMNASTICS A gymnast throws a baton up into the air during her floor routine. The height of the baton can be modeled by $y = -16x^2 + 24x + 4$, where x is the time in seconds and y is the height of the baton in feet since it was released. The ceiling of the gym can be represented by the function $y = 42$.

   a. Solve the system of equations algebraically.

   b. Will the baton reach the ceiling of the gym? Explain your reasoning.

12. DISC GOLF The height y in feet of a disc x seconds after it was thrown can be modeled by $y = -\frac{1}{30}x^2 + \frac{1}{2}x + 5\frac{2}{15}$. The height of Rodrigo's hands as he runs to catch the disc can be represented by the function $y = \frac{1}{2}x + 3$.

   a. Solve the system of equations algebraically.

   b. At what height will Rodrigo catch the disc?

13. ZOO Revenue from single-day ticket sales at a local zoo can be modeled by the function $y = -x^2 + 25x + 80$, and revenue from season pass sales can be modeled by the function $y = -4x + 200$. In both functions, x represents the number of months past January, and y represents the total revenue in thousands of dollars. Solve the system of equations algebraically to find the first month when the revenue from single-day ticket sales is equal to the revenue from season pass sales.

Mixed Exercises

Solve each system of equations.

14. $y = x^2$
$y = 2x$

15. $y = -2x^2 + 7x - 2$
$y = 3 - 4x$

16. $y = -x^2 + 4$
$y = \frac{1}{5}x + 5$

17. $y = -x^2 + 4x - 4$
$y = 2x - 3$

18. $y = -x^2 - 5x - 6$
$y = -3x - 1$

19. $y = 2x^2 - 4$
$y = 2x$

20. $y = x^2 + 7x + 12$
$y = 2x + 8$

21. $y = x^2 - x - 20$
$y = 3x + 12$

22. $y = 3x^2 - x - 2$
$y = -2x + 2$

23. $y = x^2 - x - 18$
$y = x - 3$

24. $y = x^2 - 3x + 1$
$y = x + 1$

25. $y = x^2 - 4x + 6$
$y = 2x - 3$

26. $y = -x^2 + 2x + 3$
$y = x + 2$

27. $y = x^2 + 4x - 1$
$y = 3x$

Solve each equation by using a system of equations.

28. $x^2 - 4x + 3 = x - 1$

29. $x^2 + 2x - 1 = x - 2$

30. $x^2 + 4x + 4 = 4$

31. $x^2 = -2x - 1$

32. $\frac{1}{2}x^2 - 4 = 3x + 4$

33. $x^2 + 5x + 5 = -x - 8$

34. WINGSUIT A wingsuit flyer jumps off a tall cliff. He falls freely for a few seconds before deploying the wingsuit and slowing his descent. His height during the freefall can be modeled by the function $y = -4.9x^2 + 420$, where y is the height above the ground in meters and x is the time in seconds. After deploying the wingsuit, the flyer's height is given by the function $y = -3x + 200$.

   a. To the nearest whole number, how long after jumping does the flyer deploy the wingsuit?

   b. To the nearest whole number, what was the flyer's height when he deployed the wingsuit?

35. REASONING The graph shows a quadratic function and a linear function $y = k$.

   a. How many solutions are there to the system?

   b. If the linear function were changed to $y = k + 5$, how many solutions would the system have?

   c. If the linear function were changed to $y = k - 2$, how many solutions would the system have?

36. USE A MODEL The function $R(x) = -x^2 + 5x + 14$ represents the revenue earned by a manufacturing company, where R is the revenue in millions of dollars and x is the number of items produced in millions. It cost the company $4 million to produce 3 million items.

   a. Write a function to represent the cost C in millions of dollars of producing x million items.

   b. Solve the system involving the revenue function and cost function. What does the solution represent?

37. USE TOOLS In a video game, players fling virtual rubber bands at a small target that moves around the screen. The rubber bands are all launched from the point $(-4, 0)$ and follow a path along the line $y = \frac{1}{2}x + 2$. The target moves around the screen following a parabolic path represented by the equation $y = -x^2 - 2x + 4$.

   a. Graph the path of the rubber bands and the path of the target.

   b. If a player hits the target, at approximately what point or points on the plane will this take place? Explain.

   c. Use your calculator to find the coordinates of the point or points at which the rubber bands may hit the target. Round to the nearest tenth. How do your results compare to your answer in part **b**?

38. ROCKETS Jamie is using binoculars to watch a friend set off a model rocket from a distance. Jamie's line of sight through the binoculars is a straight line from 50 meters above the launch point, decreasing $\frac{1}{2}$ a meter in height for every 1 meter of lateral distance to the point where Jamie is standing. The rocket's path follows a projectile motion formula, $4.9t^2 + vt + h$, where v is the initial velocity in meters per second and h is the launch height in meters. The rocket is launched at a velocity of 40 meters per second from the ground. At what height(s), in meters, will the rocket be in Jamie's line of sight?

   a. Write a system of equations to represent the situation.

   b. Find the solution to the system of equations graphically.

   c. Find the solution to the system of equations algebraically.

   d. Explain what the solution(s) mean in the context of the situation.

39. REGULARITY Explain a method for determining the number of solutions that exist for a system of equations from a graph of the related functions.

40. STATE YOUR ASSUMPTION Kasia found two solutions to a system of equations she used to determine whether her slinky would travel far enough to get to the next stair. One solution has a negative x-value, and one solution has a positive x-value. Which solution should Kasia use? What assumption did you make to decide?

Copyright © McGraw-Hill Education

Lesson 11-7 • Solving Systems of Linear and Quadratic Equations **661**

41. CONSTRUCT ARGUMENTS Rafaela kicked a rock off the top of a cliff to a deserted river below. The rock's path is modeled by $h(x) = -0.1x^2 + x + 600$, where $h(x)$ is the height, in meters, of the rock after x seconds. The cliff face is modeled approximately by $h(x) = 600 - 9x$. Will the rock land on the cliff or in the river? Justify your argument.

 **Higher-Order Thinking Skills**

PERSEVERE Use a graphing calculator to solve other types of systems.

42. $y = x^2 + 3x - 5$
$y = -x^2$

43. $y = \frac{3}{4}x$
$x^2 + y^2 = 1$ (*Hint*: Enter as two functions, $y = \sqrt{1 - x^2}$ and $y = -\sqrt{1 - x^2}$.)

44. ANALYZE For what value of k does the system $y = x^2 + x$ and $y = -2x + k$ have exactly one solution? Explain your reasoning.

45. ANALYZE For what value of k does the system $y = x^2 + 2$ and $y = 3x + k$ have exactly one solution? Explain your reasoning.

46. CREATE Below are a graph and a table of values that describe two equations, y_1 and y_2, one linear and one quadratic. Write a problem that could be solved using the graph and table of values, then solve your problem.

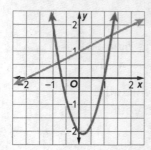

x	y_1	y_2
−2	0	12
0	1	−2
1	1.5	0

47. WRITE What is the maximum number of solutions that a system of one linear and one quadratic equation can have? What is the maximum number of solutions to a system of two linear functions? Write a paragraph to explain your reasoning.

48. FIND THE ERROR Xavier found the solution to a system of equations using the steps shown. In which step did Xavier make an error? Correct his error.

$y = x^2 - 4x - 6$
$y = 9x + 8$

Step 1: $x^2 - 4x - 6 = 9x + 8$
Step 2: $x^2 - 4x - 9x - 6 - 8 = 0$
Step 3: $x^2 - 13x - 14 = 0$
Step 4: $(x - 14)(x + 1) = 0$

Step 5: $y = 14, y = -1$
Step 6: $14 = 9x + 8; x = \frac{2}{3}$
Step 7: $-1 = 9x + 8; x = -1$

49. WHICH ONE DOESN'T BELONG? Analyze the three systems of equations given, and identify which system does not belong. Explain your reasoning.

$y = x^2 - 3x - 4$
$y = 5x + 1$

$y = -x^2 - x + 11$
$y = 2x - 10$

$y = 4x^2 + 4x - 2$
$y = x - 7$

Modeling and Curve Fitting

Explore Using Differences and Ratios to Model Data

▶ **Online Activity** Use an interactive tool to complete the Explore.

> ⊚ **INQUIRY** How can differences and ratios of
> successive *y*-values be used to write a model? ✕

Learn Modeling Real-World Situations

Different types of functions can be used to model data. You can use the equation or graph of a data set to determine the type of function that represents the data.

Differences and ratios can be used to write functions to represent data. When *x*-values are increasing at a constant rate, the differences of successive *y*-values are called *first differences*. The differences of successive first differences are called *second differences*.

Concept Summary • Linear and Nonlinear Functions

Linear	Quadratic	Exponential
Function		
$y = mx + b$	$y = ax^2 + bx + c$	$y = ab^x$, where $b > 0$
Graph		
Description		
y increases or decreases at a constant rate.	*y* increases or decreases by a square of *x*.	*y* increases or decreases by a power of *x*.
Differences/Ratios		
First differences are equal.	Second differences are equal.	Successive ratios are equal.
Uses of Difference/Ratio		
The first difference is the slope *m*.	Half of the second difference is *a*.	The common ratio is the base *b* of the function.

▶ **Go Online** You can complete an Extra Example online.

Today's Goals
- Distinguish between situations that can be modeled with linear, exponential, and quadratic functions.
- Make and evaluate predictions by fitting nonlinear functions to sets of data.

Today's Vocabulary
coefficient of determination

curve fitting

🗪 Talk About It!
When determining the best model for a function, why is it important to consider the data instead of just looking at the graph?

Copyright © McGraw-Hill Education

Think About It!

How would the process of finding the function that represents the data below differ from this example?

x	9.75	10	10.25	10.5
y	−9	−5	−1	3

Example 1 Determine a Model by Using First Differences

Look for a pattern in the data table to determine the kind of model that best describes the data. Then, write the function.

x	5	6	7	8	9
y	−1	−4	−7	−10	−13

Part A Determine the model.

Notice that the x-values are increasing by 1, so the model can be determined by examining successive differences or ratios.

$$-1 \quad -4 \quad -7 \quad -10 \quad -13$$
$$\quad -3 \quad -3 \quad -3 \quad -3$$

The _____ differences are equal. This means that the data can be represented by a _____ function.

Part B Write the function.

Substitute for m.

$y = mx + b$ Slope-intercept form of a linear function

$y = \underline{\quad}x + b$ Since the first difference is −3, the slope is −3.

Find the y-intercept.

$\underline{\quad} = -3(\underline{\quad}) + b$ $(x_1, y_1) = (5, -1)$

$-1 = -15 + b$ Simplify.

$\underline{\quad} = b$ Add 15 to each side.

Write the equation in slope-intercept form.

$y = mx + b$ Slope-intercept form

$y = \underline{\quad}x + \underline{\quad}$ Replace m with −3 and b with 14.

The data are modeled by _____.

 Go Online You can complete an Extra Example online.

Example 2 Determine a Model by Using Second Differences

Look for a pattern in the data table to determine the kind of model that best describes the data. Then, write the function.

x	−2	−1	0	1	2
y	−1	0	4	11	21

Part A Determine the model.

The *x*-values are increasing by 1, so the model can be determined by examining successive differences or ratios.

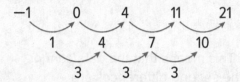

The _____ differences are equal. This means that the data can be modeled by a _____ function.

Part B Write the function.

$y = ax^2 + bx + c$ Standard form of a quadratic function

$y = \underline{\quad}x^2 + bx + \underline{\quad}$ *a* is one-half of the second
                                    difference, and *c* is the *y*-intercept
                                    from the given data point.

Find *b*.

$\underline{\quad} = 1.5(\underline{\quad})^2 + b(\underline{\quad}) + 4$ Use data points (−2, −1) for (*x*, *y*).

$-1 = 6 - 2b + 4$ Simplify.

$-1 = 10 - 2b$ Simplify.

$-11 = -2b$ Subtract 10 from each side.

$\underline{\quad} = b$ Divide each side by −2.

Write the function in standard form.

$y = ax^2 + bx + c$ Standard form of a quadratic function

$y = \underline{\quad}x^2 + \underline{\quad}x + \underline{\quad}$ $a = 1.5, b = 5.5, c = 4$

The data are modeled by _____.

 Go Online You can complete an Extra Example online.

Example 3 Determine a Model by Using Ratios

Look for a pattern in the data table to determine the kind of model that best describes the data. Then, write the function.

x	3	4	5	6	7
y	400	100	25	6.25	1.5625

Part A Determine the model.

Since the x-values are increasing by 1, the model can be determined by examining successive differences or ratios.

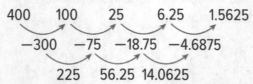

The first differences are _____. This means that the data cannot be modeled by a _____ function. The second differences are _____. This means that the data cannot be modeled by a _____ function.

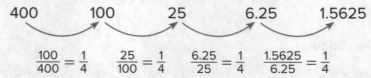

$$\frac{100}{400} = \frac{1}{4} \qquad \frac{25}{100} = \frac{1}{4} \qquad \frac{6.25}{25} = \frac{1}{4} \qquad \frac{1.5625}{6.25} = \frac{1}{4}$$

The ratios of successive y-values are equal. Therefore, the data can be modeled by an exponential function.

Part B Write the function.

$y = a(b)^x$ — Exponential function

$y = a\left(\frac{1}{4}\right)^x$ — Since the ratio is $\frac{1}{4}$, $b = \frac{1}{4}$.

Solve for a.

$\underline{\hspace{1cm}} = a\left(\frac{1}{4}\right)^3$ — $(x_1, y_1) = (3, 400)$

$400 = \left(\frac{1}{64}\right)a$ — Simplify.

$\underline{\hspace{1cm}} = a$ — Multiply each side by 64.

Write the equation.

$y = a(b)^x$ — Exponential function

$y = \underline{\hspace{1cm}}\left(\frac{1}{4}\right)^x$ — Replace a with 25,600 and b with $\frac{1}{4}$.

The data are modeled by _____.

Check

Determine the kind of model that best describes the data set. Then write the function.

x	0	1	2	3
y	−3	−6	−12	−24

Watch Out!

Constant Difference
Before checking differences and ratios, check that the x-values are increasing or decreasing by a constant value. In order to use the differences or ratios in the model function, x must be increasing by 1.

Learn Curve Fitting

You can use a graphing calculator to find a regression equation for a set of data that is approximated by a function. This process is called **curve fitting**.

The **coefficient of determination**, R^2, indicates how well the function fits the data. The closer R^2 is to 1, the better the model.

The table shows a scatter plot and graphs of linear, exponential, and quadratic functions. Which type of function is the best fit for the data?

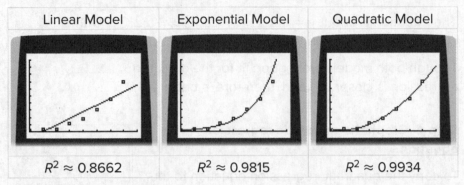

Linear Model	Exponential Model	Quadratic Model
$R^2 \approx 0.8662$	$R^2 \approx 0.9815$	$R^2 \approx 0.9934$

Based on the coefficients of determination, the data are best modeled by a _____ function.

🌐 Example 4 Find the Best Model

VIDEO STREAMING The table shows the number of households worldwide that subscribe to a video streaming service. Write a model that fits the data.

Year	2007	2008	2009	2010	2011	2012	2013	2014	2015
Households (millions)	8	9	12	19	23	33	44	57	75

Step 1 Enter the data.

Enter the data by pressing STAT and selecting the Edit option.

Let the year _____ be represented by 0.

Enter the years since 2007 into List 1 (L1).

Enter the number of households into List 2 (L2).

Step 2 Make a scatter plot.

Graph the scatter plot. Turn on Plot 1 under the STAT PLOT menu and choose the scatter plot feature.

Change the viewing window so that all data are visible by pressing ZOOM and then selecting ZoomStat.

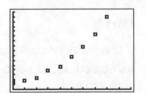

(continued on the next page)

Go Online
You can watch a video to see how to fit a curve to a set of data.

Go Online
to see how to use a graphing calculator with this example.

Use a Source
Use an outside source to find the number of subscribers or members of other services, such as ride-sharing companies, Internet service, or social media networks, over time. Then determine whether the data can be modeled by a linear, quadratic, exponential function, or none of these.

Step 3 Find the regression equation.

Exponential Regression

Select ExpReg and press ENTER.

The equation is about $y =$ _____ (_____)^x, with a coefficient of determination of approximately _____.

Quadratic Regression

Select Quadreg and press ENTER.

The equation is about $y =$ _____ $x^2 -$ _____ $x +$ _____, with a coefficient of determination of approximately _____.

Though both models are a good fit for the data, the _____ regression is closer to 1 and, therefore, a better fit.

Step 4 Graph the quadratic regression equation.

To copy the quadratic regression equation to the Y= list, press **Y=**, **VARS**, and choose **Statistics**.

From the **EQ** menu, choose **RegEQ**.

Press GRAPH.

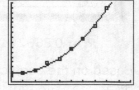

Check

RACING The table shows the speed in kilometers per hour of a racecar after x seconds.

Time x	0.5	1	1.5	2	2.5
Speed y	12	32	58	88	122

Part A

Select the model that best fits the data. ____

A. $y = -5.3x^2 + 0.2$

B. $y = 9.1x^2 + 27.8x - 4.4$

C. $y = 8.7(3.1)^x$

D. $y = 0.6(1.0)^x$

Part B

If the racecar continues to accelerate following the same trend, predict its speed at 3.5 seconds. Round to the nearest kilometer.

_____ kph

🔵 **Go Online** You can complete an Extra Example online.

🔵 **Go Online**

to learn about exponential growth patterns in Expand 11-8.

Practice

🢂 **Go Online** You can complete your homework online.

Examples 1–3

Look for a pattern in each table of values to determine which kind of model best describes the data. Then write an equation for the function that models the data.

1.

x	−3	−2	−1	0
y	−8.8	−8.6	−8.4	−8.2

2.

x	−2	−1	0	1	2
y	10	2.5	0	2.5	10

3.

x	−1	0	1	2	3
y	0.75	3	12	48	192

4.

x	−2	−1	0	1	2
y	0.008	0.04	0.2	1	5

5.

x	0	1	2	3	4
y	0	4.2	16.8	37.8	67.2

6.

x	−3	−2	−1	0	1
y	14.75	9.75	4.75	−0.25	−5.25

7.

x	−3	−2	−1	0	1	2
y	32	16	8	4	2	1

8.

x	−1	0	1	2	3
y	7	3	−1	−5	−9

9.

x	−3	−2	−1	0	1
y	−27	−12	−3	0	−3

10.

x	−2	−1	0	1	2
y	−8	−4	0	4	8

11.

x	0	1	2	3	4
y	0.5	1.5	4.5	13.5	40.5

12.

x	−1	0	1	2	3
y	27	9	3	1	$\frac{1}{3}$

13.

x	2	3	4	5	6
y	12	27	48	75	108

14.

x	0	1	2	3	4
y	80	73	66	59	52

Example 4

15. DRONES The table shows the time after a drone was launched (in seconds) x and the height of the drone, in feet, above the ground y.

x	y
1	30
2	40
3	50
4	55
5	50
6	40

   a. Make a scatter plot of the data.

   b. Which regression equation has an R^2 value closest to 1?

   c. Find an appropriate regression equation, and state the coefficient of determination. Based on the regression equation, what are the relevant domain and range?

   d. Predict the height of the drone 7 seconds after it was launched. Round to the nearest foot.

16. BAKING Alyssa baked a cake and is waiting for it to cool so she can ice it. The table shows the temperature of the cake every 5 minutes after Alyssa took it out of the oven.

Time (min)	Temperature (°F)
0	350
5	244
10	178
15	137
20	112
25	96
30	89

a. Make a scatter plot of the data.

b. Which regression equation has an R^2 value closest to 1? Is this the equation that best fits the context of the problem? Explain your reasoning.

c. Find an appropriate regression equation, and state the coefficient of determination. Based on the regression equation, what are the relevant domain and range?

d. Alyssa will ice the cake when it reaches room temperature (70°F). Use the regression equation to predict when she can ice her cake.

Mixed Exercises

USE TOOLS Use a graphing calculator to determine whether to use a *linear*, *quadratic*, or *exponential* regression equation. State the coefficient of determination.

17.

x	y
0	1.1
2	3.3
4	2.9
6	5.6
8	11.9
10	19.8

18.

x	y
1	1.67
5	2.59
9	4.37
13	6.12
17	5.48
21	3.12

19.

x	y
−2	0.2
−1	0.5
0	1
1	2
2	4
3	7.5

20. WEATHER The San Mateo weather station records the amount of rainfall since the beginning of a thunderstorm. Data for a storm is recorded as a series of ordered pairs (2, 0.3), (4, 0.6), (6, 0.9), (8, 1.2), (10, 1.5), where the x-value is the time in minutes since the start of the storm, and the y-value is the amount of rain in inches that has fallen since the start of the storm. Determine which kind of model best describes the data.

21. INVESTING The value of a certain parcel of land has been increasing in value ever since it was purchased. The table shows the value of the land parcel over time.

Year Since Purchasing	0	1	2	3	4
Land Value (thousands $)	$1.05	$2.10	$4.20	$8.40	$16.80

Look for a pattern in the table of values to determine which model best describes the data. Then write an equation for the function that models the data.

670 **Module 11** · Quadratic Functions

Copyright © McGraw-Hill Education

Name _____ Period _____ Date _____

22. BOATS The value of a boat typically depreciates over time. The table shows the value of a boat over a period of time.

Years	0	1	2	3	4
Boat Value ($)	8250	6930	5821.20	4889.81	4107.44

Write an equation for the function that models the data. Then use the equation to determine how much the boat is worth after 9 years.

23. NUCLEAR WASTE Radioactive material slowly decays over time. The amount of time needed for an amount of radioactive material to decay to half its initial quantity is known as its half-life. Consider a 20-gram sample of a radioactive isotope.

Half-Lives Elapsed	0	1	2	3	4
Amount of Isotope Remaining (grams)	20	10	5	2.5	1.25

a. Is radioactive decay a *linear* decay, a *quadratic* decay, or an *exponential* decay?

b. Write an equation to determine how many grams y of the radioactive isotope will be remaining after x half-lives.

c. How many grams of the isotope will remain after 11 half-lives?

d. Plutonium-238 is one of the most dangerous waste products of nuclear power plants. If the half-life of plutonium-238 is 87.7 years, how long would it take for a 20-gram sample of plutonium-238 to decay to 0.078 gram?

24. USE A MODEL The table shows the populations of two towns from 2015 to 2018.

a. For each town, write a function that models the town's population x years after 2015.

b. Based on the function types you used in **part a**, what can you conclude about the populations of the towns as time goes on? Explain.

Year	Population Dixon	Midville
2015	80,000	96,000
2016	84,000	96,070
2017	88,200	96,280
2018	92,610	96,630

Copyright © McGraw-Hill Education

Lesson 11-8 • Modeling and Curve Fitting **671**

25. STRUCTURE The table shows the height of an elevator above ground level at various times.

Time (s), x	0	1	2	3	4
Height (ft), y	142	124	106	88	70

a. What type of function best models the data in the table? Why?

b. The average rate of change is the change in the value of the dependent variable divided by the change in the value of the independent variable. What is the average rate of change of the function over the interval $x = 0$ to $x = 4$? What does this tell you about the motion of the elevator?

c. Write a function that models the height of the elevator as a function of time. How is the coefficient of x related to your answers to **parts a** and **b**?

26. SOCCER The table shows the total number of games a soccer team played and the number of total goals scored in a season.

Games Played	Goals Scored
1	1
2	3
3	6
4	10
5	14

a. Make a scatter plot of the data.

b. Which regression equation has an R^2 value closest to 1? What is the value?

c. Find the appropriate regression equation. What are the domain and range?

d. The team's goal is to score 25 goals for the season. Use the regression equation to predict in which game they will score their 25th goal.

27. USE A SOURCE Look up hourly temperature data for a starting time of 6:00 A.M. and every hour after that for 8 hours. Plot the points, with x representing the time in hours after 6:00 A.M. and y representing the temperature. Then, identify what type of model best describes the data.

28. PRECISION Jase used a graphing calculator to find the predicted dollar value of a used vehicle after 7 years. The calculator reported the value as 5127.45928113. How should Jase decide how many digits to report?

29. CONSTRUCT ARGUMENTS Diego claims that no function can have equal non-zero first differences and equal non-zero second differences. Is Diego correct? Use examples or counterexamples to justify your argument.

30. PERSEVERE Write a function that has constant second differences, first differences that are not constant, a y-intercept of -5, and contains the point $(2, 3)$.

31. ANALYZE What type of function will have constant third differences, but not constant second differences? Justify your argument.

32. CREATE Write a linear function that has a constant first difference of 4.

33. PERSEVERE Explain why linear functions grow by equal differences over equal intervals, and exponential functions grow by equal factors over equal intervals. (*Hint*: Let $y = ax$ represent a linear function, and let $y = a^x$ represent an exponential function.)

34. WRITE How can you determine whether a given set of data should be modeled by a *linear* function, a *quadratic* function, or an *exponential* function?

Combining Functions

Today's Goals
- Combine standard function types by using addition and subtraction.
- Combine standard function types by using multiplication.

Explore Using Graphs to Combine Functions

Online Activity Use graphing technology to complete the Explore.

> **INQUIRY** How can you use the graphs of functions to determine their sum, difference, or product?

Learn Adding and Subtracting Functions

Some situations are best modeled by the sum or difference of functions.

Consider $f(x) = x^2 - 2x + 4$ and $g(x) = -5x + 1$. To find the value of the sum of the functions when $x = -3$ or $(f + g)(-3)$, you can find $f(-3)$ and $g(-3)$ and add them.

$f(x) = x^2 - 2x + 4$	Original function	$g(x) = -5x + 1$	
$f(-3) = (-3)^2 - 2(-3) + 4$	Substitute -3 for x.	$g(-3) = -5(-3) + 1$	
$f(-3) = 19$	Simplify.	$g(-3) = 16$	

The sum is given by $f(-3) + g(-3) = 19 + 16$ or 35.

To find a function that gives the sum for all values of the domain, add the two functions and combine the like terms.

$(f + g)(x) = f(x) + g(x)$	Addition of functions
$= (x^2 - 2x + 4) + (-5x + 1)$	Substitution
$= x^2 - 2x - 5x + 4 + 1$	Combine like terms.
$= x^2 - 7x + 5$	Simplify.

Solve for $(f + g)(-3)$ to check that this method results in the same solution as $f(-3) + g(-3)$.

$(f + g)(-3) = (-3)^2 - 7(-3) + 5$	Substitute -3 for x.
$= 9 + 21 + 5$	Simplify.
$= 35$	Simplify.

So, $f(x) + g(x) = (f + g)(x)$.

Example 1 Add Functions

Given $f(x) = 3^x + 1$ and $g(x) = 6x^2 + x - 2$, find $(f + g)(x)$.

$(f + g)(x) = f(x) + g(x)$	Addition of functions
$= (\underline{}) + (6x^2 + x - 2)$	Substitution
$= 3^x + 6x^2 + x - \underline{}$	Simplify.

Example 2 Subtract Functions

Given $f(x) = -x^2 + 4x - 5$ and $g(x) = x - 7$, find each function.

a. $(f - g)(x)$

$(f - g) = $ _____ $-$ _____

$= ($_____$) - ($_____$)$

$= -x^2 + 4x - 5 - x + 7$

$= -x^2 + 3x + 2$

b. $(g - f)(x)$

$(g - f) = $ _____ $-$ _____

$= (x - 7) - ($_____$)$

$= x - 7 + x^2 - 4x + 5$

$= x^2 - 3x - 2$

Check

Given $f(x) = 5 \cdot 2^x - 1$, $g(x) = -3x^2 + 2x - 8$, and $h(x) = 4x - 5$, find each function. Write each answer in standard form.

a. $(f + h)(x)$ _____

b. $(h - g)(x)$ _____

Learn Multiplying Functions

Consider $f(x) = x^2 + x + 3$ and $g(x) = 4x$. To find $(f \cdot g)(2)$ you can find the product of $f(2)$ and $g(2)$.

$f(x) = x^2 + x + 3$	Original function	$g(x) = 4x$
$f(2) = (2)^2 + (2) + 3$	Substitute 2 for x.	$g(2) = 4(2)$
$f(2) = 9$	Simplify.	$g(2) = 8$

$f(2) \cdot g(2) = 9 \cdot 8$ or 72.

Multiply the two functions and combine like terms.

$(f \cdot g)(x) = f(x) \cdot g(x)$	Multiplication of functions
$= (x^2 + x + 3) \cdot (4x)$	Substitution
$= x^2(4x) + x(4x) + 3(4x)$	Distributive property.
$= 4x^3 + 4x^2 + 12x$	Simplify.

Solve for $(f \cdot g)(2)$.

$(f \cdot g)(2) = 4(2)^3 + 4(2)^2 + 12(2)$	Substitute 2 for x.
$= 32 + 16 + 24$ or 72	Simplify.

So, $f(x) \cdot g(x) = (f \cdot g)(x)$

🅑 **Go Online** You can complete an Extra Example online.

Example 3 Multiply Linear and Quadratic Functions

Given $f(x) = x^2 + 2x - 7$ and $g(x) = 3x - 10$, find $(f \cdot g)(x)$.

$(f \cdot g)(x) = f(x) \cdot g(x)$

$= (x^2 + 2x - 7) \cdot (\underline{\hspace{1cm}})$

$= x^2(\underline{\hspace{0.5cm}}) + x^2(\underline{\hspace{0.5cm}}) + 2x(\underline{\hspace{0.5cm}}) + 2x(\underline{\hspace{0.5cm}}) - 7(\underline{\hspace{0.5cm}}) - 7(\underline{\hspace{0.5cm}})$

$= 3x^3 - 10x^2 + 6x^2 - 20x - 21x + 70$

$= 3x^3 - 4x^2 - 41x + 70$

Think About It!

Consider the degree of the polynomials $f(x)$ and $g(x)$. Does the degree of the product meet your expectations? Explain.

Example 4 Multiply Linear and Exponential Functions

Given $g(x) = 3x - 10$ and $h(x) = \left(\frac{1}{3}\right)^x + x$, find $(h \cdot g)(x)$.

$(h \cdot g)(x) = h(x) \cdot h(x)$

$= \left[\left(\frac{1}{3}\right)^x + x\right] \cdot (\underline{\hspace{1cm}})$

$= \left(\frac{1}{3}\right)^x (\underline{\hspace{0.5cm}}) + \left(\frac{1}{3}\right)^x (\underline{\hspace{0.5cm}}) + x(\underline{\hspace{0.5cm}}) + x(\underline{\hspace{0.5cm}})$

$= 3x\left(\frac{1}{3}\right)^x - 10\left(\frac{1}{3}\right)^x + 3x^2 - 10x$

🌐 Example 5 Combine Functions

FINANCE **According to the Wall Street Journal, the average student loan debt was more than $35,000 in 2015. Hugo graduated from college with $28,500 in student loan debt. He decides to defer his payments while he is in graduate school. However, he still accrues interest on his student loans at an annual rate of 5.65%.**

Watch Out!

Percentages 5.65% = 0.0565

Part A **Write an exponential function $r(t)$ to express the amount of money Hugo owes on his student loans after time t, where t is the number of years after interest began accruing.**

$r(t) = a(1 + r)^t$ Equation for exponential growth

$= 28,500(1 + \underline{\hspace{1cm}})^t$ $a = 28,500$ and $r = 5.65\%$ or 0.0565

$= 28,500(\underline{\hspace{1cm}})^t$ Simplify.

Part B **While he is in graduate school, Hugo's parents also lend him $400 a month for rent. His parents decide not to charge him interest on this loan. Write a function $p(t)$ to represent this loan, where t is the time in years that Hugo borrows money from his parents.**

Since t is the time in years, first find the amount of money Hugo borrows from his parents each year.

$12(400) = \underline{\hspace{1.5cm}}$

So, $p(t) = \underline{\hspace{1.5cm}} t$ represents the loan from his parents as a function of time.

Go Online

You can complete an Extra Example online.

(continued on the next page)

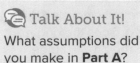

Part C Find $C(t) = r(t) + p(t)$. What does this new function represent?

$C(t) = 28{,}500(1 + 0.0565)^t +$ _____

$C(t)$ represents the total amount of money Hugo has to repay after t years.

Part D If Hugo spends 3 years in graduate school, find the total amount of money he will have to repay.

Because Hugo spends 3 years in graduate school, $t =$ ___.

$C(t) = 28{,}500(1 + 0.0565)^t + 4800t$

$C(3) = 28{,}500(1 + 0.0565)^3 + 4800($___$)$

$= 33{,}608.83 + 14{,}400$

$= \$$ _____

🌐 Example 6 Combine Two Functions

SANDWICHES **A sandwich shop charges \$7 for a large sub and sells an average of 360 large subs per day. The shop predicts that they will sell 20 fewer subs for every \$0.25 increase in the price.**

Part A Let x represent each \$0.25 price increase. Write a function $P(x)$ to represent the price of a large sub.

The price of a large sub is \$7 plus \$0.25 times each price increase.

$P(x) = 7 + 0.25x$

Part B Write a function $T(x)$ to represent the number of subs sold.

The number of subs sold is 360 minus 20 times the number of price increases.

$T(x) =$ _____

Part C Write a function $R(x)$ that can be used to maximize the revenue from sales of large subs.

The revenue from sales of large subs will be equal to the price times the number of subs sold.

$R(x) = P(x) \cdot T(x)$

$= (7 + 0.25x)($ _____ $)$

$= -5x^2 - 50x + 2520$

Part D If the sandwich shop charges \$8.50 for a large sub, find the revenue from sales of large subs. x represents each \$0.25 price increase. So, $x = 6$.

$R(x) = -5x^2 - 50x + 2520$

$R(6) = -5(6)^2 - 50(6) + 2520$

$= -180 - 300 + 2520$ or \$2040

🎥 **Go Online** You can complete an Extra Example online.

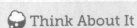
Think About It
How could you determine the amount of money the sub shop should charge for a large sub?

Practice

🔄 **Go Online** You can complete your homework online.

Examples 1 and 2

Given that $f(x) = x - 9$, $g(x) = 3x^2 - 2x + 5$, and $h(x) = -6x$, find each function.

1. $(f + g)(x)$

2. $(f - g)(x)$

3. $(f + h)(x)$

4. $(g - f)(x)$

5. $(g - h)(x)$

6. $(g + h)(x)$

7. $(h - g)(x)$

8. $(f - h)(x)$

Examples 3 and 4

Given that $f(x) = 11x$, $g(x) = x^2 - 6x + 3$, $h(x) = -x + 4$, and $j(x) = 2^x + 3$, find each function.

9. $(f \cdot g)(x)$

10. $(f \cdot h)(x)$

11. $(g \cdot h)(x)$

12. $(f \cdot f)(x)$

13. $(f \cdot j)(x)$

14. $(h \cdot j)(x)$

Examples 5 and 6

15. **TRANSPORTATION** The distance Jan typically travels after walking t seconds can be represented by the function $f(t) = 4t$. While at the airport, Jan uses the moving walkway. The distance a person travels after t seconds on a moving walkway can be represented by the function $g(t) = 3t$.

   a. Find $f(t) + g(t)$ and explain what it represents.

   b. Find $f(t) - g(t)$ and explain what it represents.

16. **CLUBS** An improvisational acting club has 32 members. The manager of the club expects its membership to increase by 4 members per year. A photography club has 60 members and is expected to grow by 10% per year.

   a. Write a function $f(t)$ to represent the number of members in the acting club after t years.

   b. Write a function $g(t)$ to represent the number of members in the photography club after t years.

   c. Find $(f + g)(t)$ and explain what this function represents.

17. MAGIC SHOW A magician currently sells tickets to his shows for $15 and averages 180 spectators per show. He estimates that he can sell 10 more tickets for each $0.75 decrease in price.

   a. Let x represent the number of $0.75 price decreases. Write a function $P(x)$ to represent the price of a ticket and a function $T(x)$ to represent the number of tickets sold.

   b. Write a function $R(x)$ that can be used to find the revenue from ticket sales.

   c. If the magician decides to sell the tickets for $12, find his revenue.

18. GOLF A golf course offers a membership plan where players pay an annual fee of $250 plus a fee for each round of golf played. The function $C(x) = 15x + 250$ represents the total cost for playing x rounds of golf. Explain how you can think of $C(x)$ as the sum of two functions. What does each of these functions represent?

Mixed Exercises

Given that $p(x) = -2x$, $q(x) = x^2 + 5$, $r(x) = x^2 - x + 2$, and $t(x) = \left(\frac{1}{4}\right)^x - 5$ find each function.

19. $(q + t)(x)$

20. $(t - r)(x)$

21. $(p \cdot q)(x)$

22. $(p \cdot r)(x)$

23. $(q \cdot r)(x)$

24. $(q \cdot q)(x)$

25. $(p \cdot t)(x)$

26. $(p + q + r)(x)$

27. STORAGE Raul is making boxes out of sheets of cardboard to hold his video game collection. He begins with a rectangular sheet that is 20 inches long and 12 inches wide, as shown. Then he cuts identical squares from each corner and bends the sides to form a box.

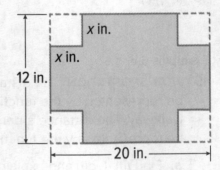

   a. If x represents the length of each side of the squares, write functions $\ell(x)$ and $w(x)$ to represent the length and width, respectively, of the resulting box.

   b. Find $(\ell \cdot w)(x)$ and explain what it represents.

   c. What is the domain of $(\ell \cdot w)(x)$? Explain.

   d. Find $(\ell \cdot w)(1.5)$ and interpret its meaning.

28. USE A MODEL Alba sells homemade granola at a farmers' market. When Alba sets the price at $4 per bag, she can sell 185 bags. She finds that for every increase of $0.25 in the price of a bag of granola, she sells 10 fewer bags.

   a. Let x represent the number of $0.25 increases in the price. Write a function $P(x)$ that gives the price of the granola as a function of x. Write a function $G(x)$ that gives the number of bags of granola Alba sells as a function of x.

   b. Explain how to combine functions to write a function $R(x)$ that gives Alba's revenue from selling granola as a function of x.

   c. Alba is considering raising the price of the granola to $4.75. Use your model to explain whether or not this is a good idea.

29. STRUCTURE Tyree makes a pattern, as shown, using shaded tiles and white tiles.

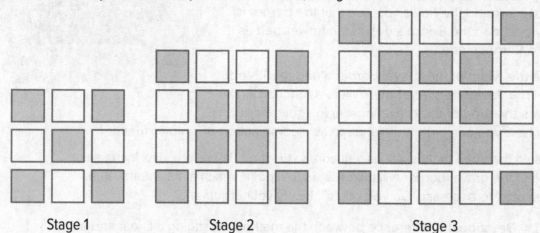

Stage 1   Stage 2   Stage 3

a. Write a function $f(n)$ that gives the number of shaded tiles needed to make stage n of the pattern and a function $g(n)$ that gives the number of white tiles needed to make stage n of the pattern.

b. Explain how to write a function $h(n)$ that gives the total number of tiles needed to make stage n of the pattern. Use the function to find the total number of tiles needed to make stage 10.

c. Explain how you know your answer to **part b** is correct.

30. USE TOOLS A chemist works in a lab that maintains a constant temperature of 22°C. She heats a saline solution and then lets the solution cool. As the solution cools, she records its temperature every 3 minutes. Her data is shown in the table.

Time (minutes)	0	3	6	9	12	15
Temperature of Solution (°C)	82.0	79.1	76.3	73.6	70.9	68.4
Temperature Above the Lab Temperature	60.0	57.1	54.3	51.6	48.9	46.4

a. Write a function $P(x)$ that models the temperature of the solution above the lab temperature in degrees Celsius x minutes after the chemist starts recording the data. Write a second function $S(x)$ that models the temperature of the solution.

b. Explain how the function you wrote for $S(x)$ in **part a** is a combination of two functions.

c. The chemist wants to know approximately how long it will take until the solution cools to a temperature of 45°C. Explain how to use your calculator to estimate this time to the nearest minute.

31. CONSTRUCT ARGUMENTS A rectangular flower bed in a garden is 12 feet long and 8 feet wide. Kazuo plans to add a gravel border around the flower bed so that the border is twice as wide along the 8-foot sides of the flower bed as it is along the 12-foot sides.

a. Let x be the width of the gravel border along the 12-foot sides of the flower bed, as shown. Write a function $A(x)$ that gives the area of the gravel border in square feet. Explain how you can write this function as a combination of simpler functions.

b. Kazuo said that the function $A(x)$ is a quadratic function. Therefore, as x increases, the area of the gravel border will increase up to a point, reach a maximum value, and then start to decrease. Do you agree? Justify your argument.

32. REGULARITY Describe the difference between the method for adding a linear and a quadratic function and the method for multiplying a linear and a quadratic function.

33. STRUCTURE Adding the linear function $j(x) = 5$ to the exponential function $k(x) = 3^x - 7$ gives a new exponential function $m(x)$, which is a translation of $k(x)$. What is the equation of the translated function, $m(x)$?

34. FRAME A picture is to be placed in a frame with a glass center that is 8 inches wide and 10 inches long. There are two frames, one that extends x inches outside the glass in each direction and one that extends $2x$ inches outside the glass in each direction. Write a function $f(x)$ in standard form to describe how much larger the area of the larger frame is than the area of the smaller frame.

35. CREATE Define $f(x)$ and $g(x)$ if $(f - g)(x) = x^2 + 14x - 12$, where $f(x)$ is a quadratic function and $g(x)$ is a linear function.

36. CREATE Define $f(x)$ and $g(x)$ if $(f \cdot g)(x) = x^3 - x^2 - 4x + 4$, where $f(x)$ is a quadratic function and $g(x)$ is a linear function.

37. ANALYZE Determine whether the following statement is *sometimes*, *always*, or *never* true. Justify your argument.

Multiplying a linear equation with two terms by a quadratic equation with three terms will result in an equation that has three terms.

38. WRITE You learned that when a linear function and a quadratic function are multiplied, the product is a third-degree polynomial. What do you think would be the degree of the product of two quadratic functions? What about the degree of the sum of two quadratic functions? Explain your reasoning.

39. FIND THE ERROR Delfina multiplied the linear function $f(x) = 6x - 8$ by the quadratic function $g(x) = 4x^2 + 7x - 19$. She found that $(f \cdot g)(x) = 24x^3 + 42x^2 + 152$. Is Delfina's solution correct? Explain your reasoning.

40. WHICH ONE DOESN'T BELONG? Consider the functions $a(x)$, $b(x)$, and $c(x)$, below, and their products, $(a \cdot b)(x)$, $(a \cdot c)(x)$, and $(b \cdot c)(x)$. Determine which of the products does not belong. Justify your conclusion.

| $a(x) = 4x + 8$ | $b(x) = -x^2 + 3x - 5$ | $c(x) = 6x - 7$ |

e Essential Question
Why is it helpful to have different methods to analyze quadratic functions and solve quadratic equations?

Module Summary

Lessons 11-1 and 11-2

Graphing Quadratic Functions

- The graph of a quadratic function is a parabola. The axis of symmetry intersects a parabola at the vertex. The vertex is either the lowest point or the highest point on a parabola.

- The standard form of a quadratic function is $f(x) = ax^2 + bx + c$, where a, b, and c are integers and $a \neq 0$.

- Quadratic functions can be translated like other functions.

- A quadratic function in the form $f(x) = a(x - h)^2 + k$ is in vertex form.

Lessons 11-3 through 11-6

Solving Quadratic Equations

- The solutions or roots of an equation can be identified by finding the x-intercepts of the graph of the related function.

- If the product of two factors is 0, then at least one of the factors must be 0.

- To complete the square for any quadratic expression of the form $x^2 + bx$, find one-half of b, the coefficient of x. Square the result. Then add the result to $x^2 + bx$.

- Quadratic Formula: $x = \frac{-b \pm \sqrt{b^2 - 4ac}}{2a}$.

Lesson 11-7

Solving Systems of Linear and Quadratic Equations

- The solution to a system of linear and quadratic equations is at the point of intersections of the graphs of each equation.

- You can use the Substitution Method or the Elimination Method to solve a system of linear and quadratic equations.

Lesson 11-8

Modeling and Curve Fitting

- The coefficient of determination, R^2, indicates how well the function fits the data.

Lesson 11-9

Combining Functions

- You can add two functions and combine the like terms.

- You can multiply two functions.

Study Organizer

Foldables

Use your Foldable to review this module. Working with a partner can be helpful. Ask for clarification of concepts as needed.

Key Features
Transformations
Solve by Factoring
Completing the Square

Name _____ Period _____ Date _____

Test Practice

1. GRAPH Graph $f(x) = x^2 + 4x - 2$. (Lesson 11-1)

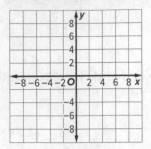

2. MULTIPLE CHOICE The height of a baseball is modeled by the function $h(x) = -16x^2 + 50x + 5$, where $h(x)$ represents the height, in feet, of the ball x seconds after being hit by a bat. Which of the following describes the most appropriate domain for this function? (Lesson 11-1)

- Ⓐ all integers
- Ⓑ positive integers
- Ⓒ all real numbers
- Ⓓ positive real numbers

3. MULTIPLE CHOICE Find the vertex of the graph of $f(x) = 4x^2 - 24x - 2$. (Lesson 11-1)

- Ⓐ $(6, -2)$
- Ⓑ $(3, -38)$
- Ⓒ $(3, -62)$
- Ⓓ $(-3, 106)$

4. MULTIPLE CHOICE Find the axis of symmetry of the graph of $f(x) = 4x^2 - 24x - 2$. (Lesson 11-1)

- Ⓐ $x = 6$
- Ⓑ $x = 3$
- Ⓒ $x = \frac{1}{3}$
- Ⓓ $x = -3$

5. OPEN RESPONSE Find the y-intercept of $f(x) = 4x^2 - 24x - 2$. (Lesson 11-1)

6. MULTIPLE CHOICE If $f(x) = -4x^2$, write a function $k(x)$ representing a reflection of $f(x)$ across the x-axis. (Lesson 11-2)

- Ⓐ $k(x) = -\frac{1}{4}x^2$
- Ⓑ $k(x) = \frac{1}{4}x^2$
- Ⓒ $k(x) = (-4x)^2$
- Ⓓ $k(x) = 4x^2$

7. GRAPH Given the parent function, $f(x) = x^2$, graph the translation 3 units to the right, $g(x)$. (Lesson 11-2)

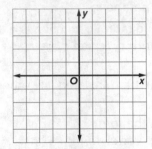

8. MULTI-SELECT Use the graph of the related function to solve $x^2 + x = 20$. Select all the solutions that apply. (Lesson 11-3)

- Ⓐ -20
- Ⓑ -5
- Ⓒ 1
- Ⓓ 4
- Ⓔ 9

9. MULTIPLE CHOICE Which quadratic function models the graph? (Lesson 11-4)

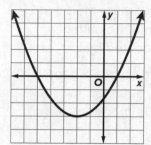

Ⓐ $y = \frac{1}{3}x^2 - \frac{4}{3}x + \frac{5}{3}$

Ⓑ $y = \frac{1}{3}x^2 - \frac{4}{3}x - \frac{5}{3}$

Ⓒ $y = \frac{1}{3}x^2 + \frac{4}{3}x - \frac{5}{3}$

Ⓓ $y = \frac{1}{3}x^2 + \frac{4}{3}x + \frac{5}{3}$

10. OPEN RESPONSE During a thunderstorm, a branch fell from a tree. Chantel estimates the branch fell from 25 feet above the ground.

The formula $h = -16t^2 + h_0$ can be used to approximate the number of seconds t it takes for the branch to reach height h from an initial height of h_0 in feet. Find the time it takes the branch to reach the ground. Round to the nearest hundredth, if necessary. (Lesson 11-4)

11. MULTIPLE CHOICE The area of the rectangle shown is 120 square inches.

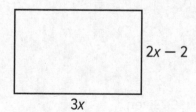

$2x - 2$

$3x$

What is the perimeter of the rectangle? (Lesson 11-4)

Ⓐ 5 inches

Ⓑ 23 inches

Ⓒ 46 inches

Ⓓ 120 inches

12. MULTIPLE CHOICE Solve $x^2 + 6x = 7$ by completing the square. (Lesson 11-5)

Ⓐ 3, 9

Ⓑ 6, 7

Ⓒ 1, −7

Ⓓ 3, 4

13. OPEN RESPONSE Write $y = 2x^2 + 4x + 6$ in vertex form. Identify the extrema, and explain whether it is a minimum or maximum. (Lesson 11-5)

14. OPEN RESPONSE Solve $x^2 - 20x + 27 = 8$ by completing the square. (Lesson 11-5)

15. OPEN RESPONSE A cat that is sitting on a boulder 5 feet high jumps off and lands on the ground. During its jump, its height h in feet is given by $h = -0.5d^2 + 2d + 5$, where d is the distance from the base of the boulder, in feet. (Lesson 11-5)

How far is the cat from the base of the boulder when it lands on the ground? Round to the nearest tenth.

What is the maximum height of the cat during its jump?

16. MULTIPLE CHOICE A local company manufactures brake drums. Based on their records, their daily profit can be approximated by the function $f(x) = x^2 + 3x - 19$, where x is the number of brake drums they manufacture. If $f(x)$ is negative, it means the company has lost money. What is the least number of brake drums that the company needs to manufacture each day in order to make a profit? (Lesson 11-6)

Ⓐ 3

Ⓑ 4

Ⓒ 7

Ⓓ 19

17. OPEN RESPONSE How many real solutions does the equation $2x^2 + 5x + 7 = 0$ have? (Lesson 11-6)

18. MULTIPLE CHOICE Find the solution(s) of the system shown.

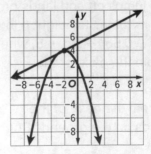

(Lesson 11-7)

Ⓐ $(-2, 4)$

Ⓑ $(-2, 4)$ and $(-4, 2)$

Ⓒ $(0, 2)$

Ⓓ $(0, 5)$

19. OPEN RESPONSE Let x represent the time in seconds and y represent the height in feet. The path of a drone can be modeled by $y = -x^2 + 19x + 1$ and the path of a projectile can be modeled by $y = 2x + 1$.

Solve a system of equations algebraically to calculate how many seconds it will take the projectile to meet the drone. (Lesson 11-7)

20. MULTIPLE CHOICE The table shows the profit earned y as it relates to the number of magazines sold x.

x	0	1	2	3	4
y	0.2	1	5	25	125

Which model best describes the data in the table? (Lesson 11-8)

Ⓐ Linear

Ⓑ Exponential

Ⓒ Quadratic

Ⓓ Square

21. MULTIPLE CHOICE Which coefficient of determination shows the best fit of the data? (Lesson 11-8)

Ⓐ $R^2 = 0.589$

Ⓑ $R^2 = 0.880$

Ⓒ $R^2 = 0.989$

Ⓓ $R^2 = 1.54$

22. OPEN RESPONSE Given $f(x) = 7x^2 + 22x - 6$ and $g(x) = 3x - 9$, find $(g - f)(x)$. (Lesson 11-9)

e Essential Question
How do you summarize and interpret data?

What will you learn?

Place a check mark (✓) in each row that corresponds with how much you already know about each topic **before** starting this module.

KEY

👎 — I don't know.   👍 — I've heard of it.   👍 — I know it!

	Before			After		
	👎	👍	👍	👎	👍	👍
find measures of center in a data set						
calculate percentiles						
represent data in dot plots, bar graphs, and histograms						
collect data and analyze bias						
represent data in box plots						
calculate standard deviation						
analyze data distributions						
transform linear data						
compare two data sets						
represent data in two-way frequency tables						
find frequencies, including marginal and conditional relative frequencies						

📖 **Foldables** Make this Foldable to help you organize your notes about statistics. Begin with 8 sheets of $8\frac{1}{2}$″ by 11″ paper.

1. **Fold** each sheet of paper in half. Cut 1 inch from the end to the fold. Then cut 1 inch along the fold.

2. **Write** the lesson number and title on each page.

3. **Label** the inside of each sheet with *Definitions* and *Examples*.

4. **Stack** the sheets. Staple along the left side. Write *Statistics* on the first page.

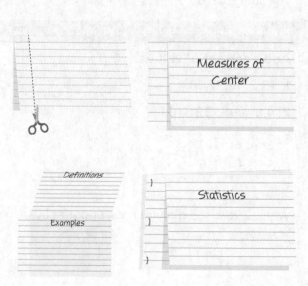

What Vocabulary Will You Learn?

Check the box next to each vocabulary term that you may already know.

- ☐ bar graph
- ☐ bias
- ☐ box plot
- ☐ categorical data
- ☐ conditional relative frequency
- ☐ distribution
- ☐ dot plot
- ☐ extreme values
- ☐ five-number summary
- ☐ histogram
- ☐ interquartile range
- ☐ joint frequencies
- ☐ linear transformation

- ☐ lower quartile
- ☐ marginal frequencies
- ☐ measurement data
- ☐ measures of center
- ☐ measures of spread
- ☐ median
- ☐ negatively skewed distribution
- ☐ outlier
- ☐ percentile
- ☐ population
- ☐ positively skewed distribution
- ☐ quartile
- ☐ range

- ☐ relative frequency
- ☐ sample
- ☐ standard deviation
- ☐ statistic
- ☐ symmetric distribution
- ☐ two-way frequency table
- ☐ two-way relative frequency table
- ☐ univariate data
- ☐ upper quartile
- ☐ variable
- ☐ variance

Are You Ready?

Complete the Quick Review to see if you are ready to start this module.
Then complete the Quick Check.

Quick Review

Example 1

Add the set of values.

12.5, 3.4, 1.75, 9

$$
\begin{array}{r}
12.5 \\
3.4 \\
1.75 \qquad \text{Align the numbers at the decimal.} \\
9 \\
\hline
26.65
\end{array}
$$

The sum is 26.65.

Example 2

Write the fraction $\frac{33}{80}$ as a percent. Round to the nearest tenth.

$\frac{33}{80} \approx 0.413$	Simplify and round.
$0.413 \cdot 100 = 41.3$	Multiply the decimal by 100.
$\frac{33}{80} \approx 41.3\%$	Write as a percent.

Quick Check

Add each set of values.

1. 13.2, 15, 17.68

2. 4.5, 1.95, 2.36, 8.1

3. $\frac{2}{3}, \frac{3}{4}, \frac{5}{6}, \frac{9}{10}$

4. −8, −4, 1, 5

Write each fraction as a percent. Round to the nearest tenth.

5. $\frac{14}{17}$

6. $\frac{7}{8}$

7. $\frac{107}{125}$

8. $\frac{625}{1024}$

How did you do?

Which exercises did you answer correctly in the Quick Check? Shade those exercise numbers below.

① ② ③ ④ ⑤ ⑥ ⑦ ⑧

Measures of Center

Copyright © McGraw-Hill Education

Learn Mean, Median, and Mode

- A **variable** is any characteristic, number, or quantity that can be counted or measured. A variable is an item of data.

- Data that have units and can be measured are called **measurement data** or quantitative data.

- Data that can be organized into different categories are called **categorical data** or qualitative data.

- Measurement data in one variable, called **univariate data**, are often summarized using a single number to represent what is average, or typical.

- Measures of what is average are called **measures of center** or central tendency. The most common measures of center are mean, median, and mode.

  Mode: the value of the elements that appear most often in a set of data

  Mean: the sum of the elements of a data set divided by the total number of elements in the set

  Median: the middle element, or the mean of the two middle elements, in a set of data when the data are arranged in numerical order

🌐 Example 1 Measures of Center

BASKETBALL **The table shows the total number of points scored in several NCAA Championship Basketball Games. Find the mean, median, and mode of the data.**

Year	Score	Year	Score
2016	150	2008	143
2015	131	2007	159
2014	114	2006	130
2013	158	2005	145
2012	126	2004	155
2011	94	2003	159
2010	120	2002	116
2009	161		

(continued on the next page)

Today's Goals
- Represent sets of data by using measures of center.
- Represent sets of data by using percentiles.

Today's Vocabulary
variable
measurement data
categorical data
univariate data
measures of center
percentile

💬 Talk About It
A set of data can have only one value for the mean and median. How many values can a set of data have for the mode? Explain your reasoning.

Study Tip
Mean When calculating the mean, your answer will always be between the least and greatest values of the data set. It can never be less than the least value or greater than the greatest value.

Think About It!

Carlos says that a set of data cannot have the same mean and mode. Do you agree or disagree? Explain your reasoning or provide a counterexample.

Mean

To find the mean, find the sum of all the points and divide by the number of years in the data set.

$$= \frac{150 + 131 + 114 + 158 + 126 + 94 + 120 + 161 + 143 + 159 + 130 + 145 + 155 + 159 + 116}{15}$$

$$= \frac{2061}{15} \text{ or about } \underline{\hspace{1cm}}.$$

The mean is about _____ points.

Median and Mode

To find the median, order the points from least to greatest and find the middle value.

94, 114, 116, 120, 126, 130, 131, 143, 145, 150, 155, 158, 159, 159, 161

_____   _____

The median is _____.

From the arrangement of data values, we can see that 159 is the only value that appears more than once. So, the mode is _____.

The mean and median are close together, so they both represent the average of the scores well. Notice that the median is greater than the mean. This indicates that the scores less than the median are more spread out than the scores greater than the median. The mode is greater than most of the scores.

Check

FOOTBALL The data show the number of interceptions thrown during one regular season for each team in the NFC. Find the mean, median, and mode. Round to the nearest whole number, if necessary.

13	17	10	12	22	14	8	11
9	12	14	18	12	8	15	11

Mean: _____

Median: _____

Mode: _____

Go Online You can complete an Extra Example online.

Explore Finding Percentiles

Online Activity Use a real-world situation to complete the Explore.

> @ **INQUIRY** How can you describe a data value based on its position in the data set?

Learn Percentiles

A **percentile** is a measure that is often used to report test data, such as standardized test scores. It tells us what percent of the total scores were below a given score.

- Percentiles measure rank from the bottom.
- There is no 0 percentile rank. The lowest score is at the 1st percentile.
- There is no 100th percentile rank. The highest score is at the 99th percentile.

Key Concept • Finding Percentiles

To find the percentile rank of an element of a data set, use these steps.

Step 1 Order the data values from greatest to least.

Step 2 Find the number of data values less than the chosen element. Divide that number by the total number of values in the data set.

Step 3 Multiply the value from Step 2 by 100.

🌐 Example 2 Find Percentiles

FIGURE SKATING **The table shows the total points scored by each country in the team figure skating event in the 2014 Olympic Winter Games. Find the United States' percentile rank.**

Country	Score
Canada	65
China	20
France	22
Germany	17
Great Britain	8
Italy	52
Japan	51
Russia	75
Ukraine	10
United States	60

(continued on the next page)

Go Online You can complete an Extra Example online.

Study Tip

Percent vs Percentile *Percent* and *percentile* mean two different things. For example, a score at the 40th percentile means that 40% of the scores are either the same as the score at the 40th percentile or less than the score at that rank. It does not mean that the person scored 40% of the possible points.

Step 1 Order the data.

Order the data values from greatest to least.

Country	Score
Russia	75
Canada	65
United States	60
Italy	52
Japan	51
France	22
China	20
Germany	17
Ukraine	10
Great Britain	8

Study Tip

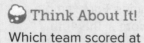

Percentiles The team with the highest score is at the 99th percentile rank, and the team with the lowest score is at the 1st percentile rank.

Step 2 Divide.

Divide the number of teams with scores lower than the United States by the total number of teams.

$$\frac{\text{number of teams below the United States}}{\text{Total number of teams}} = \_\_\_\_$$

Step 3 Multiply by 100.

$\frac{7}{10} \cdot 100$ or _____

The United States figure skating team scored at the ____th percentile in the 2014 Olympics.

Think About It!

Which team scored at the 40th percentile?

Check

DRUM CORPS The table shows the scores of the corps that competed in the Drum Corps International World Championship World Class Finals in 2015.

Corps	Score
Bluecoats	96.925
Blue Devils	97.650
Blue Knights	91.850
Blue Stars	85.150
Boston Crusaders	86.800
Carolina Crown	97.075
Crossmen	85.025
Madison Scouts	88.750
Phantom Regiment	90.325
Santa Clara Vanguard	93.850
The Cadets	95.900
The Cavaliers	88.325

The Cavaliers scored at the _____th percentile. The _____ scored at the 75th percentile.

 Go Online You can complete an Extra Example online.

Practice

Go Online You can complete your homework online.

Copyright © McGraw-Hill Education

Example 1

Find the mean, median, and mode for each data set.

1. {17, 11, 8, 15, 28, 20, 10, 16}

2. {2.5, 6.4, 7.0, 5.3, 1.1, 6.4, 3.5, 6.2, 3.9, 4.0}

3.

2	1	1	5	7
3	2	4	6	2

4.

50	30	40	10
20	80	60	90
10	30	110	70

5. number of students helping at a booth each hour: 3, 5, 8, 1, 4, 11, 3

6. weight in pounds of boxes loaded onto a semi-truck: 201, 201, 200, 199, 199

7. car speeds in miles per hour observed by a highway patrol officer: 60, 53, 53, 52, 53, 55, 55, 57

8. number of songs downloaded by students last week in Ms. Turner's class: 3, 7, 21, 23, 63, 27, 29, 95, 23

9. ratings of an online video: 2, 5, 3.5, 4, 4.5, 1, 1, 4, 2, 1.5, 2.5, 2, 3, 3.5

Example 2

MARCHING BAND A competition was recently held for 12 high school marching bands. Each band received a score from 0 through 100, with 100 being the highest.

10. Find Hamilton High School's percentile rank.

11. Find Monmouth High School's percentile rank.

12. Find Freeport High School's percentile rank.

Band	Score	Band	Score
Freeport	78	Madison	69
Ross	85	Monmouth	67
Hamilton	88	Carlisle	65
Groveport	94	Dupont	48
Lakehurst	56	Cave City	90
Benton	77	Monroe	80

Mixed Exercises

13. **REASONING** The mean number of people at the movies on Saturday nights throughout the year is 425, and the median is 412. Explain why the mean could be slightly higher.

14. **REASONING** The mode length of time it takes to fly from New York City to Chicago is 2 hours 35 minutes, and the mean is 3 hours 15 minutes. Explain why the mode could be slightly lower.

15. FOOTBALL Find the mean, median, and mode for the data set. The weights in pounds of 5 offensive linemen of a football team: 217, 212, 285, 245, 301.

16. WEB SITES The ratings for a new recipe Web site varied from very low, 1 point, to very high, 10 points, with half of the scores receiving a rating of 7. If a new rating of 7 were added to the data set, how would the mode be affected? Explain.

17. Find a mean of {16, 19, 22, 27, 33, 19, 25}.

18. SINGING In a singing competition that involved 50 contestants, Reina's score ranked higher than 40 of the contestants. In what percentile did Reina score?

19. SPORTS The table shows the number of points scored by a basketball team during their first several games. Find the mean, median, and mode of the number of points scored.

Game	1	2	3	4	5	6	7	8
Points	43	50	52	47	55	61	48	56

20. PERFORMANCE At a bodybuilding competition, Shawnte earned a score of 42 points. There were 19 competitors who received a lower score than Shawnte and 5 competitors who earned a higher score. What was Shawnte's percentile rank in the bodybuilding competition?

21. GRADES On her first four quizzes, Rachael has earned scores of 21, 24, 23, and 17 points. What score must Rachael earn on her fifth and final quiz so that both the mean and median of her quiz scores is 21?

22. QUIZZES Sequon scored 95, 86, 81, 83, and 95 on his math quizzes this quarter. Find the mode of his quiz scores.

23. BAND Out of the 30 bands at the competition, Coastal High School's band scored higher than 27 others. Find the percentile rank for Coastal High School's band.

24. SHOPPING The table shows the prices of comparable laptop computers at different retailers.

   a. Find the mean, median, and mode of the prices.

   b. Why are the mean and median much lower than the mode?

   c. After deliberation, Nikki is interested in buying a laptop from either retailer C, F, or J. What are the percentile ranks for the laptop at each of these retailers?

Retailer	Price ($)
A	389
B	425
C	350
D	499
E	475
F	360
G	319
H	425
I	299
J	379

25. NOVELS The table shows the lengths (in words) of the seven novels on the required reading list of Miguel's language arts class.

   a. Find the mean, median, and mode for the data set.

   b. Predict which novels will be lower than the 50th percentile in length. Verify your prediction.

   c. If *Lord of the Flies* comes with an additional 10,020-word online reading assignment, then how will the median length be affected? How will the mean be affected?

Novel	Number of Words
Old Yeller	35,968
Lord of the Flies	59,900
Moby Dick	206,052
Jane Eyre	183,858
Great Expectations	183,349
Call of the Wild	31,750
The Color Purple	66,556

26. VOLUNTEERING The table shows the number of hours different students spent volunteering as part of a community outreach program. Find the mean, median, and mode of the data set.

Volunteer Hours				
25	30	35	40	35
25	50	45	25	90

27. USE A SOURCE Research the total medal counts for Canada, France, Japan, Russia, Brazil, and Great Britain at the 2016 Rio de Janeiro Olympics. Make a table of the data you collect. Then find the percentile rank of each country.

28. CONSTRUCT ARGUMENTS The table shows the number of pet adoptions each week for a shelter over a two-month period. If there are 65 pets adopted next week, how are the mean, median, and mode affected?

Pet Adoptions			
7	19	26	20
23	21	24	20

29. **STATE YOUR ASSUMPTION** A data set has a mean of 37, a median of 36.5, and a mode of 37. What assumption(s) can you make about the dataset?

30. **BOWLING** The table shows Lucinda's score for each of her last ten bowling games.
    a. Find the mean, median, and mode of the scores. Round to the nearest whole number.
    b. Why is the mean slightly higher than the median?

Game	Score
1	220
2	235
3	255
4	210
5	240
6	220
7	225
8	220
9	250
10	210

31. **DANCE COMPETITION** At a dance competition, Pascal earned a score of 73 points. There were 12 competitors who received a lower score than Pascal and 3 competitors who earned a higher score. What was Pascal's percentile rank in the dance competition?

32. **CREATE** Create a data set that has a mean of 11, a median of 10, and a mode of 8.

33. **WRITE** Describe how an outlier value that is greater than the numbers in the data set affects each measure of center.

34. **ANALYZE** Determine whether the statement is *true* or *false*. If it is false, explain how to make the statement true.

    To find percentile rank, divide the selected value by the total of all the values.

35. **PERSEVERE** Describe the effect on the mean, median, and mode of a set when all the items in the set are multiplied by the same number.

36. **WHICH ONE DOESN'T BELONG?** Analyze each situation. Which situation is NOT best described by the median of the data? Explain.

An art gallery has many items for sale that are reasonably priced, but it also carries luxury priced paintings.	Most of the students volunteered 2 hours each week, but James volunteered 8 hours per week.	The amusement park had about the same number of attendees each day. On the annual bring-a-friend-for-free day, the number of attendees tripled.

37. **FIND THE ERROR** Julio is studying botany and has been tracking the growth of 10 tomato plants each week. The first week, the plants measured the following growth: 1 in., 1.5 in., 2.2 in., 0.5 in., 1 in., 1.25 in., 1.4 in., 2 in., 2.1 in., 1.9 in. In his research paper, Julio includes the median growth value for the week. Has Julio chosen the best measure of center to describe the plant growth? Explain.

38. **STRUCTURE** Explain how you determine that a data set is best described by the mean.

39. **WRITE** Explain in your own words the process for finding a percentile rank.

Representing Data

Learn Dot Plots

One way to represent data is by using a **dot plot**, which is a diagram that shows the frequency of data on a number line.

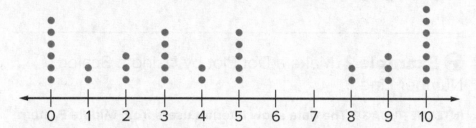

Copyright © McGraw-Hill Education

Key Concept • Making Dot Plots

Step 1 Write the data points in order from least to greatest.

Step 2 Make a number line that starts at the least data point and ends at the greatest data point. Choose an appropriate scale.

Step 3 Plot the dots on the number line. Stack the points when there is more than one data point with the same number.

Step 4 If appropriate, include a label for the number line and title for the dot plot.

Example 1 Make a Dot Plot

Represent the data as a dot plot.

11, 12, 14, 15, 12, 13, 15, 13, 9, 15, 12, 13, 15, 15, 11

Step 1 Write the data points in order from least to greatest.

9, ___, 11, ___, 12, 12, 13, 13, ___, 14, 15, 15, ___, 15, 15

Step 2 Make a number line.

The data are whole numbers ranging from ___ to ___. So, make

a number line starting at 9 with intervals of ___.

Step 3 Plot the dots on the number line.

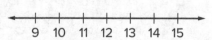

Step 4 If appropriate, include a label for the number line and title for the dot plot.

Because no information is given regarding what these data represent, no title is needed for this dot plot.

⬤ Go Online You can complete an Extra Example online.

Today's Goals
- Represent sets of data by using dot plots.
- Determine whether discrete or continuous graphical representations are appropriate, and represent sets of data by using bar graphs or histograms.

Today's Vocabulary
dot plot
bar graph
histogram

🍩 Think About It!

What would be the benefit of representing data in a dot plot?

Study Tip

Accuracy Count the listed data and check that the number of data points matches the sum of the frequency table. Missing just one or two pieces of data can change the values of statistics.

Check

Represent the data as a dot plot.

8, 6, 0, 2, 7, 1, 8, 1, 4, 8, 0, 1, 2, 8, 4, 7, 1, 5, 9, 1

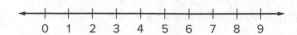

0 1 2 3 4 5 6 7 8 9

🌐 **Example 2** Make a Dot Plot by Using a Scaled Number Line

INTERNET USAGE **The data show Internet users from Middle Eastern countries as a percentage of their total population. Represent the data as a dot plot.**

96.4	**57.2**	**33.0**	**74.7**	**86.1**	**78.7**	**80.4**
78.6	**64.6**	**91.9**	**65.9**	**28.1**	**93.2**	**22.6**

Step 1 Write the data points in order from least to greatest.

_____, 28.1, _____, 57.2, 64.6, 65.9, 74.7, _____, 78.7, 80.4, 86.1, 91.9, _____, 96.4

Step 2 Make a number line.

The data range from _____ to _____. Since these data represent a broad range with specific values, it is unlikely that any data point is represented more than once. To represent the data in a meaningful way, _____ the number line.

Step 3 Plot the dots on the number line.

Internet Usage from Middle Eastern Countries

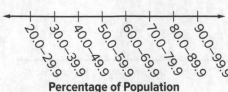

20.0–29.9 30.0–39.9 40.0–49.9 50.0–59.9 60.0–69.9 70.0–79.9 80.0–89.9 90.0–99.9

Percentage of Population

Step 4 If appropriate, include a label for the number line and title for the dot plot.

🐦 **Go Online** You can complete an Extra Example online.

💬 **Talk About It!**

How would these data appear if the number line were scaled by 1 instead of 10? Do you think a scale of 1 would be a good way to represent the data? Explain.

Check

MOUNTAINS The data give the elevation of the highest mountain peaks in the United States. Create a dot plot that best represents the data.

| 20,308 | 18,009 | 17,402 | 16,421 | 16,391 |
| 16,237 | 15,325 | 14,951 | 14,829 | 14,573 |

Highest Mountain Peaks in the U.S.

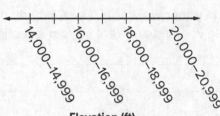

Elevation (ft)

Learn Bar Graphs and Histograms

A **bar graph** is a graphical display that compares categories of data using bars of different heights. Bar graphs are used when the data are discrete. To indicate this, there is a space between each of the bars.

A **histogram** is a graphical display that uses bars to display numerical data that have been organized in equal intervals. A histogram represents continuous data, so the bins have no spaces between them.

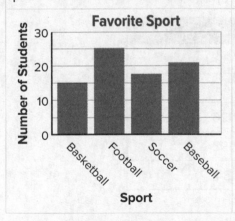

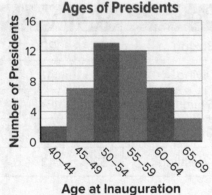

Key Concept • Making a Bar Graph or Histogram

Step 1 Determine whether the data should be represented as a bar graph or histogram.

Step 2 Determine appropriate categories or bins, and tally the data, if necessary.

Step 3 Draw bars to represent each category or bin.

Step 4 Label the axes. If appropriate, include a title for the graph.

🌐 Apply Example 3 Determine an Appropriate Graph for Discrete Data

OLYMPICS **The table shows the total number of Olympic medals won by U.S. athletes competing in selected events from the first Summer Olympics in 1896 through 2012. Make a graph of the data to show the total medals won for each sport.**

1 What is the task?

Describe the task in your own words. Then list any questions that you may have. How can you find answers to your questions?

_____

_____

_____

2 How will you approach the task? What have you learned that you can use to help you complete the task?

_____

_____

_____

3 What is your solution?

Use your strategy to solve the problem.

Event	Gold	Silver	Bronze	Total
Boxing	49	23	39	
Diving	48	41	43	
Swimming	230	164	126	
Track & Field	319	247	193	
Wrestling	52	43	34	

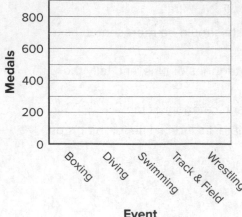

Total Medals Won by U.S. Athletes

4 How can you know that your solution is reasonable?

 **Write About It!** Write an argument that can be used to defend your solution.

_____

_____

_____

Check

VIDEO GAMES The table shows the number of active video game players in each country. Make a graph that best displays the data.

Country	Australia	Brazil	France	Germany	Italy	Poland	Spain	Turkey	UK	US
Players (millions)	9.5	40.2	25.3	38.5	18.6	11.8	17	21.8	33.6	157

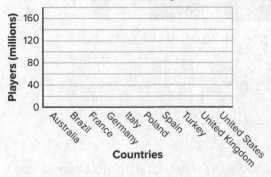

🌐 Example 4 Determine an Appropriate Graph for Continuous Data

MARATHON **The results of the top finishers of the 2015 New York City Marathon, wheelchair division, are given below. Determine whether the data are _discrete_ or _continuous_. Then make a graph.**

1:30:54   1:30:55   1:34:05   1:35:19   1:35:21   1:35:37   1:35:38   1:36:45
1:36:59   1:38:39   1:39:22   1:39:22   1:39:27   1:39:27   1:40:36   1:43:04

Step 1 Because racers can finish with any time, the data are

_____ and you can use a _____.

(_continued on the next page_)

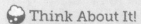

Think About It!

Describe the histogram. What does it show you about the racers' times? What do the gaps in the graph represent?

Step 2 Because the data are spread over several minutes, group the data by the minute. Then, tally each interval.

Time (h:m:s)	Frequency	Time (h:m:s)	Frequency
1:30:00–1:30:59		1:37:00-1:37:59	
1:31:00–1:31:59		1:38:00-1:38:59	
1:32:00–1:32:59		1:39:00-1:39:59	
1:33:00–1:33:59		1:40:00-1:40:59	
1:34:00–1:34:59		1:41:00-1:41:59	
1:35:00–1:35:59		1:42:00-1:42:59	
1:36:00–1:36:59		1:43:00-1:43:59	

Steps 3 and 4 Draw a bar to represent each bin. Label the axes. Include a title for the graph.

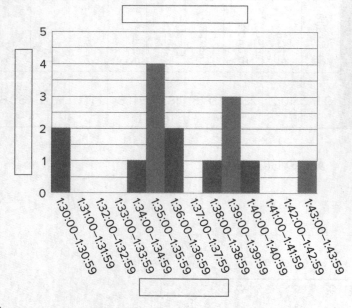

Check

PHOTO SHARING The table shows the users of a photo sharing app by age group. Make a graph that best displays the data.

Age	18-24	25-34	35-44	45-54	55-64	65+
Users (%)	45	26	13	10	6	1

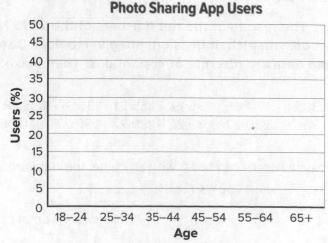

🧭 **Go Online** You can complete an Extra Example online.

Practice

Go Online You can complete your homework online.

Examples 1 and 2

1. **READING** The table shows the number of books read by students in a summer reading program. Make a dot plot of the data.

Number of Books Read				
3	8	5	6	5
4	5	5	4	5
5	6	8	8	4

2. **QUIZ SCORES** Represent the quiz scores as a dot plot. Scale the number line as needed.

   50, 45, 24, 28, 27, 38, 21, 22, 23, 42, 41, 35, 37, 25, 43

Examples 3 and 4

3. **SURVEY** A survey was conducted among students in Mr. Dalton's science class to determine a field trip destination. The results are shown in the table at the right. Make a graph to display the data.

Destination	Number of Votes
zoo	6
museum	4
observatory	11
state park	7

4. **MOVIES** In a survey, students were asked to name their favorite type of movie. Of those surveyed, 8 chose action movies, 6 chose comedies, 5 chose horror movies, 3 chose dramas, and 7 chose science fiction movies (sci-fi). Determine whether the data are discrete or continuous. Then make a graph.

5. **CONCERT** The table shows the number of attendees by age at a concert. Determine whether the data should be shown in a *bar graph* or *histogram*. Then make an appropriate graph for the data.

Concert Attendees	
0–9	400
10–19	1440
20–29	2400
30–39	2000
40–49	960
50–59	560
60–69	240

Mixed Exercises

6. **PRIZES** The table shows the number of prizes won by customers at a carnival game each of the past several days. Determine whether the data are discrete or continuous. Then make an appropriate graph for the data.

Prizes Won				
37	29	53	32	42
21	41	45	17	27
44	34	24	34	31
19	51	48	35	54
46	38	39	49	25

7. **JOGGING** The number of miles Lisa jogged each of the last 10 days are 3, 4, 6, 2, 5, 8, 7, 6, 4, and 5.

   a. Choose the most appropriate type of data display and graph the data.

   b. How many days did Lisa jog at least 4 miles?

   c. What was the greatest number of miles she jogged in a day?

8. **MOVIES** The number of movies that are released theatrically each year are shown in the table.

   a. Select an appropriate display for the data. Explain your reasoning.

Year	2010	2011	2012	2013	2014	2015	2016
Number of Movies Released	563	609	678	661	709	708	718

Source: comScore, MPAA

   b. Make a graph of the data.

9. **RUNNING** The ages of the participants in a 10K race at Masonville are 65, 47, 23, 70, 41, 55, 32, 29, 56, 39, 12, 57, 25, 33, 15, 18, 35, 22, 63, 49, 23, 30, 37, 40, and 50.

   a. Construct an appropriate data display for the data.

   b. How many participants are less than 30 years old?

   c. In what interval is the most frequent age?

10. **ORCHESTRA** The ages of the members of an orchestra are 39, 43, 31, 53, 41, 25, 35, 46, 27, 34, 37, 26, 51, 29, 36, 40, 33, 28, 48, 26, 42, and 38 years. Make a graph of the data.

11. **PRECISION** A scientific research study tracks the growth of an insect in millimeters. The growth data for each insect in the study during week 1 are 1.1, 1.25, 1.3, 1.67, 1.9, 2.35, 2.1, 2.3, 1.5, 1.7, 2.25, 2.1, 2.45, 1.37, 1.83. The scientist is preparing a histogram to show the distribution of growth across the population. How should the scientist break down his data into categories?

12. **PETS** The pets owned by Liza's classmates are rabbit: 2, dog: 6, cat: 3, horse: 2, bird: 5, mouse: 1, fish: 3, and other: 1.

   a. Make a dot plot of the data.

   b. How many types of pets are represented by the dot plot?

   c. Which pet is the most popular?

13. **ANALYZE** Make two conclusions about a product that received the ratings shown in the dot plot. Justify your conclusions.

14. **REGULARITY** Explain when a histogram is the best model for data, and describe the process of creating a histogram.

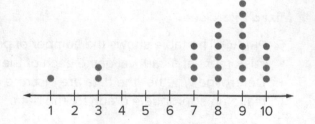

15. **WRITE** Explain why it may be necessary to scale the number line of a dot plot.

16. **PERSEVERE** Using the data provided in the double bar graph about peanut butter, what are two conclusions the grocery store could infer?

17. **STRUCTURE** How is a bar graph similar to a histogram? How is it different?

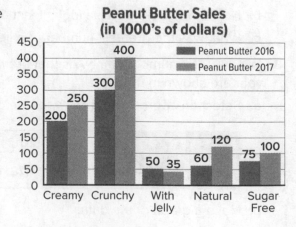

Using Data

Copyright © McGraw-Hill Education

Today's Goals
- Identify potential bias in sampling methods and questions.
- Identify potential bias in statistics and representations of data.

Today's Vocabulary
population
sample
bias
statistic

Explore Phrasing Questions

Online Activity Use a real-world situation to complete the Explore.

INQUIRY How can the way you collect data affect the results?

Learn Collecting Data

A **population** consists of all the members of a group of interest about which data will be collected. Since it may be impractical to examine every member of a population, a subset of the group, called a **sample**, is sometimes selected to represent the population. The sample can then be analyzed to draw conclusions about the entire population.

Sample data are often used to estimate a characteristic of a population. Therefore, a sample should be selected so that it closely represents the entire population. Also, the larger the sample size, or the more samples taken, the better it represents the population.

A **bias** is an error that results in a misrepresentation of a population. If a sample favors one conclusion over another, the sample is biased and the data are invalid.

Talk About It!
Some polls use both landlines and cell phones. How might this alleviate the issue of bias with the landline-only sampling method?

Example 1 Sample Bias

POLLS Before the 2010 elections for members of the U.S. House of Representatives, pollsters called American households on their landline phones to see how they planned to vote. What kind of sample bias might have affected the poll?

Step 1 Identify the intended population.

The population is _____.

Step 2 Identify the sample method.

The data for this poll were collected over landline phones, so the sample consists of likely voters who have a landline.

Step 3 Determine potential bias.

Because not all likely voters have landline phones, the results could be _____ because not all likely voters are available for this sample.

Go Online You can complete an Extra Example online.

Check

SOCIAL MEDIA Shia wants to determine the age of the average internet user. He posts a poll to his friends on a social media site, asking their age. Is this a good sample? If not, what kind of sample bias might have affected the poll? _____

A. This is a good sample.

B. This is not a good sample. He asked only people on the Internet.

C. This is not a good sample. He asked only about users' ages.

D. This is not a good sample. He asked only his friends on a specific Web site.

Example 2 Question Bias

SOFT DRINKS A survey organization wants to see what percent of New York City citizens support a ban on soft drinks. The question posed is, "Do you support a ban on soft drinks, which contribute to heart disease and tooth decay?"

Part A Identify bias.

Step 1 Identify the purpose of the question: To find the _____ of New York citizens who _____ on soft drinks.

Step 2 Identify potential bias in the question: The question lists some of the health risks of soft drinks. This might make respondents _____ to respond that they do support a ban.

Part B Identify interests.

The bias in the question might make respondents more likely to support a ban. This bias could serve the interests of health groups who want to ban soft drinks or companies who sell competing drinks, like juices.

Check

FILM One of your friends wants to determine whether people in your class prefer to watch movies or television. She asks, "Do you prefer to watch movies or television?"

Part A Does this question potentially bias the results? _____

A. No; the question is as neutral as possible.

B. Yes; your friend asked only people in your class.

C. Yes; your friend provided only two options.

D. Yes; the framing of the question influences the respondent to choose *television*.

Part B Whose interests might be served by asking the question in this way?

Go Online You can complete an Extra Example online.

Learn Using Statistics and Representations

A **statistic** is a measure that describes a characteristic of a sample. Like data, statistics and representations of data are nonneutral. When the average of a set of data is discussed, it uses a measure of center: mean, median, or mode. However, depending on the data set and what information is being conveyed, one measure of center might not give the whole picture of the data. Even if the data is being discussed in whole, it can also be misrepresented. For example, a person might manipulate the scales of the axes of a graph or how the data are represented graphically to misrepresent the data.

Example 3 Data Summaries

TEACHING **A teacher wants to tell his students how the average student did on an exam, so he looks at the scores in his gradebook. Two students scored a 0 because they stopped showing up for class in the last month and did not take the exam. He uses the mean, 71, as the measure of center. Does the mean accurately represent these data?**

0, 0, 82, 83, 85, 87, 88, 88, 91, 91, 91

Step 1 Identify the other measures of center. Round your answer to the nearest unit.

median = _____

mode = _____

Step 2 Analyze the measures of center and how they align with the information the teacher wants to convey.

Mean: The mean, 71, _____ by the two 0 scores. However, no one who showed up for the exam scored below an 82, so the mean does not do a good job of indicating the performance of the students who took the exam.

Median: The median, 87, _____ by the extreme values. It provides a more accurate average for how students performed on the test because it includes the scores of the two students who did not take the exam at all.

Mode: The mode, 91, is both the score most students received and the highest score received on the exam, but it _____ accurately portray how students performed on average.

Because the teacher wants to discuss the performance of students who took the exam, the mean _____ the best measure of center. It indicates that all students who took the exam performed _____ than their actual scores.

Math History Minute

With M. A. Girschick, **David Blackwell** (1919–2010) authored the classic book *Theory of Games and Statistical Decisions*. In 1965, he became the first African American president of the American Statistical Society.

Go Online You can complete an Extra Example online.

Check

READING Karen writes down the number of books she has read for each of her classes so far: 5, 6, 4, 5, 5, 6, 5, 4, 5. Using the mean, 5, she says that she reads around 5 books on average for an English course. Does the mean accurately represent the data for the situation? Explain. _____

A. No; the mean is overly influenced by the low numbers.

B. No; the mean is overly influenced by the high numbers.

C. No; the mean doesn't tell how many pages are in the average book.

D. Yes; the mean accurately represents the data for the situation.

🌐 Example 4 Data Representation

SOCCER A group compares two soccer players who play the same position for different teams. They make a graph of the number of goals scored throughout the season for each player. Do the graphs misrepresent the data? Whose interests might be served by the representation?

Player 1

Total Goals / Months Since Start of Season

Player 2

Total Goals / Months Since Start of Season

Part A Identify misleading representations.

Step 1 Identify the purpose of the graphs. The purpose of the graphs is to compare the number of _____.

Step 2 Identify differences in the graphs. The graphs appear to be the same in terms of the data being represented, but the *y*-axis for the second player goes up to 90 in increments of ___, whereas the first goes up to 45 in increments of ___.

Step 3 Identify how this affects the representation of the data. Although the numbers are the same, the scale for Player 1 makes it look like he is scoring more goals than Player 2.

Part B Identify interests.

The misleading representation of the data makes it appear that Player 1 scores more goals than Player 2. This bias could serve the interests of the team, sponsors of Player 1's team, or Player 1's agent.

🅝 Go Online You can complete an Extra Example online.

Think About It!

If the two graphs had the same scales but were comparing a goalkeeper and a forward, how might the data be misleading when comparing the skill of the two players?

Use a Source

Find a graph or set of statistics online. Ask yourself, are the data accurately represented? Whose interests are served by the graph or statistics?

Practice

Example 1

1. **SPORTS** Awan wants to know what the favorite sport is among students. To find out, he asks everyone he sees leaving school after basketball practice. Identify the intended population and determine the potential sample bias.

2. **STORES** Raya wants to conduct a survey at a nearby mall to determine which are the mall's most popular stores. How could she choose a sample that is unbiased?

Example 2

3. **MUSIC** Shea is shopping online, and a survey question pops up that says, "Music education enriches student learning. Do you support music education in schools?"

   a. Identify potential bias in the question.

   b. Identify whose interests may be served by the question.

4. **CANDIDATES** There are three candidates for mayor. To investigate how the townspeople feel about the candidates, a newspaper posts a poll that lists the three candidates and asks which candidate people support. The poll appears on the same page as an opinion piece in support of one of the candidates.

   a. Identify potential bias in the question.

   b. Identify whose interests may be served by the question.

Example 3

5. **BUTTERFLIES** Tania recorded the number of butterflies she saw on her daily runs each day for a week. The numbers are: 1, 8, 2, 2, 5, 6, and 4. Find the mean, median, and mode of the data. Which measure(s) are appropriate to accurately summarize the data?

6. **OUTLIERS** In a data set with an outlier, which measure of center, mean or median, is the better measure to use to describe the center of the data? Explain your reasoning.

Example 4

7. **SALES** The graphs show the number of T-shirts sold at a baseball tournament for two years by two different vendors. The tournament director wants to compare the vendors. Do the graphs misrepresent the data? How does that difference affect the interpretation?

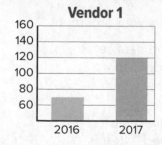

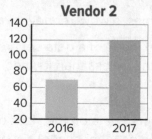

8. **SCALE** If the same set of data is graphed with a scale of 0 to 10 on the *y*-axis and then with a scale of 0–100 on the *y*-axis, what effect does that have on the representation of the data?

Mixed Exercises

9. SCIENCE A school wants to know which area of science; physics, biology, or chemistry, is most interesting to its students. Would it be better to survey students in a class that is an elective or required to get a sample with the least bias? Explain your reasoning.

10. TAX Before surveying people about whether they favor or oppose a proposed tax, the surveyors want to present information about the tax. Suppose the surveyors give facts about the tax without giving opinions. How could the facts given by the surveyors introduce bias?

11. CONSTRUCT ARGUMENTS The weights, in pounds, of several dolphins at a sea animal care facility are 185, 222, 755, 801, 835, 990, and 1104. Which measure of center best represents the data? Justify your conclusion.

12. FOOD DRIVE The chart shows the number of canned goods collected by Valley High School in 2012 and 2017. Is the graph misleading? Explain.

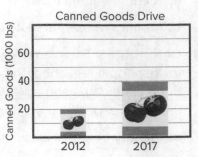

13. REASONING The number of participants at reading club for six weeks are 11, 12, 10, 13, 10, and 10. Without calculating the measures of center, how would adding an outlier of 24 participants affect which measure of center most appropriately represents the data?

14. PRECISION A community garden has 8 tomato plants with heights ranging from 0.4 to 0.9 meters. Regina found the median to be 0.7 meters, which she rounded and reported as 1 meter. Is Regina's report of the median accurate? Explain your reasoning.

15. REGULARITY Describe a general method for assessing a sample for bias.

16. STRUCTURE How are the median and mean scores affected if all data values in a set are increased by a specific value, such as 10?

17. CREATE Create two sets of data and display them in a graph or chart that shows bias toward one of the sets of data.

18. WRITE Write two scenarios that have different examples of sample bias. Have a classmate rewrite your statements without bias.

19. CREATE Think of a topic about which you can survey the teachers at your school. Conduct the survey. Explain whether your survey question(s) introduce bias.

20. ANALYZE Is a biased sample *sometimes*, *always*, or *never* valid? Justify your argument.

21. PERSEVERE If the mean, median, and mode of a data set are equal, the data set is symmetric. If a data set has a mean that is less than its median, what does that tell you about the data set?

22. FIND THE ERROR Two students collected data on the sizes of box turtle shells. Olivia measured 8 turtles from a pond near her school. Caleb measured 2 turtles from each of 4 ponds around town. Which is more likely to be free of sample bias? Explain your reasoning.

Measures of Spread

Explore Using Measures of Spread to Describe Data

Online Activity Use a real-world situation to complete the Explore.

> **INQUIRY** Why might you describe a data set with more than the mean?

Learn Range and Interquartile Range

Statisticians use **measures of spread** or variation to describe how widely data values vary. One such measure is the **range**, which is the difference between the greatest and least values in a set of data.

Quartiles divide a data set arranged in ascending order into four groups, each containing about one-fourth, or 25%, of the data. A **five-number summary** contains the minimum, quartiles, and maximum of a data set.

The **median** marks the second quartile, Q_2, and separates the data into upper and lower halves.

The **lower quartile**, Q_1, is the median of the lower half.

The **upper quartile**, Q_3, is the median of the upper half.

A **box plot**, or box-and-whisker plot, is a graphical representation of the five-number summary of a data set. A box is drawn from Q_1 to Q_3 with a vertical line at the median. This box represents the **interquartile range**, or IQR, which is the difference between the upper and lower quartiles. The whiskers of the box plot are drawn from Q_1 to the minimum and from Q_3 to the maximum.

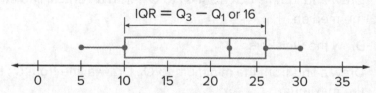

IQR = $Q_3 - Q_1$ or 16

Example 1 Range

GRADES **What is the range of the scores?**

79, 83, 88, 62, 91, 99, 70

Step 1 Arrange the data in ascending order. ___, 70, 79, 83, 88, 91, ___

Step 2 Determine the range.

range = greatest value − least value

99 − 62 or ___

Go Online You can complete an Extra Example online.

Copyright © McGraw-Hill Education

Today's Goals
- Determine measures of spread, including the range and interquartile range, of a set of data.
- Determine the standard deviation of a data set.

Today's Vocabulary
measures of spread
range
quartiles
five-number summary
median
lower quartile
upper quartile
box plot
interquartile range
standard deviation

Study Tip

Ordering Because the range involves only the greatest and least values in a data set, it can be determined without ordering the numbers. However, it is often useful to order the data to avoid missing a number.

💭 Think About It!

If Q_1 were located between two numbers, how would you determine its value?

🌎 Example 2 Make a Box Plot

BOX OFFICE **A financial analyst for a movie studio wants to determine how much most of the top-earning movies have grossed to compare his studio's recent grosses. The worldwide grosses, in millions of dollars, for the top 10 highest-grossing films of all time are given. Determine the five-number summary and draw a box plot of the data to see the spread of the data.**

2788	2187	2060	1670	1520
1516	1405	1342	1277	1215

Part A Determine the five-number summary.

Step 1 Arrange the data in ascending order.

1215, 1277, 1342, 1405, 1516, 1520, 1670, 2060, 2187, 2788

Step 2 Determine the five-number summary of the data.

1215, 1277, 1342, 1405, 1516, 1520, 1670, 2060, 2187, 2788

min $Q_1 = 1342$ $Q_2 = \dfrac{1516 + 1520}{2}$ $Q_2 = 2060$ max

or 1518

Part B Construct a box plot.

Step 1 Construct a number line.

Because the minimum is 1215 and the maximum is 2788, your number line must include those values.

Step 2 Draw the box.

Draw and label a box from Q_1 to Q_3, with a vertical line at the median.

Step 3 Draw the whiskers.

Draw a line from the minimum to Q_1. Draw a line from Q_3 to the maximum.

$Q_1 = 1342$
$Q_2 = 1518$ $Q_3 = 2060$

1200 1400 1600 1800 2000 2200 2400 2600 2800

🌎 **Go Online** You can complete an Extra Example online.

Check

MARRIAGE The average age at which women first get married differs by country. The ages for eight countries are shown. Select the box plot for the data. ___

$$25, 26, 21, 27, 31, 31, 29, 20$$

A. **Average Age of Women at First Marriage**

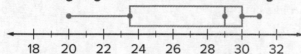

B. **Average Age of Women at First Marriage**

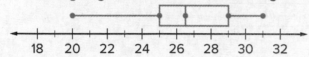

C. **Average Age of Women at First Marriage**

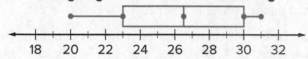

D. **Average Age of Women at First Marriage**

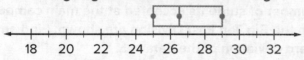

🌐 Example 3 Interquartile Range

AUDIO Sarah wants to upload a song that she recorded and share it with her friends. She wants to know whether her song, which is currently −12 decibels on average, is the right volume compared to other songs online so listeners do not have to adjust their volume. She writes down the average volume of seven songs. Find the interquartile range of these average volumes.

$$-18, -20, -8, -13, -14, -15, -18$$

Step 1 Order the data: $-20, -18, -18, -15, -14, -13, -8$

Step 2 Determine Q_1 and Q_3.

Step 3 Determine the *IQR*.

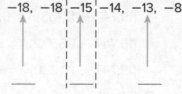

$IQR = Q_3 - Q_1 = -13 - ($___$)$ or ___

Check

WRITING Cora is writing a novel and tracks the number of pages she writes each day for a week. The number of pages she wrote each day for a week is shown. Find the interquartile range of the data set. ___

$$6, 5, 0, 3, 8, 1, 4$$

🔘 **Go Online** You can complete an Extra Example online.

Think About It!

If all the data in a set were the same value, then what would the standard deviation be?

Learn Standard Deviation

In a data set, the **standard deviation** shows how the data deviate from the mean. A majority of the data in a set, approximately two-thirds, is contained within 1 standard deviation below and above the mean. So, if two data sets have the same mean, but one has a greater standard deviation, then the data in that set is more spread out from the mean.

Key Concept · Standard Deviation

Step 1 Find the mean μ.

Step 2 Find the square of the difference between each data value x_n and the mean, $(\mu - x_n)^2$.

Step 3 Find the sum of all the values in Step 2.

Step 4 Divide the sum by the number of values in the set of data n. This value is the variance.

Step 5 Take the square root of the variance.

Formula $\sigma = \sqrt{\dfrac{(\mu - x_1)^2 + (\mu - x_2)^2 + \cdots + (\mu - x_n)^2}{n}}$

🌐 Example 4 Calculate Standard Deviation

UNIVERSITY **The number of students accepted at the main campus of each university in the Big Ten East Division are shown. Find and interpret the standard deviation of the data set.**

27,300	**12,333**	**15,570**	**21,610**
17,413	**25,772**	**18,230**	

Step 1 Find the mean μ.

$$\mu = \frac{27,300 + 12,333 + 15,570 + 21,610 + 17,413 + 25,772 + 18,230}{7} \text{ or } \underline{\hspace{1cm}}$$

Step 2 Find the square of the differences, $(\mu - x_n)^2$.

$(19,747 - 27,300)^2 = 57,047,809$

$(19,747 - 12,333)^2 = 54,967,396$

$(19,747 - 15,570)^2 = 17,447,329$

$(19,747 - 21,610)^2 = 3,470,769$

$(19,747 - 17,413)^2 = 5,447,556$

$(19,747 - 25,772)^2 = \underline{\hspace{2cm}}$

$(19,747 - 18,230)^2 = \underline{\hspace{2cm}}$

Step 3 Find the sum.

$57,047,809 + 54,967,396 + \ldots + 2,301,289 = \underline{\hspace{3cm}}$

Step 4 Divide by the number of values. This value is the variance.

$$\frac{176,982,773}{7} \approx \underline{\hspace{2cm}}$$

Step 5 Find the standard deviation.

$$\sqrt{25,283,253} \approx \underline{\hspace{2cm}}$$

Think About It!

If another data point less than 14,719 or greater than 24,775 is added to the data set, how would that change the standard deviation?

 Go Online

You can complete an Extra Example online.

Practice

⏺ **Go Online** You can complete your homework online.

Example 1

Find the range of each data set.

1. 12, 27, 43, 52, 43, 18, 45, 53, 26

2. 132, 127, 129, 130, 141, 125, 138, 129

3. 56, 101, 78, 49, 55, 108, 111, 64

4. 5.9, 6.2, 3.9, 3.7, 8.5, 6.2, 9.0, 8.7, 4.5, 9.3

5. **EXERCISE** Kent tracked his daily number of minutes of exercise. Find the range of the data set.

Number of Minutes of Exercise					
30	35	25	28	40	38
36	29	34	45	42	39

Example 2

Determine the five-number summary and draw a box plot of the data.

6. prices in dollars of smartphones: 311, 309, 312, 314, 399, 312

7. attendance at an event for the last nine years: 68, 99, 73, 65, 67, 62, 80, 81, 83

8. books a student checks out of the library: 17, 9, 10, 17, 18, 5, 2

9. ounces of soda dispensed into 36-ounce cups: 36.1, 35.8, 35.2, 36.5, 36.0, 36.2, 35.7, 35.8, 35.9, 36.4, 35.6

10. ages of riders on a roller coaster: 45, 17, 16, 22, 25, 19, 20, 21, 32, 37, 19, 21, 24, 20, 18, 22, 23, 19

Example 3

Find the interquartile range of each data set.

11. 43, 36, 51, 68, 50, 27, 38, 81, 33

12. 201, 225, 217, 240, 232, 252, 228, 231

13. 94, 87, 105, 99, 118, 97, 102, 85

14. 8.4, 7.1, 6.3, 6.8, 9.2, 7.3, 8.8, 7.9, 5.3, 8.2

15. **HEART RATE** A nurse tracked the heart rates of several patients. Find the interquartile range (IQR) of the data set: 108, 88, 119, 75, 96, 88, 100, 99, 125, 81.

Example 4

Find the standard deviation.

16. {10, 9, 11, 6, 9}

17. {6, 8, 2, 3, 2, 9}

18. {23, 18, 28, 36, 15}

19. {44, 35, 40, 37, 43, 38, 40}

20. **PARKING** A city councilor wants to know how much revenue the city would earn by installing parking meters on Main Street. He counts the number of cars parked on Main Street each weekday: {64, 79, 81, 53, 63}. Find the standard deviation.

Mixed Exercises

21. **REASONING** A hockey team keeps track of how many goals it scores each game: {2, 4, 0, 3, 7, 2}. Find and interpret the standard deviation of the data.

22. FOOTBALL The table shows information about the number of carries a running back had over a number of years. Find and interpret the standard deviation of the number of carries.

Year	Number of Carries
2006	31
2007	90
2008	105
2009	115
2010	162

23. MOVIES The manager at a movie theater kept track of the age of each person in a matinee movie: 67, 62, 65, 38, 69, 67, 59, 41, 43, 36, 45, 22, 69, 68, 18, 15, 9, 60, 64.

   a. Determine the five-number summary for the data set.

   b. Draw a box plot of the data.

24. GAS PRICES Renee is planning a road trip to her aunt's house. To estimate how much the trip will cost, she goes online and finds the price of a gallon of gasoline for 5 randomly selected gas stations along the route: $2.09, $2.19, $3.99, $2.39, $2.29.

   a. Determine the five-number summary for the data set.

   b. Draw a box plot of the data.

Find the range, five-number summary, interquartile range, and standard deviation for each data set. Then draw a box plot of the data.

25. SEASHELLS Jorja collected the following number of seashells for the last nine trips to the beach: 5, 11, 7, 12, 13, 17, 3, 15, 14.

26. SHOE SIZE The following shoe sizes of students at a high school were randomly recorded for one hour: 6, 8, 8.5, 10, 12, 6.5, 7, 8, 8.5, 7.5, 9, 11.5, 10, 13, 5.5, 6.5, 5, 9.5.

27. FIND THE ERROR Jennifer and Megan are determining one way to decrease the size of the standard deviation of a set of data. Is either correct? Explain your reasoning.

Jennifer	Megan
Remove the outliers from the data set.	Add data values to the data set that are equal to the mean.

28. ANALYZE Determine whether the statement *Two random samples taken from the same population will have the same mean and standard deviation* is *sometimes, always,* or *never* true. Justify your argument.

29. CREATE Write your own survey question and collect data about your question from 8 classmates. Use that data to find the range, five-number summary, interquartile range, and standard deviation for the data set. Then draw a box plot of the data.

30. WRITE What does the interquartile range tell you about how data clusters around the median of the data?

Distributions of Data

Learn Shapes of Distributions

Analyzing the shape of a **distribution** can help you learn a lot about the data it represents. When data are graphed, the shape of the distribution can be seen.

Today's Goals
- Interpret differences in the shapes of distributions.
- Account for the possible effects of extreme data points.

Today's Vocabulary
distribution

symmetric distribution

negatively skewed distribution

positively skewed distribution

outlier

Key Concept • Symmetric and Skewed Distributions

Histograms	Box Plots	Dot Plots

In a **symmetric distribution**, the mean and median are approximately equal.

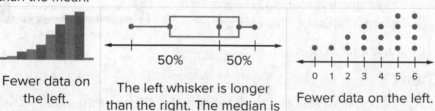

The data are evenly distributed.	The whiskers are the same length. The median is in the center of the data.	The data are evenly distributed.

A **negatively skewed distribution** typically has a median greater than the mean.

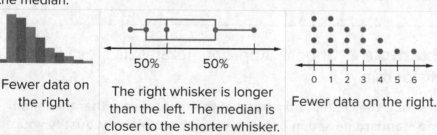

Fewer data on the left.	The left whisker is longer than the right. The median is closer to the shorter whisker.	Fewer data on the left.

A **positively skewed distribution** typically has a mean greater than the median.

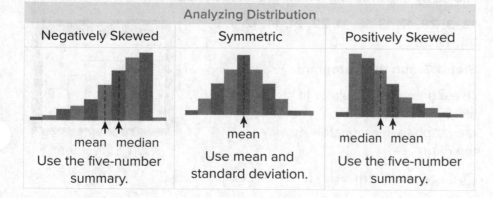

Fewer data on the right.	The right whisker is longer than the left. The median is closer to the shorter whisker.	Fewer data on the right.

Analyzing Distribution

Negatively Skewed	Symmetric	Positively Skewed
mean median	mean	median mean
Use the five-number summary.	Use mean and standard deviation.	Use the five-number summary.

Go Online
You may want to complete the Concept Check to check your understanding.

Study Tip

Distribution Shape
For a histogram, drawing a curve over the data bars may help you see the distribution.

Talk About It!

How could different bin widths of a histogram affect the shape of the distribution? Explain your reasoning.

Study Tip:

Window Settings On a TI-84, use **ZoomStat** from the **Zoom** menu to get a basic fitting view window. Then, adjust the window parameters and bin width.

Go Online

to see how to use a graphing calculator with these examples.

Example 1 Analyze Distribution by Using Technology

Use a graphing calculator to construct a histogram and box plot for the data. Then describe the shape of the distribution.

78, 53, 24, 75, 76, 83, 78, 60, 64, 53, 36, 47, 32, 75, 54, 68, 68, 74, 85, 42

Method 1 Histogram

Steps 1 and 2 Enter the data and then graph the histogram.

Step 3 Analyze the histogram.

The histogram is higher on the right and has a tail on the left. Therefore, the distribution is

_____.

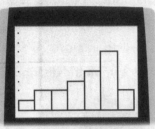

[20, 90] scl: 10 by [0, 8] scl: 1

Method 2 Box Plot

Steps 1 and 2 Enter the data and graph the box plot.

Step 3 Analyze the box plot.

The left whisker is longer than the right and the median is closer to the shorter whisker. Therefore, the distribution is _____.

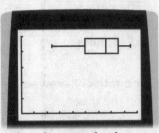

[0, 90] scl: 10 by [0, 5] scl: 1

Check

Use a histogram or box plot to determine the shape of the data.

61, 135, 217, 388, 354, 459, 512, 243, 440, 307

The shape of the distribution is _____.

Example 2 Choose Appropriate Statistics by Using a Histogram

Describe the center and spread of the data using either the mean and standard deviation or the five-number summary. Justify your choice by constructing a histogram for the data.

18, 3, 28, 17, 13, 18, 11, 22, 21, 14, 12, 7, 9, 24, 17, 28

Step 1 Graph the histogram.

Use a graphing calculator to create a histogram. Adjust the parameters of the graph to appropriately display the data.

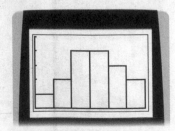

[0, 30] scl: 5 by [0, 5] scl: 1

Go Online You can complete an Extra Example online.

This graph shows that the frequency of the data in the middle is high while frequency of data to the left and right are low. Therefore, the distribution is symmetric.

Step 2 Calculate statistics.

The distribution is symmetric, so the mean and standard deviation are good statistics to represent the data.

To display the statistics, press **STAT**, access the **CALC** menu, select **1-VAR Stats**, and press **ENTER**.

The mean $\bar{x}$ is about 16.4 with a standard deviation σ of about 7.0.

💬 **Think About It!**
Why is mean an appropriate statistic to represent the center for these data?

Example 3 Choose Appropriate Statistics by Using a Box Plot

Describe the center and spread of the data using either the mean and standard deviation or the five-number summary. Justify your choice by constructing a box plot for the data.

202, 148, 21, 60, 74, 140, 462, 157, 225, 23, 88, 241, 59, 139, 351

Step 1 Graph the box plot.

Use a graphing calculator to create a box plot. Adjust the parameters of the graph to appropriately display the data.

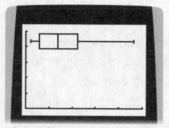

The right whisker is longer than the left and the median is slightly closer to the right whisker. So, this distribution is _____.

[0, 500] scl: 100 by [0, 5] scl: 1

Step 2 Calculate statistics.

The distribution is positively skewed, so use the

_____.

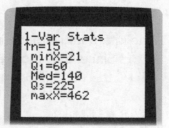

To display the statistics, press [**stat**], access the **CALC** menu, select **1-VAR Stats**, and press [**enter**]. Use the down arrow key to display more statistics.

Maximum: _____

Minimum: _____

Median: _____

Lower Quartile: _____

Upper Quartile: _____

 Go Online You can complete an Extra Example online.

Learn Extreme Data Points

The least and greatest values in a set of data are called **extreme values**. An **outlier** is a value that is more than 1.5 times the interquartile range above the third quartile or below the first quartile. Outliers can significantly skew the mean and standard deviation.

🌐 Example 4 Choose Appropriate Statistics with Extreme Data Points

SHARKS The lengths, in feet, of adult sharks of various species are shown. Describe the center and spread of the data using appropriate statistics, and identify the effect of extreme data points.

33	5.2	12.5	11.5	12	0.6	11	20	23	10
6.5	18	13	5	12	4	1	20	12	46

Step 1 Make a box plot.

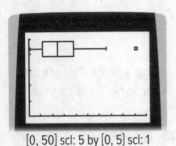

[0, 50] scl: 5 by [0, 5] scl: 1

Step 2 Analyze the graph.

Notice that the right whisker is longer than the left. So, this distribution is _____.

The plot also shows that there is an _____.

Step 3 Calculate statistics.

Include the mean with the five-number summary to see the effect of the outlier.

Mean: _____   Median: _____

Max: _____   Min: _____

Lower Quartile: _____

Upper Quartile: _____

The interquartile range is _____. Since 46 is more than 19 + _____ (13.15), 46 is an outlier.

Step 4 Describe the effect of the outlier.

Since 46 is an outlier, it has affected the mean. To see how much, remove 46 from the set of data and display the statistics again.

Mean: _____   Median: _____

Notice that the median did not change when the extreme data point was removed, but the mean did. Without the outlier, the mean and median are closer to the same value.

Check

PRECIPITATION The table shows the annual rainfall in Death Valley, CA.

Year	2006	2007	2008	2009	2010	2011	2012	2013
Rainfall (in.)	0.85	0.18	1.04	0.26	2.41	0.98	0.40	0.51

Part A What year(s) represent outlier(s)? _____

Part B Because of the outlier, the mean is _____.

🐺 **Go Online** You can complete an Extra Example online.

Think About It!

Suppose a new species of shark is discovered that has an average length of 50 feet. How would two extreme data points affect the measures of center?

🐺 Go Online

to see how to use a graphing calculator with this example.

Practice

Go Online You can complete your homework online.

Example 1

Use a graphing calculator to construct a histogram and a box plot for the data. Then describe the shape of the distribution.

1. 55, 65, 70, 73, 25, 36, 33, 47, 52, 54, 55, 60, 45, 39, 48, 55, 46, 38
   50, 54, 63, 31, 49, 54, 68, 35, 27, 45, 53, 62, 47, 41, 50, 76, 67, 49

2. 42, 48, 51, 39, 47, 50, 48, 51, 54, 46, 49, 36, 50, 55, 51, 43, 46, 37
   50, 52, 43, 40, 33, 51, 45, 53, 44, 40, 52, 54, 48, 51, 47, 43, 50, 46

Example 2

Describe the center and spread of the data using either the mean and standard deviation or the five-number summary. Justify your choice by constructing a histogram for the data.

3. 32, 44, 50, 49, 21, 12, 27, 41, 48, 30, 50, 23, 37, 16, 49, 53, 33, 25
   35, 40, 48, 39, 50, 24, 15, 29, 37, 50, 36, 43, 49, 44, 46, 27, 42, 47

4. 82, 86, 74, 90, 70, 81, 89, 88, 75, 72, 69, 91, 96, 82, 80, 78, 74, 94
   85, 77, 80, 67, 76, 84, 80, 83, 88, 92, 87, 79, 84, 96, 85, 73, 82, 83

Example 3

Describe the center and spread of the data using either the mean and standard deviation or the five-number summary. Justify your choice by constructing a box plot for the data.

5. 47, 16, 70, 80, 28, 33, 91, 55, 60, 45, 86, 54, 30, 98, 34, 87, 44, 35
   64, 58, 27, 67, 72, 68, 31, 95, 37, 41, 97, 56, 49, 71, 84, 66, 45, 93

6. 64, 36, 32, 65, 41, 38, 50, 44, 39, 34, 47, 35, 46, 36, 53, 35, 68, 40
   36, 62, 34, 38, 59, 46, 63, 38, 67, 39, 59, 43, 39, 36, 60, 47, 52, 45

Example 4

7. **FLYING** The various prices of a flight from Los Angeles to New York are shown.
   $182, $234, $264, $271, $277, $314, $317, $455
   a. Make a box plot of the data.
   b. Calculate the statistics that best represent the data.
   c. Describe the effect of the outlier.

8. **EXERCISE** Yoshiko tracked her minutes of exercise each day for 10 days as shown. 57, 60, 53, 59, 57, 61, 61, 54, 62, 10
   a. Make a box plot of the data.
   b. Calculate the statistics that best represent the data.
   c. Describe the effect of the outlier.

Mixed Exercises

USE TOOLS **Use a graphing calculator to construct a histogram and a box plot for the data. Then describe the shape of the distribution.**

9. 14, 71, 63, 42, 24, 76, 34, 77, 37, 69, 54, 64, 47, 74, 59, 43, 76, 56
   78, 52, 18, 54, 39, 28, 56, 74, 68, 36, 20, 49, 67, 47, 69, 68, 72, 69

10. 53, 34, 36, 38, 43, 49, 52, 36, 39, 37, 58, 45, 37, 38, 46, 52, 45, 39
55, 39, 40, 55, 38, 40, 42, 38, 45, 36, 46, 39, 35, 41, 49, 43, 52, 34

11. 51, 19, 46, 64, 29, 51, 58, 30, 55, 31, 34, 31, 50, 37, 40, 39, 40, 41
42, 32, 24, 48, 43, 45, 38, 43, 58, 47, 34, 36, 50, 54, 46, 28, 60, 22

12. TRACK Daryn recorded the number of laps he walked around the track each week. Use a graphing calculator to construct a histogram for the data, and describe the shape of the distribution.

17, 21, 23, 26, 27, 28, 28, 27, 33, 34, 33, 27, 29, 22, 19, 28, 35

13. GOLF Mr. Swatsky's geometry class's miniature golf scores are shown below. Use a graphing calculator to construct a box plot for the data, and describe the shape of the distribution.

Scores
36, 38, 38, 39, 40, 42, 44, 46, 46, 47, 48, 48, 50,
52, 52, 53, 54, 55, 56, 56, 56, 60, 57, 58, 63

14. HAIR LENGTH Ruth recorded the lengths, in centimeters, of hair of students in her school. Describe the center and spread of the data using either the mean and standard deviation or the five-number summary. Justify your choice by creating a box plot for the data.

40, 39, 37, 26, 25, 40, 35, 34, 26, 39, 42, 33, 26, 25, 34, 38, 41, 34
37, 39, 32, 30, 22, 38, 36, 28, 27, 39, 34, 26, 36, 38, 25, 39, 23, 8

15. PRESIDENTS The ages of the presidents of the United States at the time of their inaugurations are shown. Describe the center and spread of the data using either the mean and standard deviation or the five-number summary. Justify your choice by creating a box plot for the data.

Ages of Presidents
57, 61, 57, 57, 58, 57, 61, 54, 68, 51, 49, 64, 50, 48, 65,
52, 56, 46, 54, 49, 51, 47, 55, 55, 54, 42, 51, 56, 55, 51,
54, 51, 60, 62, 43, 55, 56, 61, 52, 69, 64, 46, 54, 47

16. AUTOMOTIVE A service station tracks the number of cars they service per day.

Cars Serviced
40, 47, 37, 42, 46, 31, 50, 41, 17, 43, 36, 45, 21, 43, 45, 23, 49, 50,
48, 26, 42, 46, 35, 52, 27, 51, 31, 44, 35, 27, 46, 39, 33, 50, 45, 50

a. Use a graphing calculator to construct a histogram for the data, and describe the shape of the distribution.

b. Describe the center and spread of the data using either the mean and standard deviation or the five-number summary. Justify your choice.

17. COMMUTE The number of miles that Armando drove each week during a 15-week period is shown.

Distance (miles)
62, 110, 92, 430, 73, 84, 525,
123, 86, 290, 114, 98, 103, 312, 71

    a. Use a graphing calculator to construct a box plot. Describe the center and spread of the data.

    b. Armando visited four colleges during this period, and these visits account for the four highest weekly totals. Remove these four values from the data set. Use a graphing calculator to construct a box plot that reflects this change. Then describe the center and spread of the new data set.

    c. Calculate and compare the mean and median for the original data set to the mean and median for the data set from **part b**.

18. ELEVATION The table contains data about 10 elevations in the United States.

    a. Use a graphing calculator to construct a box plot for the data, and describe the shape of the distribution.

    b. Describe the center and spread of the data using either the mean and standard deviation or the five-number summary. Justify your choice.

    c. If there is an outlier, describe its effect on the statistics.

Elevations in the US	
Mt McKinley, AK	20,237
Mt Whitney, CA	14,494
Mt Elbert, CO	14,433
Mt Rainier, WA	14,410
Gannett Peak, WY	13,804
Mauna Kea, HI	13,796
Kings Peak, UT	13,528
Wheeler Peak NM	13,161
Boundary Peak, NV	13,140
Granite Peak, MT	12,799

19. USE A MODEL The histograms show the weight of sample boxes of two brands of pasta.

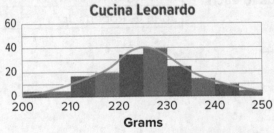

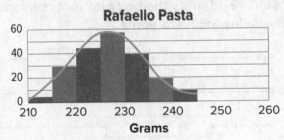

    a. Do the two packages of pasta likely have the same advertised weight? Which manufacturer's quantity control appears better? Explain your answers based on the distributions.

    b. Infer the two population distribution shapes by analyzing the smooth curves across the tops of the histograms. Describe the shapes you observe.

20. STRUCTURE The United States has been sending astronauts up in the Space Shuttle since 1981. The table provides data regarding the duration of Space Shuttle flights from 1981 to 1985, and then from 2005 to 2011.

Length of Flights from 1981–1985 (days)

Days: 2, 2, 8, 7, 5, 5, 6, 6, 10, 8, 7, 6, 8, 8, 3, 7,
7, 7, 8, 7, 4, 7, 7

Length of Flights from 2005–2011 (days)

Days: 14, 13, 12, 13, 14, 13, 15, 13, 16, 14, 15,
13, 13, 16, 14, 11, 14, 15, 12, 13, 16, 13

Choose and calculate the statistics appropriate for the distribution of the data sets. Use the statistics to compare the two sets.

21. REASONING Gerardo live streams with 15 of his friends. Most of his streams have lasted 10-15 days so far, however he has two streams that have lasted 93 days. Describe what Gerardo's data distribution would look like currently and how it would be affected if he lost his longest streams.

22. CONSTRUCT ARGUMENTS Examine the two box plots shown. Without knowing the data points but assuming the same scale, what conclusion can be made? Justify your argument.

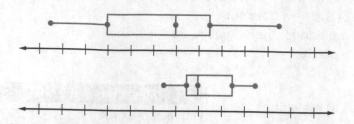

23. SUPREME COURT The table gives the ages of the Supreme Court Justices in 2017.

a. Use a graphing calculator to construct a histogram for the data, and describe the shape of the distribution.

b. Describe the center and spread of the data using appropriate statistics. Justify your choice.

c. If there is an outlier, describe its effect on the statistics.

Supreme Court Justices	
Neil Gorsuch	49
Elena Kagan	57
Sonia Sotomayor	62
Samuel Anthony Alito	67
Stephen G. Breyer	78
Ruth Bader Ginsburg	84
Clarence Thomas	68
Anthony M Kennedy	80
John G. Roberts Jr.	62

24. PERSEVERE Identify the box plot that corresponds to each of the following histograms.

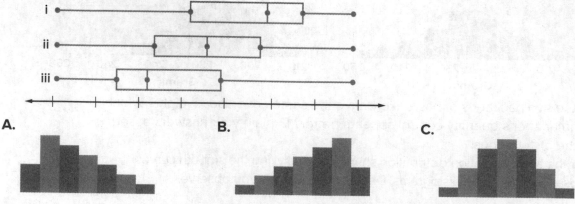

25. ANALYZE Research and write a definition for a *bimodal distribution*. How can the measures of center and spread of a bimodal distribution be described?

26. CREATE Give an example of a set of real-world data with a distribution that is symmetric and one with a distribution that is not symmetric.

27. WRITE Explain why the mean and standard deviation are used to describe the center and spread of a symmetrical distribution and the five-number summary is used to describe the center and spread of a skewed distribution.

Comparing Sets of Data

Today's Goal
• Describe the effects that linear transformations have on measures of center and spread.

Today's Vocabulary
linear transformation

Explore Transforming Sets of Data by Using Addition

Online Activity Use graphing technology to complete the Explore.

> **INQUIRY** How can you find the measures of center and spread of a set of data that has been transformed using addition?

Explore Transforming Sets of Data by Using Multiplication

Online Activity Use graphing technology to complete the Explore.

> **INQUIRY** How can you find the measures of center and spread of a set of data that has been transformed using multiplication?

Learn Linear Transformations of Data

A **linear transformation** is one or more operations performed on a set of data that can be written as a linear function. Common linear transformations are adding a constant to or multiplying a constant by every value in the set of data.

Key Concept • Linear Transformations of Data	
Transformations Using Addition	**Transformations Using Multiplication**
A real number k is added to every value in a set of data, $k \neq 0$.	Every value in a set of data is multiplied by a constant k, $k > 0$.
Measures of Center	
The mean, median, and mode of the new set of data can be found by adding k to the mean, median, and mode of the original set of data.	The mean, median, and mode of the new set of data can be found by multiplying each original statistic by k.
Measures of Spread	
The range and standard deviation of the new set of data will be unchanged.	The range and standard deviation of the new set of data can be found by multiplying each original statistic by k.

🐸 **Think About It!**

Compare and contrast the two methods.

🐸 **Think About It!**

Use a calculator and the data set to examine the effect of multiplying by a constant k, $k < 0$. How can the mean, median, mode, range, and standard deviation of the new data set be found?

Example 1 Transformations Using Addition

Find the mean, median, mode, range, and standard deviation of the data set obtained after adding 6 to each value.

$$8, 11, 3, 6, 15, 3, 5, 7, 14, 3, 5, 4$$

Method 1 Add k to the measures of center and spread of the original set of data.

Find the mean, median, mode, range, and standard deviation of the original data set.

Mean _____	Mode _____	Standard Deviation _____
Median _____	Range _____	

Add 6 to the mean, median, and mode. The range and standard deviation are unchanged.

Mean _____	Mode _____	Standard Deviation _____
Median _____	Range _____	

Method 2 Add k to each data value of the original set of data.
Add 6 to each data value.

$$\text{\_\_\_\_}, 17, 9, 12, 21, \text{\_\_\_\_}, 11, 13, \text{\_\_\_\_}, 9, 11, \text{\_\_\_\_}$$

Mean _____	Mode _____	Standard Deviation _____
Median _____	Range _____	

Example 2 Transformations Using Multiplication

Find the mean, median, mode, range, and standard deviation of the data set obtained after multiplying each value by 4.

$$12, 18, 20, 12, 14, 18, 11, 21, 13, 18, 11, 24$$

Find the measures of center and spread for the original data set.

Mean	Median	Mode	Range	Standard Deviation
16	16	18	13	4.2

Multiply the measures of center and spread by 4.

Mean	Median	Mode	Range	Standard Deviation
_____	64	72	_____	_____

Check

Find the mean, median, mode, range, and standard deviation of the data set obtained after multiplying each value by 0.6. Round to the nearest tenth, if necessary.

$$45, 33, 43, 51, 39, 48, 34, 39, 30, 39, 47, 44$$

🔵 **Go Online** You can complete an Extra Example online.

🌐 Example 3 Compare Symmetric Distributions of Data

RESTAURANTS The numbers of customers eating at a restaurant during breakfast, lunch, and dinner each day are shown below.

Breakfast: 71, 58, 65, 48, 44, 56, 68, 64, 51, 67, 74, 62, 59, 53, 62, 73, 54, 49, 63, 55

Lunch: 115, 105, 87, 108, 117, 110, 92, 101, 114, 91, 109, 96, 100, 98, 103, 111, 95, 94, 102, 106

Dinner: 76, 62, 91, 76, 79, 68, 65, 89, 81, 76, 90, 82, 79, 74, 71, 73, 84, 87, 81, 64

Part A Construct a histogram or box plot for each set of data. Then describe the shape of each distribution.

Method 1 Histogram

Enter the data in **L1**, **L2**, and **L3**. From the **STAT PLOT** menu, enter **L1** as the **Xlist** for Plot 1, **L2** for Plot 2, and **L3** for Plot 3. Select 📊 as the plot type for each Plot. View each histogram by turning on Plot 1, Plot 2, and then Plot 3. Use the same window dimensions and bin width for each graph.

Breakfast

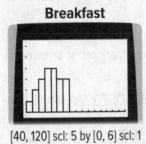

[40, 120] scl: 5 by [0, 6] scl: 1

Lunch

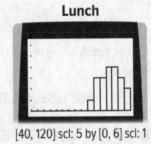

[40, 120] scl: 5 by [0, 6] scl: 1

Dinner

[40, 120] scl: 5 by [0, 6] scl: 1

For each time of day, the distribution is high in the middle and low on the left and right. Therefore, all of the distributions are _____.

Method 2 Box Plot

Enter the data using the same process. Select 📦 as the plot type for each set of data. To view all of the box plots at once, turn on Plot 1, Plot 2, and Plot 3 and graph.

For each time of day, the lengths of the whiskers are approximately equal, and the median is in the middle of the data. The left and right sides are approximately mirror images of one another. Therefore, all of the distributions are _____.

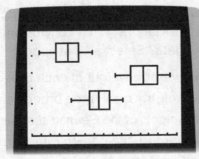

[40, 120] scl: 5 by [0, 6] scl: 1

(continued on the next page)

🌐 **Go Online** You can complete an Extra Example online.

Study Tip

Window and Bin Settings When setting the window dimensions for multiple sets of data, try setting the minimum and maximum as the least and greatest values of all the sets. When selecting a bin width, consider the context of the situation. For example, if the data does not include fractional numbers, as would be the case with number of people, use a whole number as the bin width.

Copyright © McGraw-Hill Education

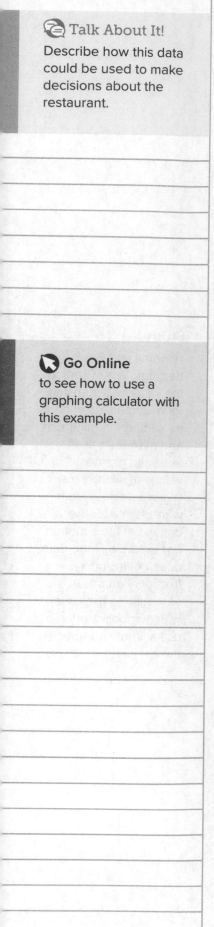

Part B Compare the data sets using the means and standard deviations.

All of the distributions are symmetric, so use the _____

and _____ to describe the centers and spreads.

Breakfast

```
1-Var Stats
 x̄=59.8
 Σx=1196
 Σx²=72910
 Sx=8.550777127
 σx=8.334266614
↓n=20
```

Lunch

```
1-Var Stats
 x̄=102.7
 Σx=2054
 Σx²=212346
 Sx=8.58456387
 σx=8.367197858
↓n=20
```

Dinner

```
1-Var Stats
 x̄=77.4
 Σx=1548
 Σx²=121218
 Sx=8.592530416
 σx=8.374962686
↓n=20
```

The means vary, with breakfast having the lowest average number of

customers and lunch having the highest average number of customers.

However, the standard deviations are approximately _____.

This means that, while the _____ for each time of

day is very different, the _____

generally _____ from day to day.

Check

DOGS The weights, in pounds, for a sample of the three most popular breeds of dogs are shown below.

Labrador Retriever: 75, 59, 63, 68, 67, 59, 69, 63, 60, 76, 70, 74, 67, 68, 71, 65, 62, 74, 66, 78

German Shepherd: 53, 61, 58, 74, 85, 80, 72, 57, 64, 69, 81, 75, 73, 64, 76, 68, 66, 51, 67, 73

Golden Retriever: 62, 59, 67, 72, 64, 67, 69, 76, 63, 64, 73, 69, 71, 75, 59, 64, 69, 59, 74, 68

Part A Use a graphing calculator to construct a histogram or box plot for each set of data. Then complete the statement about the shape of each distribution.

All of the distributions are _____.

Part B Compare the data sets using the means and standard deviations. What conclusion(s) can you make about the sets of data? Select all that apply. _____

A. The average weight of each breed is about the same.

B. The weights of all three breeds are very close to their means.

C. The weights of the German shepherds vary more than the other breeds.

D. On average, the golden retrievers weigh much more than the other breeds.

E. The means of the weights differ by less than 1.5 pounds.

F. The weights of the Labrador retrievers and golden retrievers are generally closer to their means than the German shepherds' weights are to their mean.

🌐 Example 4 Compare Skewed Distributions of Data

SPORTS **The numbers of high school boys and girls, in hundred thousands, participating in tennis from 2001–2015 are shown below.**

Boys (hundred thousands)	Girls (hundred thousands)
144, 139, 145, 153, 149, 153, 157, 156, 157, 163, 161, 160, 157, 161, 157	164, 160, 163, 168, 169, 174, 177, 172, 178, 182, 182, 181, 181, 184, 183

Part A Construct a histogram or box plot for each set of data. Then describe the shape of each distribution.

Method 1 Histogram

Enter the data in **L1** and **L2**. From the **STATPLOT** menu, enter **L1** as the **Xlist** for Plot 1 and **L2** for Plot 2. Select 📊 as the plot type for each Plot. View each histogram by turning on Plot 1, and then Plot 2. Use the same window dimensions and bin width for each graph.

Boys

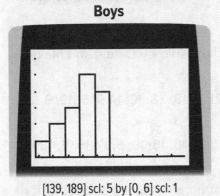

[139, 189] scl: 5 by [0, 6] scl: 1

Girls

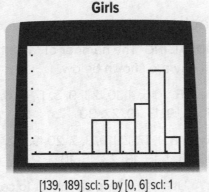

[139, 189] scl: 5 by [0, 6] scl: 1

Both distributions are _____ on the _____ and have _____ on the _____. Therefore, both distributions are _____.

Method 2 Box Plot

Enter the data using the same process. Select 📦 as the plot type for each set of data. To view both box plots at once, turn on Plot 1 and Plot 2 and graph.

For each distribution, the left whisker is longer than the right, and the median is closer to the right whisker. Therefore, both distributions are

_____.

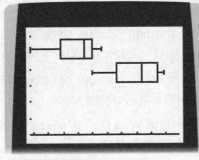

[139, 189] scl: 5 by [0, 6] scl: 1

Part B Compare the data sets using the five-number summaries.

Both distributions are skewed, so use the five-number summary to compare the data.

(continued on the next page)

🌐 **Go Online** You can complete an Extra Example online.

Go Online
to see how to use a
graphing calculator with
this example.

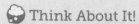 **Think About It!**
Compare two other
statistics from the five-
number summaries.
What does this tell you
about the number of
girls and boys that
participated in tennis?

The upper quartile for the number of
boys that participated in tennis is 160,
while the minimum number of girls that
participated is 160. This means there
were only 160,000 or more boys
participating in tennis for 25% of the
years, while at least 160,000 girls
participated every year.

We can conclude that many more girls
participated in tennis from 2001 to 2015
than boys.

Boys

1-Var Stats
↑n=15
 minX=139
 Q₁=149
 Med=157
 Q₃=160
 maxX=163

Girls

1-Var Stats
↑n=15
 minX=160
 Q₁=168
 Med=177
 Q₃=182
 maxX=184

Check

FUNDRAISING The number of raffle tickets sold by Darius and Makya
each day are shown below.

Darius: 5, 1, 15, 4, 10, 23, 9, 3, 17, 2, 6, 21, 5, 13, 28, 10, 14, 7, 5, 19, 9, 22,
10, 8, 15, 9, 13, 19, 22, 30

Makya: 18, 1, 17, 10, 19, 3, 7, 20, 9, 22, 12, 13, 16, 18, 16, 5, 17, 15, 6, 11, 18,
14, 16, 18, 1, 16, 18, 23, 15, 10

Part A Use a graphing calculator to construct a histogram or box plot for
each set of data. Then complete the statement about the shape of
each distribution.

The distribution of Darius' raffle ticket sales is _____.

The distribution of Makya's raffle ticket sales is _____.

Part B Compare the data sets using the five-number summaries. What
conclusion(s) can you make about the sets of data? Select all
that apply. _____

A. The median number of tickets Darius sold is much higher than the
median number of tickets Makya sold.

B. The median number of tickets Darius sold is the same as the lower
quartile of Makya's sales.

C. The data from Darius' sales is spread over a wider range than the
data from Makya's sales.

D. The median number of tickets each student sold was the same.

E. The fewest number of tickets each student sold in a day was 1.

F. The upper 50% of Darius' data spans from 10 to 30, while the upper
75% of Makya's data spans from 10 to 23.

 Go Online You can complete an Extra Example online.

Practice

🢅 **Go Online** You can complete your homework online.

Example 1

Find the mean, median, mode, range, and standard deviation of each data set that is obtained after adding the given constant to each value.

1. 52, 53, 49, 61, 57, 52, 48, 60, 50, 47; +8

2. 101, 99, 97, 88, 92, 100, 97, 89, 94, 90; +(−13)

3. 27, 21, 34, 42, 20, 19, 18, 26, 25, 33; +(−4)

4. 72, 56, 71, 63, 68, 59, 77, 74, 76, 66; +16

Example 2

Find the mean, median, mode, range, and standard deviation of each data set that is obtained after multiplying each value by the given constant.

5. 11, 7, 3, 13, 16, 8, 3, 11, 17, 3; ×4

6. 64, 42, 58, 40, 61, 67, 58, 52, 51, 49; ×0.2

7. 33, 37, 38, 29, 35, 37, 27, 40, 28, 31; ×0.8

8. 1, 5, 4, 2, 1, 3, 6, 2, 5, 1; ×6.5

Examples 3 and 4

9. BASEBALL The total wins per season for the first 17 seasons of the Marlins are shown. The total wins over the same time period for the Cubs are also shown.

Marlins
64, 51, 67, 80, 92, 54, 64, 79, 76, 79, 91, 83, 83, 78, 71, 84, 87

Cubs
84, 49, 73, 76, 68, 90, 67, 65, 88, 67, 88, 89, 79, 66, 85, 97, 83

a. Use a graphing calculator to construct a box plot for each set of data. Then describe the shape of each distribution.

b. Compare the data sets using either the means and standard deviations or the five-number summaries. Justify your choice.

10. HEALTH CLUBS To plan their future equipment purchases, the Northville Health Club randomly chooses 8 patrons and tracks how many minutes they spend on the treadmill.

a. Use a graphing calculator to construct a histogram for each set of data. Then describe the shape of each distribution.

b. Compare the data sets using either the means and standard deviations or the five-number summaries. Justify your choice.

Minutes on Treadmill Last Week	Minutes on Treadmill This Week
30	20
30	30
45	45
20	45
60	30
30	60
30	50
45	45

Mixed Exercises

Find the mean, median, mode, range, and standard deviation of each data set that is obtained after adding or multiplying each value by the given constant(s).

11. 98, 95, 97, 89, 88, 95, 90, 81, 87, 95; +2

12. 32, 30, 27, 29, 25, 33, 38, 26, 23, 31; ×1.6

13. 14, 17, 13, 9, 15, 7, 12, 16, 8, 9; ×5

14. 5, 12, 7, 3, 8, 5, 7, 1, 4, 7, 3, 9; +22

15. 12, 15, 16, 12, 12, 15, 17, 19, 22, 27, 42, 42; +5

16. 49, 43, 26, 39, 40, 30, 33, 64, 26, 45, 23, 26; ×3, +(−8)

17. 71, 72, 68, 70, 72, 67, 68, 72, 65, 70; ×0.2

18. 112, 91, 108, 129, 80, 99, 78, 80; +(−15)

19. 57, 38, 42, 51, 39, 44, 33, 55; +(−7), ×2

20. 55, 50, 58, 52, 56, 57, 50, 55, 50; ×2, +5

21. BOWLING The scores of 15 bowlers are shown in the table.

Score
211, 123, 183, 176, 224, 115, 109, 136, 152, 177, 127, 196, 143, 166, 170

a. Find the mean, median, mode, range, and standard deviation of the scores.

b. The handicap of the bowling team will add 56 points to each score. Find the statistics of the scores while including the handicap.

22. COMPETITION The distances that 18 participants threw a football are shown in the table.

Distance (feet)
96, 94, 114, 85, 96, 109, 90, 109, 67, 82, 98, 79, 69, 70, 106, 96, 112, 84

a. Find the mean, median, mode, range, and standard deviation of the participants' distances.

b. Find the statistics of the participants' distances in yards.

23. TEMPERATURE The monthly average high temperatures for Lexington, Kentucky, are shown in the table.

Temperature (°F)
40, 45, 55, 65, 74, 82, 86, 85, 78, 67, 55, 44

a. Find the mean, median, mode, range, and standard deviation of the temperatures.

b. Find the statistics of the temperatures in degrees Celsius. Recall that $C = \frac{5}{9}(F - 32)$.

24. FANTASY SPORTS The weekly total points of Scott's and Azumi's fantasy baseball teams are shown in the tables.

Scott's Team
109, 99, 121, 137, 131, 141, 77,
83, 139, 92, 42, 133, 98, 153, 124,
102, 113, 117, 112, 128, 107, 147

Azumi's Team
113, 121, 98, 104, 106, 123, 175,
141, 109, 129, 49, 110, 112, 144,
106, 119, 127, 88, 132, 93, 137, 123

a. Use a graphing calculator to construct a box plot for each set of data. Then describe the shape of each distribution.

b. Compare the data sets using either the means and standard deviations or the five-number summaries. Justify your choice.

c. How does eliminating the outliers of each data set affect the statistics and comparison from **part b**?

25. BUSINESS Saeed owns an electronics store. He is revising his pricing for phone accessories. His current prices for an assortment of accessories are listed at the right. He has also determined that the mean price for the same assortment of accessories at a rival store is $10.99.

Saeed's Price Data ($)		
14.99	4.49	9.99
18.49	12.99	6.99
8.49	21.99	13.49
13.99	9.99	10.99
12.49	4.49	12.99

a. Saeed wants to match his rival's prices. Make a table to list the new prices. Explain.

b. Compare the mean and standard deviation of the current prices to the new prices.

26. REASONING Two different samples on the shell diameter of a species of snail are shown.

Sample A (mm)		
45	35	37
40	42	40
28	38	31

Sample B (mm)		
26	44	40
27	35	28
26	39	31

a. Use the median and interquartile range to compare the samples.

b. Based on your findings and on the data points in each sample, which sample appears to be more representative? Explain your reasoning.

27. STRUCTURE Height data samples of 17-year-old male and female students are shown. Use the mean and standard deviation to compare the samples.

Heights of Male Students (inches)		
71	69	67
68	69	70
72	74	68
71	69	72

Heights of Female Students (inches)		
67	62	69
65	71	66
63	65	68
66	63	70

Copyright © McGraw-Hill Education

28. CONSTRUCT ARGUMENTS Francisca is planning a two-week vacation to one of two cities and wants to base her decision on the weather history for the same dates as her vacation. She has collected the number of days that it has rained during this two-week period for each city over the past 10 years. The results are shown.

City A		City B	
5	0	4	4
7	6	6	5
5	6	3	7
6	6	4	3
3	2	5	7

a. Determine the shape of each distribution, and use the appropriate statistics to find the center and spread for each set of data.

b. Which city do you think Francisca should visit on her vacation? Justify your argument.

29. WRITE Compare and contrast the benefits of displaying data using histograms and box plots.

30. ANALYZE If every value in a set of data is multiplied by a constant k, $k < 0$, then how can the mean, median, mode, range, and standard deviation of the new data set be found?

31. PERSEVERE A salesperson has 15 SUVs priced between $33,000 and $37,000 and 5 luxury cars priced between $44,000 and $48,000. The average price for all of the vehicles is $39,250. The salesperson decides to reduce the prices of the SUVs by $2000 per vehicle. What is the new average price for all of the vehicles?

32. ANALYZE If k is added to every value in a set of data, and then each resulting value is multiplied by a constant m, $m > 0$, how can the mean, median, mode, range, and standard deviation of the new data set be found? Justify your argument.

33. WRITE Explain why the mean and standard deviation are used to compare the center and spread of two symmetrical distributions, and the five-number summary is used to compare the center and spread of two skewed distributions or a symmetric distribution and a skewed distribution.

Summarizing Categorical Data

Explore Categorical Data

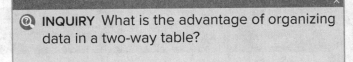 **Online Activity** Use a real-world situation to complete the Explore.

> **@ INQUIRY** What is the advantage of organizing data in a two-way table?

Learn Two-Way Frequency Tables

A **two-way frequency table** or *contingency table* is used to show the frequencies of data from a survey or experiment classified according to two categories, with the rows indicating one category and the columns indicating the other.

Suppose you are constructing a two-way frequency table based on two categories, grade level and employment. The table is constructed below for sample values.

Grade	Employed	Unemployed	Totals
Junior	8	12	20
Senior	15	10	25
Totals	23	22	45

Subcategories: The subcategories are the column and row headers that represent the two different types of categories. In this case, Employed, Unemployed, Junior, and Senior are the subcategories.

Joint frequencies: **Joint frequencies** are the values for every combination of subcategories. So, 8 is a joint frequency that represents the number of students who are employed and juniors.

Marginal frequencies: **Marginal frequencies** are the totals of each subcategory. So, 20 is a marginal frequency that represents the total number of juniors.

Today's Goals
- Organize categorical data in a two-way frequency table.
- Determine and interpret the values in a two-way relative frequency table.

Today's Vocabulary
two-way frequency table

joint frequencies

marginal frequencies

relative frequency

two-way relative frequency table

conditional relative frequency

Use a Source

Create your own two-way frequency table. Find data online that divides a group of subjects into two categories, with each subject fitting into one subcategory of each. For example, in the data shown the categories are whether each person is male or female and whether each person's name is Casey or Riley. Determine the subcategories, enter the given data, and fill in any cells for which values are not provided.

🌐 **Example 1** Use a Two-Way Frequency Table

NAMES **Unisex names are names often used for both males and females. At one point, the most common unisex names in the U.S. were Casey and Riley, with 176,544 Caseys and 154,861 Rileys. During that time, there were 104,161 males with the name Casey and 75,882 females with the name Riley. Organize the data in a two-way frequency table.**

Steps 1 and 2 Enter the given data in a table. Then use the information given to fill in the rest of the cells.

Top Unisex Names in the U.S.			
	Casey	**Riley**	**Totals**
Male	104,161		
Female		75,882	
Totals	176,544	154,861	

Male Rileys: 154,861 − 75,882 = 78,979

Total Males: 104,161 + 78,979 = 183,140

Female Caseys: 176,544 − 104,161 = 72,383

Total Females: 72,383 + 75,882 = 148,265

Totals: W176,544 + 154,861 = 331,405

Check

TECHNOLOGY Pew Research Center released a survey that asked whether participants thought technological advancements in the future will make people's lives better or worse. Of the people interviewed, 423 earned less than $50,000 per year and 328 earned $50,000 or more. Of those earning less than $50,000 per year, 262 thought that people's lives would get better, and 240 of those who earned $50,000 or more thought the same. Organize the data in a two-way frequency table.

Will technological advancements in the future make people's lives better or worse?			
	Better	**Worse**	**Totals**
< $50,000			
≥ $50,000			
Totals			

🌐 **Go Online** You can complete an Extra Example online.

Learn Two-Way Relative Frequency Tables

A **relative frequency** is the ratio of the number of observations in a category to the total number of observations. A **two-way relative frequency table** can help you see patterns of association in the data. To create a two-way relative frequency table, divide each of the values by the total number of observations and replace them with their corresponding decimals or percents.

A **conditional relative frequency** is the ratio of the joint frequency to the marginal frequency. Because each two-way frequency table has two categories, each two-way relative frequency table can provide two different conditional relative frequency tables.

🌐 Example 2 Use a Two-Way Relative Frequency Table

PARENTING Many parents monitor their teenagers' Internet usage. The Pew Research Center conducted a survey of whether parents do or do not check what sites their teens had visited and whether they are the parent of a teen between the ages of 13 and 14 or between the ages of 15 and 17. The results of the survey are shown. Organize the data in a relative frequency table by age group, and interpret the data.

How Parents Monitor Teenagers' Internet Usage			
	Does Check	Does Not Check	Totals
13 to 14	299	140	439
15 to 17	348	273	621
Totals	647	413	1060

Part A Organize the data in a relative frequency table.

How Parents Monitor Teenagers' Internet Usage			
	Does Check	Does Not Check	Totals
13 to 14	$\frac{299}{1060} \approx 28.2\%$	$\frac{140}{1060} \approx 13.2\%$	$\frac{439}{1060} \approx 41.4\%$
15 to 17	$\frac{348}{1060} \approx 32.8\%$	$\frac{273}{1060} \approx 25.8\%$	$\frac{621}{1060} \approx 58.6\%$
Totals	$\frac{647}{1060} \approx 61.0\%$	$\frac{413}{1060} \approx 39.0\%$	$\frac{1060}{1060} = 100\%$

Part B Interpret the data.

Do more parents check what sites their teens have visited, or do more parents not check?

____% of parents do check the sites their teens have visited compared to ____% who do not.

> 🧠 **Think About It!**
>
> Based on the data, do you think there is an association between a teen's age and whether their parents check their Internet usage? Explain.

Copyright © McGraw-Hill Education

If the condition for the relative frequency table were whether a person had voted or not rather than age group, the joint frequency for people between the ages of 18 and 24 who voted would be 52.5%. If someone claims that this indicates 52.5% of all people who voted in the 2012 election were between the ages of 18 and 24, would they be correct? Justify your argument.

Example 3 Use a Two-Way Conditional Relative Frequency Table

VOTING According to the U.S. Census Bureau, voter turnout describes how many eligible voters show up to vote in an election. The table shows the number of eligible voters who did and did not vote in 2012 for the oldest and youngest eligible age groups. Organize the data in a conditional relative frequency table by age group, and interpret the data.

Voter Turnout			
	Voted	Did Not Vote	Totals
18 to 24	12,515	13,275	25,790
75 and over	11, 344	5380	16,724
Totals	23,859	18,655	42,514

Part A Organize the data in a conditional relative frequency table by age group.

Step 1 Determine which marginal frequencies to use.

The conditional relative frequency relates the number of voters or nonvoters to the age group, so the relevant marginal frequencies are the total numbers of voters for each age group.

Step 2 Determine the ratios of the joint frequencies to the marginal frequencies.

Voter Turnout			
	Voted	Did Not Vote	Totals
18 to 24	$\frac{12,515}{25,790} \approx 48.5\%$	$\frac{13,275}{25,790} \approx$ ____ %	$\frac{25,790}{25,790} = 100\%$
75 and over	$\frac{11,344}{16,724} \approx$ ____ %	$\frac{5380}{16,724} \approx 32.2\%$	$\frac{16,724}{16,724} = 100\%$

Part B Interpret the data.

Which age group has the higher voter turnout?

The percent of eligible voters aged 18 to 24 that voted is ____%, and the percent for those aged 75 and over is ____%. Based on the data, there is an association between age and whether a person voted. People aged 18 to 24 were more likely to not have voted than people aged 75 and over.

Go Online You can complete an Extra Example online.

Practice

Example 1

TREATS **The owner of a snow cone stand keeps track of the sizes and flavors sold one afternoon. He sold 125 snow cones in all. Of these, 40% were large snow cones, 32% were grape, and 12% were small watermelon snow cones. The stand sold 15 more cherry snow cones than grape. The most popular snow cone of the day was small cherry, with a total of 35 sales.**

1. Construct a two-way frequency table to organize the data.

2. How many large grape snow cones were sold?

3. How many watermelon snow cones were sold in all?

4. How many more small snow cones were sold than large snow cones?

Example 2

FOREIGN LANGUAGE **Christy surveyed several students at her school and asked each person what foreign language he or she is studying. The results are shown in the table.**

	Male	Female	Total
Spanish	18	20	38
French	16	12	28
German	6	8	14
Total	40	40	80

5. Construct a relative frequency table by converting the data in the table to percentages. Round to the nearest tenth, if necessary.

6. Find the joint relative frequency of a female student who is studying French.

7. Interpret the data.

Example 3

CLASS PRESIDENT **In a poll for senior class president, 68 of the 145 male students said they planned to vote for Santiago. Out of 139 female students, 89 planned to vote for his opponent, Measha.**

8. Construct a conditional relative frequency table based on voter preference. Show your calculations.

9. What does each conditional relative frequency represent?

10. What is the probability that a vote for Measha will come from a female student? How is this different from the probability that a female student intends to vote for Measha?

Mixed Exercises

VETERINARIAN The two-way frequency table shows the number of dogs and cats that were seen at a veterinarian's office and the primary purpose of their visit.

	Dog	Cat	Total
Exam	12	5	17
Shots	6	3	9
Grooming	7	2	9
Total	25	10	35

11. How many dogs were seen for an exam today?

12. How many more dogs than cats were seen at the veterinarian's office?

BIRD WATCHING A group of bird-watchers has been tracking the number of tree swallows, cardinals, and goldfinches in a region. Over the weekend, a total of 40 birds were observed. Of those, 45% were male, 37.5% were cardinals, and 12.5% were male tree swallows. Twice as many female cardinals were observed as male cardinals. There were 5 female goldfinches spotted.

13. Construct a two-way frequency table to organize the data.

14. How many more female tree swallows were seen than male cardinals?

15. How many male goldfinches and female cardinals were seen?

16. How many more female birds were seen than male birds?

SCHOOL ACTIVITIES The two-way frequency table shows the number of students who participate in school sports or clubs at Monroe High School.

	Sports or Clubs	No Sports or Clubs	Total
Freshmen	48	60	108
Sophomores	60	72	132
Juniors	51	69	120
Seniors	57	63	120
Total	216	264	480

17. Construct a relative frequency table by converting the data in the table to percentages. Round to the nearest tenth, if necessary.

18. Find the joint relative frequency of a sophomore who participates in school sports or clubs.

19. What percentage of freshmen do not participate in school sports or clubs? Round to the nearest tenth percent, if necessary.

20. What percentage of seniors participate in school sports or clubs? Round to the nearest tenth percent, if necessary.

SCHOOL MASCOT **The freshmen and sophomores at Lakeview High School are tasked with adopting a new school mascot next school year. The district asked a representative group of students to vote for one of the three mascot finalists and to indicate to which grade they belong. The results are shown in the table.**

School Mascot Vote Results			
	Freshmen	Sophomores	Total
Panthers	30	36	66
Hornets	17	33	50
Lions	28	31	59
Total	75	100	175

21. How many students voted for Panthers?

22. How many students voted for Lions?

23. How many sophomores were in the representative group?

24. Of the students who voted for Hornets, how many of them are freshmen?

25. Of the students who voted for Lions, how many of them are sophomores?

26. To the nearest whole, what percent of all the students voted for Lions?

27. To the nearest whole, what percent of all the students voted for Panthers?

28. To the nearest whole, what percent of all the students who voted were freshmen?

THANKSGIVING PIE **An online poll collected a sample of Thanksgiving pie preferences for different U.S. regions.**

	Apple	Sweet Potato	Pumpkin	Totals
West	77	4	13	
Midwest	32		54	
South		63	24	
Northeast	92	2		
Total	213	75	117	

29. PRECISION Complete the table. Then find each relative frequency to the nearest tenth of a percent.

30. USE A MODEL Assuming the poll is representative of the whole population, what is a reasonable estimate of the probability that a family will be from the northeast and will be eating pumpkin pie on Thanksgiving?

31. STRUCTURE Construct a table of conditional relative frequencies based on pie preference. Round each percent to the nearest tenth. Interpret the meaning of the probabilities in the context of the problem.

32. REGULARITY If we had found the conditional relative frequencies by dividing by the total replies from each region, what would be the meaning of the probability in each cell?

VEHICLES The table shows the relative frequencies of drive systems for different vehicle types in a school parking lot. There are 215 vehicles in the lot.

	2WD	AWD	Totals
Hatchbacks	42%	4%	
Sedans	28%	6%	
SUVs	1%	19%	
Total			215

33. **USE TOOLS** Construct a table to show the joint and marginal frequencies.

34. **REASONING** Without calculating individual frequencies, how many times greater will the conditional relative frequencies based on drive systems for AWD be than the relative frequencies for AWD, and why?

35. **PERSEVERE** Len conducted a survey among a random group of 1000 families in his home state of California. He wanted to determine whether there is an association between gasoline prices and distances traveled on family vacations. He collected the following information. According to Len's two-way frequency table, does there appear to be an association between gasoline prices and vacation distances traveled? Explain.

	$1.75–$3.24 per gallon	$3.25-$4.74 per gallon	Total
Less than 250 miles	109	255	364
More than 250 miles	329	86	415
No vacation travel	34	187	221
Total	472	528	1000

36. **CREATE** Select your own data for a two-way frequency table, write a question related to the data in the table, and provide the solution.

37. **WRITE** Compare two-way relative frequency tables and two-way conditional relative frequency tables.

38. **FIND THE ERROR** Magdalena took a survey of students in her school to find out what snack was most popular.

Favorite Snack Vote Results			
	Freshmen	Sophomores	Total
Fruit Snack	65	61	126
Granola	27	21	48
Yogurt	21	18	39
Total	113	100	213

a. Interpret the data based on the conditional relative frequency related to age groups.

b. Magdalena claims that fruit snack is the most popular snack for freshmen and sophomores, and Ben claims that a higher percentage of sophomores prefer fruit snack than do freshmen. Is either correct? Explain your reasoning.

ⓔ Essential Question
How do you summarize and interpret data?

Module Summary

Lessons 12-1 and 12-4

Measures of Center and Spread

- The mean of a data set is the sum of the elements of the data set divided by the total number of elements in the set.

- The median of a data set is the middle element or the mean of the two middle elements in the set of data when the data are arranged in numerical order.

- The mode of a data set is the value of the elements that appear most often in the set of data.

- The formula for standard deviation, with mean $\overline{x}$ and n terms is

$$\sigma = \sqrt{\frac{(\overline{x} - x_1)^2 + (\overline{x} - x_2)^2 + \ldots + (\overline{x} - x_n)^2}{n}}.$$

Lessons 12-2 and 12-3

Representing and Using Data

- Dot plots, bar graphs, and histograms are commonly used to represent data.

- Bar graphs are used with discrete data, and histograms are used with continuous data.

- A population is all members of a group of interest about which data will be collected. A sample is a subset of the population.

- A bias is an error that results in a misrepresentation of a population.

Lesson 12-5

Distributions of Data

- In a symmetric distribution, the mean and median are approximately equal.

- A negatively skewed distribution typically has a median greater than the mean. A positively skewed distribution typically has a mean greater than the median.

- An outlier is a value that is more than 1.5 times the interquartile range above the third quartile or below the first quartile.

Lesson 12-6

Comparing Sets of Data

- A linear transformation is one or more operations performed on a set of data that can be written as a linear function.

- Common linear transformations are adding a constant to or multiplying a constant by every value in the set of data.

Lesson 12-7

Two-Way Frequency Tables

- A two-way frequency table shows the frequencies of data classified according to two or more categories.

Study Organizer

📖 Foldables

Use your Foldable to review this module. Working with a partner can be helpful. Ask for clarification of concepts as needed.

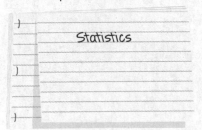

Name _____   Period _____   Date _____

Test Practice

1. GRAPH Make a dot plot of the quiz scores of Ms. Perez's third period class.

Quiz Scores			
85	88	75	100
90	90	88	72
72	79	88	85

(Lesson 12-2)

2. OPEN RESPONSE When is it a good idea to scale the number line when making a dot plot?

(Lesson 12-2)

3. MULTI-SELECT Which of the statements are true regarding dot plots, bar graphs, and histograms? Select all that apply.

Lesson 12-2

(A) Dot plots use a number line and dots to represent very large amounts of data.

(B) Bar graphs are used to represent data that is continuous.

(C) Histograms are used to represent data that is continuous.

(D) Bar graphs are used to represent data that is discrete.

(E) Histograms are used to represent data that is discrete.

4. MULTIPLE CHOICE Which dot plot correctly models these data values?

36, 38, 42, 36, 36, 40, 42, 38, 38, 39, 40, 38, 38, 38, 40 (Lesson 12-2)

(A)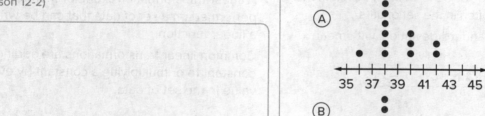

(B)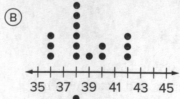

(C)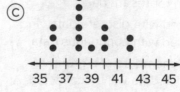

(D)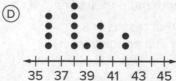

5. OPEN RESPONSE Given the set of data in the table, describe what size intervals could be used when making a histogram. (Lesson 12-2)

Ages of Guests at a Picnic
74, 26, 32, 4, 61, 56, 16, 15, 17, 28, 39, 42, 47, 72, 66, 12, 16, 38, 35, 8, 16, 11, 10, 41, 47, 5, 13, 77, 24, 30, 9, 62

6. GRAPH A survey was conducted among students in Mr. Sadiq's history class to determine their favorite major topic covered in class this semester. The results are shown in the table. Make a bar graph to display the data. (Lesson 12-2)

Topic	Number of Votes
Civil War	12
Revolutionary War	5
The Industrial Revolution	8
Westward Expansion	10

7. OPEN RESPONSE Akeem wants to determine how long it took students in his class to complete a 1-mile run.

Running Time (min)
18.5, 8.4, 10.2, 27.1, 9.5, 10.9, 17.0, 5.3, 6.1, 8.4, 8.4, 9.9, 10.0, 7.4, 8.4

State two types of displays Akeem could use to appropriately display his data. (Lesson 12-2)

8. MULTIPLE CHOICE Which box plot correctly models these data values? 95, 72, 84, 98, 87, 75, 100, 86, 90, 81, 93, 90 . (Lesson 12-4)

Ⓐ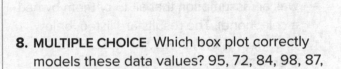
72 75 78 81 84 87 90 93 96 99 102

Ⓑ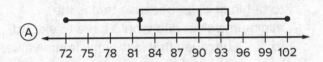
72 75 78 81 84 87 90 93 96 99 102

Ⓒ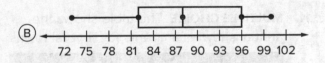
72 75 78 81 84 87 90 93 96 99 102

Ⓓ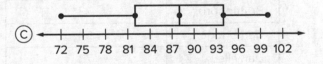
72 75 78 81 84 87 90 93 96 99 102

9. **GRAPH** The table shows the number of pages each student read in one night.

Pages Read
13, 15, 8, 22, 11, 17, 15, 9, 14, 16, 13

Create a box plot to represent the set of data. (Lesson 12-4)

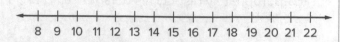

8 9 10 11 12 13 14 15 16 17 18 19 20 21 22

10. **OPEN RESPONSE** Fifteen people in their fifties were surveyed about the number of apps they have on their cell phone. (There was an assumption that all 15 of them owned a cell phone). The results are listed, below.
6, 0, 11, 8, 9, 6, 7, 3, 1, 2, 10, 7, 22, 5, 13
(Lesson 12-4)

A box plot to represent this data would have to begin at _____ because that is the minimum value, and would have to extend to _____ because that is the maximum value.

The most appropriate scale to display the data in the box plot should be _____ or 2.

11. **MULTIPLE CHOICE** The table shows the annual snowfall amounts for several towns.

Town	Snowfall (in.)
Westfield	246
Brattleboro	73
Cambridge	54
Danville	73
Shelburne	86
Lowell	67

Which measure(s) of center and measure(s) of spread best describe the set of data? (Lesson 12-5)

Ⓐ mean

Ⓑ median

Ⓒ standard deviation

Ⓓ five-number summary

12. **OPEN RESPONSE** *True* or *false*: Histogram B has more variability than Histogram A. (Lesson 12-6)

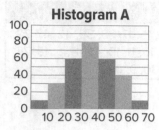
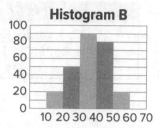

13. **MULTIPLE CHOICE** The junior varsity dance team is selecting the color of their new uniforms. The team consists of 28 freshmen and sophomores. Of the 16 freshmen, 7 want red uniforms and 9 want black uniforms. Only 4 of the sophomores want black uniforms. How many total team members want red uniforms? (Lesson 12-7)

Ⓐ 7

Ⓑ 13

Ⓒ 15

Ⓓ 21

14. **OPEN RESPONSE** The table shows the frequencies of positions for different offensive players on a school football team. There are 38 offensive players on the team.

	Senior	Junior	Sophomore
Quarterback	1	1	0
Running Back	2	1	1
Receiver	3	2	2
Lineman	13	6	6

Suppose a junior player was picked at random. What is the probability that player is a receiver? (Lesson 12-7)

Module 7

Quick Check

1. (4, 0)   **3.** (0, 0)   **5.** $x = 6 - 2y$ **7.** $m = 2n + 6$

Lesson 7-1

1. 1; consistent; independent

3. 0; inconsistent

5. 1; consistent; independent

7. 1; consistent; independent

9. 1; consistent; independent

11. 1 solution; (0, −3)

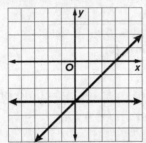

13. no solution

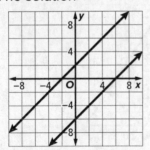

15. infinitely many solutions

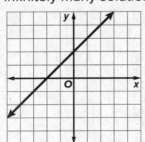

17a. $y = 400x + 1000$; $y = 5900 - 300x$

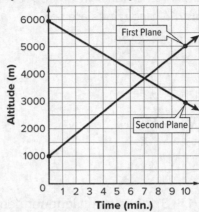

17b. After 7 minutes the planes will be at the same altitude.

19. $y = 3x + 6$ and $y = 6$; (0, 6)

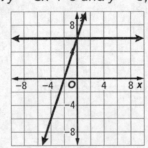

21. $y = -12x + 90$ and $y = 30$; (5, 30)

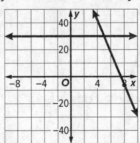

23. $y = 2x + 5$ and $y = 2x + 5$; infinitely many solutions

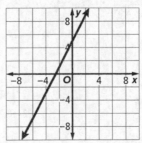

25. approximately (2.68, 1.01)

27. approximately (2.67, −0.88)

29. Sample answer: $x + y = 260$; $2.5x + 0.75y = 450$; approximately (145.71, 114.29); The bookstore will make a weekly profit of $450 with total weekly sales of 260 items when about 146 books and about 114 magazines are sold.

31. no solution; inconsistent

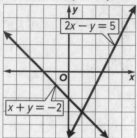

33. 1 solution; (1, −3); consistent; independent

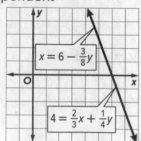

35. infinitely many solutions; consistent; dependent

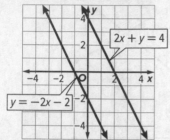

37. infinitely many solutions; consistent; dependent

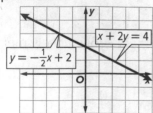

39. 1 solution; (2, 1); consistent; independent

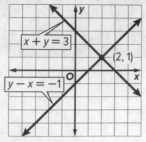

41a. x = time, in seconds, y = distance from where Olivia started to the finish line, in feet; $y = 20x$; $y = 15x + 150$

41b.

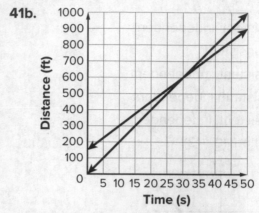

41c. 600 ft

43a. x = time walking in minutes, y = time on bike in minutes; $3x + 2y = 70$, $x = y + 15$

43b.

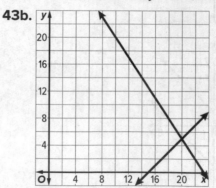

43c. 20 minutes

45. (−2, 3)

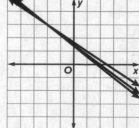

47. Sample answer: $4x + 2y = 14$, $12x + 6y = 18$; This system is inconsistent, while the others are consistent and independent.

49. Graphing clearly shows whether a system of equations has one solution, no solution, or infinitely many solutions. However, finding the exact value of x and y from a graph can be difficult.

51. Francisca; If the item is less than $100, then $10 off is better. Of the item is more than $100, then the 10% is better.

Lesson 7-2

1. (1, 6)   **3.** (29, 53)   **5.** (1, 1)   **7.** infinitely many

9. no solution   **11.** (0, 1)   **13.** (2, 5)

15. infinitely many

17a. Sample answer: $a + b = 5$; $0.7a + 0.2b = 0.65(5)$

17b. 4.5 mL from Beaker A and 0.5 mL from Beaker B

19. $\left(\frac{1}{2}, -\frac{3}{8}\right)$

21. Sample answer: In 2011, the population of Ecuador was about 15,180,000 and the population of Chile was about 17,150,000. The population of Ecuador increased by 1,210,000 and the population of Chile increased by 760,000 from 2011 to 2016. Let $x =$ the number of 5-year periods and $y =$ population. The system is $y = 15,180,000 + 1,210,000x$ and $y = 17,150,000 + 760,000x$. Solve by substitution to find that $x \approx 4.4$, or $4.4 \times 5 = 22$ years. So, the population of Ecuador and Chile will be equal in about $2011 + 22 = 2033$. (Source: World Bank)

23. Let $x =$ tens digit and $y =$ units digit of the original number; $10y + x = 10x + y - 45$; $x = 3y + 1$; (7, 2); The original number is 72.

25. Neither; Guillermo substituted incorrectly for b. Cara solved correctly for b, but misinterpreted the pounds of apples bought.

27. Sample answer: The solutions found by each of these methods should be the same. However, it may be necessary to estimate when using a graph. So, when a precise solution is needed, you should use substitution.

29. An equation containing a variable with a coefficient of 1 can easily be solved for the variable. That expression can then be substituted into the second equation for the variable.

Lesson 7-3

1. (−3, 4)   **3.** (−3, 1)   **5.** (4, −2)   **7.** (8, −7)

9. (4, 7)   **11.** (4, 1.5)   **13.** (2, 1)   **15.** (11, 0)

17. (−3, 7)   **19.** (2, −1)   **21.** (−3, −5)

23. (10, 4)   **25.** (7, 5)   **27.** (2, −3)

29. −2 and −4

31a. $r + s = 181$ and $r - s = 119$

31b. 31 state senators and 150 state representatives

33. (4, −1)   **35.** $\left(-1, 3\frac{1}{3}\right)$

37. (−36, −4)   **39.** 34 games

41a. Sample answer: $4p + 2n = 18.50$, $7p + 2n = 26.75$, where p is the price of a bag of popcorn and n is the price of a plate of nachos

41b. (2.75, 3.75); A bag of popcorn costs $2.75 and a plate of nachos costs $3.75.

43a. Add the equations because this will eliminate the variable y, and then you can solve for x.

43b. (5, −2)

43c.
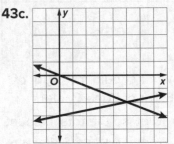

Sample answer: The point of intersection on the graph will match the solution (5, −2).

43d. The equations would be equivalent. There would be infinitely many solutions, all real numbers x and y satisfying the equation $x - 5y = 15$.

43e. There would be no solution because the lines would be parallel and would never intersect.

45. Sample answer: $x + y = 1$ and $-x - y = 1$; This system of equations has no solutions.

47. Sample answer: $-x + y = 5$, I used the solution to create another equation with the coefficient of the x-term being opposite of its corresponding coefficient.

49. Sample answer: It would be most beneficial to use elimination to solve a system of equations when one variable has either the same coefficient in both equations or one variable has coefficients that are additive inverses in the equations.

Lesson 7-4

1. $(-1, 3)$ **3.** $(-3, 4)$ **5.** $(-2, 3)$ **7.** $(3, 5)$

9. $(1, -5)$ **11.** $(0, 1)$

13a. $2x + y = 592.30$ and $x + 2y = 691.31$, where x is the number of MLB games and y is the number of NBA games

13b. MLB: $164.43, NBA: $263.44

15. 8 and -1

17. wash: $6, vacuum: $2

19a.

	Tropical Breeze	Kona Cooler	Total
Amount of Juice (qt)	t	k	10
Amount of Pineapple Juice (qt)	$0.2t$	$0.5k$	4

19b. $\left(3\frac{1}{3}, 6\frac{2}{3}\right)$; The owner should mix $3\frac{1}{3}$ qt of Tropical Breeze and $6\frac{2}{3}$ qt of Kona Cooler.

19c. $3\frac{1}{3}$ qt $+ 6\frac{2}{3}$ qt $= 10$ qt, so the total amount is correct, and $0.2\left(3\frac{1}{3}\text{ qt}\right) + 0.5\left(6\frac{2}{3}\text{ qt}\right) = 4$ qt, so the amount of pineapple juice in the new drink is correct.

21. Jason; In order to eliminate the t-terms, you can multiply the second equation by 2 and then subtract, or multiply the equation by -2 and then add. Daniela did not subtract the equations correctly.

23. Sample answer: $2x + 3y = 6$ and $4x + 9y = 5$

25. Sample answer: It is more helpful to use substitution when one of the variables has a coefficient of 1 or if a coefficient can be reduced to 1 without turning other coefficients into fractions. Otherwise, elimination is more helpful because it will avoid the use of fractions when solving the system.

Lesson 7-5

1.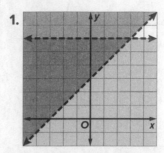

3.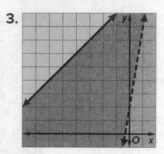

5. no solution

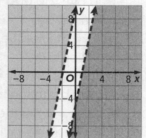

7.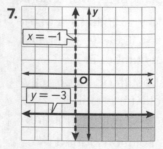

9.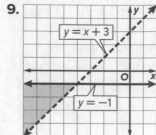

11. no solution

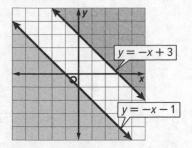

13a. Sample answer: Let x = miles of walking and y = hours at gym; $x \geq 9$, $x \leq 12$, $y \geq 4.5$, $y \leq 6$

13b.

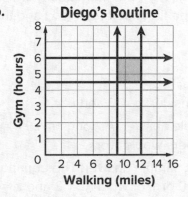

Diego's Routine

13c. Sample answers: gym 5 h, walk 9 mi; gym 6 h, walk 10 mi, gym 5.5 h, walk 11 mi

15. $y \leq x + 2$, $y \geq x - 3$

17. $y \geq x + 1$, $y < 1$

19. The solution set is the region where the graphs of the inequalities overlap. The point (2.5, 1) is not in the overlapping region, so it is not a solution. A solution must make all of the inequalities in the system true statements: $4x - 5y \geq 2 \rightarrow 4(2.5) - 5(1) \geq 2 \rightarrow 10 - 5 \geq 2 \rightarrow 5 \geq 2$; $2x + 3y > 8 \rightarrow 2(2.5) + 3(1) > 8 \rightarrow 5 + 3 > 8 \rightarrow 8 > 8$; The first inequality is true, but the second inequality is false. So, (2.5, 1) is not a solution.

21. Let x = tins of popcorn and y = tins of peanuts; $x + y \leq 200$; $x \geq y$; $3x + 4y \leq 900$; $x \geq 0$ and $y \geq 0$.

23. Sample answer: (3, 3)

25. Sometimes; sample answer: $y > 3$, $y < -3$ will have no solution, but $y < -3$, $y < 3$ will have solutions.

27. 9 units^2

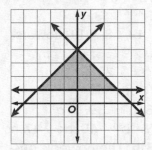

Module 7 Review

1. B, C, D   **3.** B

5. 4 wooden frames and 3 plastic frames

7. one solution; $(-2, 7)$

9. (10, 4)

11. A, B   **13.** $r = 6$, $t = 5$ **15.** D   **17.** C

Module 8

Quick Check

1. 4^5 **3.** $m^3 p^5$ **5.** 32   **7.** $\frac{1}{16}$

Lesson 8-1

1. $2q^6$ **3.** $9w^8 x^{12}$ **5.** $7b^{14} c^8 d^6$ **7.** $j^{20} k^{28}$
9. 2^8 or 256   **11.** $4096 r^{12} t^6$ **13.** $y^3 z^3$
15. $-15m^{11}$ **17.** $9p^2 r^4$ **19.** 1.8×10^5 flowers
21. $14{,}850{,}000°C$ **23.** 2.9465×10^{12} ml
25. $25x^6 y^{10}$ **27.** $64m^{24} n^{12}$ **29.** $9c^{16}$
31. $32{,}000 k^{10} m^{19}$ **33.** $800 x^8 y^{12} z^4$
35. $30x^5 y^{11} z^6$ **37.** $0.064 h^{15}$ **39.** $\frac{16}{25} a^4$
41. $8m^3 p^6$ **43.** $288 a^{31} b^{26} c^{30}$ **45.** no   **47.** no
49. yes   **51a.** $20x^7$ **51b.** The power is one-fourth the previous power: $\frac{1}{4}(20x^7)$, or $5x^7$.
53. $12x$ ft^2 **55a.** If the process of raising a number to an exponent were commutative, then $a^b = b^a$ for all numbers a and b. This is not true, because $2^3 = 8$, whereas $3^2 = 9$.
55b. If the process of raising a number to an exponent were associative, then $(a^b)^c = a^{(b^c)}$ for all numbers a, b, and c. This is not true, because $(4^3)^2 = 64^2 = 4096$, whereas $4^{(3^2)} = 4^9 = 262{,}144$.   **55c.** If the process of raising a number to an exponent were to distribute over addition, then $(a + b)^c = a^c + b^c$ for all numbers a, b, and c. This is not true, because $(1 + 2)^3 = 3^3 = 27$, but $1^3 + 2^3 = 1 + 8 = 9$.   **55d.** If the process of raising a number to an exponent were to distribute over multiplication, then $(ab)^c = a^c b^c$ for all numbers a, b, and c. This is true. It is a Power of a Product Property.
57. Sample answer: Use the Power of a Power Property to simplify the expression to $\frac{a^{2tm}}{b^{2t^2}}$.
59. Sample answer: $x^4 \cdot x^2$; $x^5 \cdot x$; $(x^3)^2$
61. Jade is correct; Sample answer: Only the h is raised to the power of 6. The exponent of the second g is 1.

Lesson 8-2

1. $m^2 p$ **3.** c^2 **5.** $\frac{p^6 t^{21}}{1000}$ **7.** $\frac{9n^2 p^6}{49q^4}$ **9.** $\frac{81 m^{20} r^{12}}{256 p^{32}}$
11. $k^2 m p^2$ **13.** $-4x^2 y z^3$ **15.** $3d$ **17.** x^2

19. $25^3 = 15{,}625$ **21.** $26^2 = 676$ **23.** $-\frac{w^9}{3}$
25. $\frac{4a^8 c^9}{25 b^6 d^6}$
27. $\frac{64 a^{15} b^{30}}{27}$ **29.** $\frac{8x^6}{3y^3}$ **31.** $\frac{8x^6 y^3 z^{12}}{27}$
33. $\frac{c^{24}}{64 d^{12}}$ **35.** 10^7 **37.** $3x^2 y$
39a. $n(1.04^t)$
39b. 26.5%; $\frac{n(1.04^8)}{n(1.04^2)} = 1.04^{(8-2)} = 1.04^6 \approx 1.2653$; $126.5\% - 100\% = 26.5\%$
41. 8
43. No; sample answer: The denominator of both expressions is y^6, but the numerators are different. The first numerator is x^3 and the second is x^2.
45. C
47. a^x
49a. Sample answer: $h = 3$ and $k = 1$
49b. Yes; sample answer: As long as $h - k = 2$, then any numbers would work.
51. Sometimes; sample answer: The equation is true when $x = 0$ or 1, or when $y = 2$ and $z = 2$, but it is false in all other cases.
53. Sample answer: The Quotient of Powers Property is used when dividing two powers with the same base. The exponents are subtracted. The Power of a Quotient Property is used to find the power of a quotient. You find the power of the numerator and the power of the denominator.
55. $\frac{(-3)^8}{(-2)^4}$; The numerator and denominator do not have the same base.

Lesson 8-3

1. $\frac{r^2}{n^9}$ **3.** $\frac{1}{f^{11}}$ **5.** $\frac{g^4 h^2}{f^6}$ **7.** $-\frac{3}{u^4}$ **9.** $\frac{5m^6 y^3 r^5}{7}$
11. $\frac{r^4 p^2}{4m^3 t^4}$ **13.** $\frac{y^9 z^2}{x^4}$ **15.** $\frac{p^4 r^2}{t^3}$ **17.** $\frac{-f}{4}$
19. 5   **21.** 6   **23.** 1
25. $1600 k^{13}$ **27.** $\frac{5q}{r^6 t^3}$ **29.** $\frac{4g^{12}}{h^4}$ **31.** $\frac{4x^8 y^4}{z^6}$
33. $\frac{16 z^2}{y^8}$ **35.** 3   **37a.** 12,500
37b. 6.25 petabytes
37c. $\frac{1}{10^9}$
39. No; sample answer: The exponent does not affect the coefficient in $4n^{-5}$, so the two expressions simplify to $4n^5$ and $\frac{4}{n^5}$.

41. $\frac{1}{p^{-3}} = \frac{1}{(p^3)^{-1}} = \frac{1}{\left(\frac{1}{p^3}\right)} = 1 \cdot \frac{p^3}{1} = p^3$

43. 750

45a. $\left(\frac{4a^3}{2a^{-2}}\right)^4 = (2a^5)^4 = 16a^{20}$

45b. $\left(\frac{4a^3}{2a^{-2}}\right)^4 = \frac{(4a^3)^4}{(2a^{-2})^4} = \frac{256a^{12}}{16a^{-8}} = 16a^{20}$

45c. Sample answer: When simplifying monomials, the order of applying the Quotient of Powers Property and Power of a Quotient Property does not matter.

47a. Both; sample answer: Colleen applied the Power of a Power Property first, while Tyler applied the Quotient of a Power Property first. Their answers are equivalent.

47b. Sample answer: Colleen's answer is in simplest form. Tyler's answer uses negative exponents, so it is not considered simplest form.   **49.** Sample answer: Lola is correct. If we are asked to simplify $\frac{a^7}{a^2}$, instead of using the Division Property of Exponents ($a^{7-2} = a^5$), we could rewrite the fraction as a^7a^{-2}, and then use the Multiplication Property of Exponents to get $a^{7+(-2)} = a^5$.   **51.** Sample answer: Students measure the width of their classroom and the width of a pencil, then write and solve a question such as: How many times wider is the classroom than the pencil?

53. Each power is 1 less than the previous power. Each value is half the previous value. So, each time 1 is subtracted from the power, divide the value by 2. Since $2^1 = 2$, then 2^{1-1}, or $2^0 = 2 \div 2$, or 1.

Lesson 8-4

1. $\sqrt{15}$ **3.** $4\sqrt{k}$ **5.** $26^{\frac{1}{2}}$ **7.** $2(ab)^{\frac{1}{2}}$ **9.** $\frac{1}{2}$ **11.** 9
13. 4   **15.** $\frac{2}{5}$ **17.** 7   **19.** $\frac{1}{3}$ **21.** 243   **23.** 625
25. $\frac{27}{1000}$ **27.** 96 ft/s
29. 9 astronomical units   **31.** \$51,000   **33.** $\sqrt[3]{17}$
35. $7\sqrt[3]{b}$ **37.** $29^{\frac{1}{3}}$ **39.** $2a^{\frac{1}{3}}$ **41.** 0.3   **43.** a
45. 16   **47.** $\frac{1}{3}$ **49.** $\frac{1}{27}$ **51.** $\frac{1}{\sqrt{k}}$ **53a.** 107
53b. 236   **55.** 4.1 m   **57.** Sample answer: The expressions in a, c, and d are equivalent; The expression in b is not equivalent to the others. Let $p = 4$, $f = 3$, and $k = 2$, then $p^{\frac{f}{k}} = 8$, $p^{\frac{f}{5}} \approx 2.30$, $\sqrt[k]{p^f} = 8$, and $(\sqrt[k]{p})^f = 8$.
59. 112π cm²   **61.** Sample answer: $2^{\frac{1}{2}}$ and $4^{\frac{1}{4}}$

63. $-1, 0, 1$ **65.** Sample answer: Both Sachi and Makayla are correct. Sachi addressed the numerator of the rational exponent first and then the denominator. Makayla addressed the denominator of the rational exponent first and then the numerator. Both methods will lead to the correct answer, 9.

Lesson 8-5

1. $2\sqrt{13}$ **3.** $6\sqrt{2}$ **5.** $9\sqrt{3}$ **7.** $5\sqrt{2}$ **9.** $2\sqrt{5}$
11. $4\sqrt{6}$ **13.** $\frac{5\sqrt{3}}{7}$ **15.** $\frac{2\sqrt{2}}{3}$ **17.** $\frac{2\sqrt{21}}{11}$ **19.** $4b^2$
21. $9a^6d^2$ **23.** $2a^2b\sqrt{7b}$ **25.** $40\sqrt{3}$ ft/s
27. 0.9 second   **29.** 2   **31.** 6   **33.** 3   **35.** $3\sqrt[3]{9}$
37. $3\sqrt[3]{6}$ **39.** $2\sqrt[3]{10}$ **41.** $2\sqrt[3]{12}$ **43.** $2\sqrt[3]{9}$ **45.** 3
47. $\frac{3\sqrt[3]{2}}{5}$ **49.** $\frac{\sqrt[3]{9}}{4}$ **51.** 4   **53.** $3b^2\sqrt[3]{2b^2}$ **55.** $\frac{4}{7x}$
57. $3x^2y^4\sqrt[3]{10x}$ **59.** $7m^4n^2\sqrt{5mn}$ **61.** $\frac{4df^4\sqrt{2d}}{5e^2}$
63. $4x^4y^5z^6\sqrt[3]{5x^2y^2z^2}$ **65.** $\frac{2y^3\sqrt{11y}}{x}$ **67.** $18\sqrt{3}$ m²
69. $8\sqrt{7}$ **71.** 27.6 in.
73a. The Product Property of Square Roots states that the square root of a product is equal to the product of the square roots of the factors. For this property, $a \geq 0$ and $b \geq 0$.
73b. The Quotient Property of Square Roots states that the square root of a quotient is equal to the quotient of the square roots of the numerator and denominator. For this property, $a \geq 0$ and $b > 0$.   **73c.** Both properties state that finding a square root can be done before or after certain other operations.
75. Sample answer: $\sqrt{5b} \cdot \sqrt{15b^4} = 5b^2\sqrt{3b}$
77. Sample answer: 0.25. Any number between 0 and 1 will have a square root larger than itself.
79. Sample answer: Because $4\sqrt{12} \times 6\sqrt{15} = 144\sqrt{5}$, $4\sqrt{12}$ in. is a possible length and $6\sqrt{15}$ in. is a possible width.

Lesson 8-6

1. $5\sqrt{7}$ **3.** $11\sqrt{15}$ **5.** $3\sqrt{r}$ **7.** $-\sqrt{5}$
9. $-3\sqrt{13} + 5\sqrt{2}$ **11.** $12\sqrt{3} + \sqrt{2}$
13. $115\sqrt{149}$ m   **15.** $26\sqrt{37} + 10\sqrt{41}$ in.
17. $\frac{20 - 10\sqrt{2}}{\sqrt{\pi}}$ cm or $\frac{20\sqrt{\pi} - 10\sqrt{2\pi}}{\pi}$ cm
19. $18\sqrt{5}$ **21.** $24\sqrt{14}$ **23.** $198\sqrt{2}$
25. $4 + 2\sqrt{3}$ **27.** $5\sqrt{10}$ **29.** $60 + 32\sqrt{10}$

31. $\dfrac{1-5\sqrt{5}}{5}$ **33.** 2 **35.** $2\sqrt{6} - 13\sqrt{3} + 4\sqrt{2}$
37. $15\sqrt{15}$ **39.** $\dfrac{3\sqrt{2}}{2} + 2\sqrt{15}$
41. $2ac\sqrt{6} + ad\sqrt{15} + 2bc\sqrt{10} + 5bd$
43. $63\sqrt{15}$ ft^2 **45a.** $2\sqrt{2}$ s **45b.** about 2.8 s
47. $11.25\sqrt{2}$ ft **49.** $\dfrac{5}{6}\sqrt{5}$ in.
51. True; $(x + y)^2 > \left(\sqrt{x^2 + y^2}\right)^2 \rightarrow x^2 +$
$2xy + y^2 > x^2 + y^2 \rightarrow 2xy > 0$; Because
$x > 0$ and $y > 0$, $2xy > 0$ is always true.
So, $(x + y)^2 > \left(\sqrt{x^2 + y^2}\right)^2$ is always true for all
$x > 0$ and $y > 0$.
53. Sample answer: $\sqrt{12} + \sqrt{27} = 5\sqrt{3}$;
Simplify $\sqrt{12}$ to get $2\sqrt{3}$. Simplify $\sqrt{27}$ to get
$3\sqrt{3}$.
Because $2\sqrt{3}$ and $3\sqrt{3}$ have the same
radicand, they can be combined.
55. $6\sqrt{12}$; Sample answer: $\sqrt{12}$ is the only
radical that does not simplify to $k\sqrt{5}$.

Lesson 8-7

1. 9 **3.** 6 **5.** 8 **7.** 8 **9.** 5 **11.** 2 **13.** 16 ohms
15. 20 months **17.** 3 **19.** $-\dfrac{1}{4}$ **21.** 11 **23.** 2
25. 2 **27.** 37.52 ft **29.** Use 4 as the common
base. Replace 8 with $4^{\frac{3}{2}}$ and 16 with 4^2 because
$4^{\frac{3}{2}} = (\sqrt{4})^3$, or 8 and $4^2 = 16$. So, $4^{\frac{3}{2}(2 + x)} =$
$4^{2(2.5 - 0.5x)}$. Distribute and get $4^{(3 + 1.5x)} = 4^{(5 - x)}$.
Set the exponents equal to each other: $3 +$
$1.5x = 5 - x$. Solve for x: $x = 0.8$, or $\dfrac{4}{5}$.
31. Sample answer: There is no way to make
a common base between 2 and 5. **33.** $x = 0$
and $x = -4$ **35.** $3^{2x + 1} = 243$; This equation
has a solution of $x = 2$. The other three
equations have a solution of $x = 3$.

Module 8 Review

1. A, B **3.** B **5.** 9 **7.** C **9.** 0.000001
11. Sample answer: Because $b^{\frac{1}{n}} = \sqrt[n]{b}$, rewrite
$729^{\frac{1}{6}}$ as $\sqrt[6]{729}$. Because $729 = 3 \cdot 3 \cdot 3 \cdot 3 \cdot 3 \cdot$
3, you can write $\sqrt[6]{729}$ as $\sqrt[6]{3 \cdot 3 \cdot 3 \cdot 3 \cdot 3 \cdot 3}$,
which equals 3. **13.** B **15.** 27
17. A **19.** $2\sqrt{5} + 6\sqrt{7}$ **21.** B **23.** 6

Module 9

Quick Check

1. -196

3. 0.25

5. 32

7. -3

Lesson 9-1

1. No; the domain values are at regular intervals and the range values have a common difference 3.

3. Yes; the domain values are at regular intervals and the range values have a common ratio 2.

5. No; there is no common ratio between the picture areas.

7a. y-intercept $= 50$; D $= \{$all real numbers$\}$, R $= \{y \mid y > 0\}$

7b.

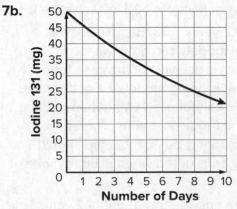

7c. Because time cannot be negative, the relevant domain is $\{x \mid x \geq 0\}$. Because the amount of Iodine 131 cannot be negative, and the amount when $x = 0$ is 50 mg, the relevant range is $\{y \mid 0 < y \leq 50\}$.

9.

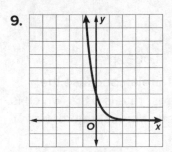

2; D $= \{$all real numbers$\}$, R $= \{y \mid y > 0\}$; $y = 0$

11.

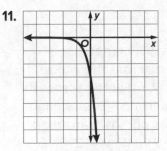

-3; D $= \{$all real numbers$\}$, R $= \{y \mid y < 0\}$; $y = 0$

13.

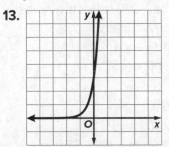

3; D $= \{$all real numbers$\}$, R $= \{y \mid y > 0\}$; $y = 0$

15a. 1038 millibars

15b. about 794 millibars

15c. It decreases.

17. $f(x) = 3(2^x)$

19. Sample answer: The number of teams competing in a basketball tournament can be represented by $y = 2^x$, where the number of teams competing is y and the number of rounds is x. The y-intercept of the graph is 1. The graph increases rapidly for $x > 0$. With an exponential model, each team that joins the tournament will play all of the other teams. If the scenario were modeled with a linear function, each team that joined would play a fixed number of teams.

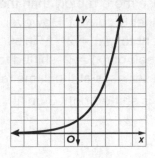

Lesson 9-2

1. translated up 8 units   **3.** compressed horizontally **5.** reflected across the x-axis; translated 1 unit right   **7.** reflected across the y-axis; translated   4 units up   **9.** stretched vertically; shifted 2 units up   **11.** translated right 3 units   **13.** $y = -2^x$ **15.** $y = 2^{-x} + 5$
17. stretched vertically by a factor of 2000
19. stretched vertically by a factor of 20
21. translated up 6 units   **23.** reflected across the x-axis; compressed vertically
25. reflected across the y-axis
27. $g(x) = 2^x + 3$ **29.** $g(x) = 5^{x-2}$
31. $g(x) = 6^x + 5$
33. $g(x) = \frac{1}{2}(4^x)$ **35.** $g(x) = 2^{3x}$
37. $g(x) = 5^x - 2$ **39.** $g(x) = 5^{x-4}$
41a. translated up 500 units   **41b.** $500
43. The graph has been reflected over the x-axis and reflected over the y-axis. It has been stretched vertically by a factor of 3 and shifted up 1 unit.

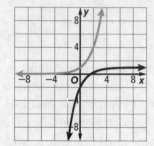

45. The graphs of these two exponential functions are the same. $f(x) = 4^{x+2} = 4^x \cdot 4^2 = 16 \cdot 4^x = g(x)$.

47. Jennifer is correct. Sample answer: As it is written, the function is multiplied by 2, which causes the graph to rise more rapidly than the parent graph, so Jennifer is correct. However, $g(x) = 2(2^x)$ is equivalent to $g(x) = 2^{x+1}$. This graph is the parent graph of $f(x) = 2^x$ shifted to the left one unit, but it still rises at the same rate.

49. The first pair; $g(x)$ is shifted right 3 units instead of left 3 units.

Lesson 9-3

1. $y = 4 \cdot 2^x$ **3.** $y = 10 \cdot 3^x$ **5.** $y = 3 \cdot 4^x$
7. $y = 3 \cdot 2^x$ **9.** $y = \left(\frac{1}{4}\right)^x$ **11.** $f(x) = 50 \cdot 2^x$, where x is the number of 30-minute time periods

13. $f(x) = 43{,}000{,}000 \cdot (1.23)^x$, where x is the number of years since 2010.
15a. $P = 8{,}192{,}426(1.009)^t$ **15b.** about 9,370,872
17a. $Z = 60{,}000(0.90)^t$ **17b.** about $31,886
19. $2200   **21.** 360 million
23. $y = 2.6 \cdot 4^x$ **25.** about 77,529
27. Sample answer: The equation can be rewritten in the form $y = a(1 + r)^x$ to find the amount of original investment, a, and the rate of increase or decrease. Because $a = 2400$, he invested $2400. Because $1 + r = 0.95$ and is less than 1, his investment is decreasing in value. A graphing calculator can be used to find that the investment will be worth $1200 in about 13.5 years.

29a. $P(t) = 128(1.25)^t$

29b. an increase of approximately 41 deer per year   **29c.** No; the amount of increase is exponential, not linear.   **29d.** There is no common difference over equal intervals (differences are 32, 40 and 50). There is a common factor (factor is 1.25 in each case.)

31. about 9.2 years   **33.** Sample answer: Exponential models can grow without bound, which is usually not the case for the situation that is being modeled. For instance, a population cannot grow without bound due to space and food constraints. Therefore, the situation that is being modeled should be carefully considered when used to make decisions.

35a. Sample answer: 5%; about 14.2 years

35b. Sample answer: 10%; about 6.6 years

35c. Sample answer: about 10.4 years; about $8320

Lesson 9-4

1a. $A(t) = (1.021)^t$; $A(t) = (1.0052)^{4t}$

1b. Bank B has the better plan because the effective quarterly interest rate is 0.8%, which is greater than the quarterly interest rate of about 0.52% for Bank A.

1c. About 3.2%; sample answer: This confirms the result of part **b** because 3.2% is greater than the annual interest rate at Bank A, so Bank B has the better plan.

3. Bank A; Bank A has a quarterly interest rate of 0.95%. Bank B has a quarterly interest rate of about 0.92%. Bank A's quarterly interest rate is higher.

5. Species B; the population of Species A is decreasing at a rate of about 0.25% per quarter. The population of Species B is decreasing at a rate of about 0.34% per quarter. The population of Species B is decreasing at a faster rate.

7. Plan A

9. Account A; Account A has a semi-annual interest rate of 2.3%. Account B has a semi-annual interest rate of about 2.1%. Account A's semi-annual interest rate is greater.

11. Account A; Account A has a monthly interest rate of 0.5%. Account B has a monthly interest rate of about 0.21%. Account A's monthly interest rate is greater.

13. $T(t) = 72 + 140(0.67)^t$

15. Sample answer: Bank A offers a savings account with a 0.6% interest rate compounded quarterly. Bank B offers a savings account with a 2% interest rate compounded annually. Bank A offers the better interest rate because it has a higher effective annual interest rate of about 2.4%.

Lesson 9-5

1. The ratios are not the same, so the sequence is not geometric.

3. Since the ratio is the same for all of the terms, 5, the sequence is geometric.

5. The ratios are not the same, so the sequence is not geometric.

7. Because the ratio is the same for all of the terms, $\frac{1}{2}$, the sequence is geometric.

9. The ratios are not the same, so the sequence is not geometric.

11. The ratios are not the same, so the sequence is not geometric.

13. −250, 1250, −6250

15. 108, 324, 972

17. −2058; −14,406; −100,842

19. 54, 162, 486

21. $\frac{1}{10}, \frac{1}{20}, \frac{1}{40}$

23. $\frac{1}{3}, \frac{1}{18}, \frac{1}{108}$

25. 387,420,489

27. 177,147

29. $a_n = 4 \cdot \left(\frac{3}{2}\right)^{n-1}$

31. $1310.72

33a. $a_n = P \cdot 1.005^n$

33b. $538.84

35. $a_n = \frac{9}{16}\left(\frac{2}{3}\right)^{n-1}; \frac{4}{81}$

37. $a_n = -8\left(\frac{1}{4}\right)^{n-1}; -\frac{1}{2048}$

39. Sample answer: The average annual salary is about $39,416, and the average annual rate of increase is about 3%. $a_n = 39,416(1.03)^{n-1}$; ≈ 69,116.19; This means that after 20 years of employment the average annual salary will be about $69,116.19.

41. −3, −12, −48

43a. The first method provides a starting salary of $100 and an $8 per month raise. The second method provides a starting salary of $0.01 and doubles it each month.

43b. The first situation is linear because there is a common difference of $8. The equation is $y = 8x + 92$. The second situation is exponential because it is a geometric sequence with a common ratio of 2. The equation is $y = 0.01(2)^{x-1}$.

43c. Sample answer: As long as I do not need money immediately, I would use the second method. In the last month, I would make $y = 0.01(2)^{23} = $83,886.08$ due to the fact that the payment is growing exponentially. In the last month, in the first method I would make $y = 8(24) + 92 = 284.

45. If the values fit a geometric sequence, then $r = \sqrt{\frac{540}{180}} = \sqrt{3}$. This would mean that the interior angles of a square would have a sum of $180\sqrt{3} \approx 312°$. Since the sum of the angles in a square is 360°, this is not a geometric sequence.

47. Neither; Haro calculated the exponent incorrectly. Matthew did not calculate $(-2)^8$ correctly.

49. Sample answer: When graphed, the terms of a geometric sequence lie on a curve that can be represented by an exponential function. They are different in that the domain of a geometric sequence is the set of natural numbers, while the domain of an exponential function is all real numbers. Thus, geometric sequences are discrete, while exponential functions are continuous.

51. Sample answer: In the geometric sequence 6, 3, 1.5, ..., the value of r is 0.5 and the absolute value of a_{n+1} will be closer to zero than the value of a_n.

Lesson 9-6

1. 23, 30, 37, 44, 51

3. 8, 20, 50, 125, 312.5

5. 13, −29, 55, −113, 223

7. $a_1 = 12$, $a_n = a_{n-1} - 13$, $n \geq 2$

9. $a_1 = 2$, $a_n = a_{n-1} + 9$, $n \geq 2$

11. $a_1 = 40$, $a_n = -1.5a_{n-1}$, $n \geq 2$

13. $a_1 = 3$, $a_n = a_{n-1} - 1$, $n \geq 2$

15. $a_1 = 2$, $a_n = a_{n-1} + 1$, $n \geq 2$

17. $a_1 = \frac{5}{2}$, $a_n = a_{n-1} - 1$, $n \geq 2$

19a. 875, 1050, 1225, 1400, 1575

19b. $a_1 = 175$, $a_n = a_{n-1} + 175$, $n \geq 2$

19c. $a_n = 175n$

21a. $a_1 = 6$, $a_n = 0.9a_{n-1}$, $n \geq 2$

21b. $a_n = 6(0.9)^{n-1}$

23. $a_n = -12n + 10$

25. $a_1 = 45$, $a_n = a_{n-1} - 7$, $n \geq 2$

27. $a_1 = -11$, $a_n = a_{n-1} + 5$, $n \geq 2$

29. $a_n = 16(4)^{n-1}$

31. $a_n = 500(1.05)^{n-1}$

33. Ramon has 2 parents, 4 grandparents, 8 great-grandparents, and so on. We can write a geometric sequence to count the number of ancestors in a given generation. The recursive formula is $a_1 = 2$, $a_n = 2_{n-1}$, $n \geq 2$. The explicit formula is $a_n = 2^n$. Ramon's claim is about the 8th generation back: $a_8 = 2^8 = 256$. Ramon is correct.

35a. Sample answer: B3 = B2 + B1 and C2 = B2 ÷ B1

35b. The ratio approaches a constant value of 1.618034.... For larger values of n, the Fibonacci numbers behave like a geometric sequence with a common ratio of 1.618034....

37. Both; Sample answer: The sequence can be written as the recursive formula $a_1 = 2$, $a_n = (-1)a_{n-1}$, $n \geq 2$. The sequence can also be written as the explicit formula $a_n = 2(-1)^{n-1}$.

39. False; sample answer: A recursive formula for the sequence 1, 2, 3, ... can be written as $a_1 = 1$, $a_n = a_{n-1} + 1$, $n \geq 2$ or as $a_1 = 1$, $a_2 = 2$, $a_n = a_{n-2} + 2$, $n \geq 3$.

41. Sample answer: In an explicit formula, the nth term a_n is given as a function of n. In a recursive formula, the nth term a_n is found by performing operations to one or more of the terms that precede it.

Module 9 Review

1.

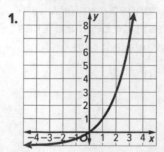

3. As x increases, y increases; and, as x decreases, y approaches 0.

5. A

7. A

9. D

11. Local Credit Union; sample answer: The monthly interest rate is 0.12% higher than at First & Loan, and the annual interest rate is 1.6% higher than at First & Loan.

13. 146 people

15. $a_1 = 20$, $a_n = a_{n-1} + 15$, $n \geq 2$

17. row 2: 12; row 3: 12, 48; row 4: 48, 192

Module 10

Quick Check

1. $a^2 + 5a$ **3.** $n^2 - 3n^3 + 2n$ **5.** $13x$
7. simplified

Lesson 10-1

1. no   **3.** yes; 4; trinomial   **5.** yes; 2; binomial
7. $5x^2 + 3x - 2$; 5   **9.** $-5c^2 - 3c + 4$; -5
11. $t^5 + 2t^2 + 11t - 3$; 1   **13.** $-3x^4 + \frac{1}{2}x + 7$; -3
15. $6x + 12y$ **17.** $3a + 5b$ **19.** $2m^2 + m$
21. $d^2 - 3d$ **23.** $3f + g + 1$ **25.** $7c^2 + 6c - 3$
27. $3c^3 - c^2 - 3c + 3$ **29.** $-2x - 5y + 1$
31. $-x^2y - 3x^2 + 4y$ **33.** $-6p^2 + 2np + n$
35a. $D = 0.5x + 2$ **35b.** \$7,000,000
37. quadratic trinomial   **39.** quartic binomial
41. quintic polynomial   **43.** $9x + 4y - 17z$
45. $2c^2 - c + 8$ **47a.** $4\pi rh + 4\pi r^2$
47b. 15.7 m²   **49.** Both $0x^2 + 0x$ and 0 are
polynomials.   **51a.** $-13x^2 - x + 10$;
$(5x^2 - 3x + 7) - (18x^2 - 2x - 3) = (5x^2 - 18x^2)$
$+ (-3x + 2x) + (7 + 3)$ **51b.** $13x^2 + x - 10$;
$(18x^2 - 2x - 3) - (5x^2 - 3x + 7) = (18x^2 - 5x^2)$
$+ (-2x + 3x) + (-3 - 7)$ **51c.** The second
result is the negative of the first. Reversing the
order of subtraction with integers yields the
negative of the original difference. For a and b
integers, $(a - b) = -(b - a)$.   **53.** No; neither
of them found the additive inverse correctly.
All terms should have been multiplied by
-1.   **55.** $6n + 9$ **57.** Sample answer: To add
polynomials in a horizontal method, combine
like terms. For the vertical method, write the
polynomials in standard form, align like terms
in columns, and combine like terms. To subtract
polynomials in a horizontal method, find the
additive inverse of the polynomial that is being
subtracted, and then combine like terms. For
the vertical method, write the polynomials in
standard form, align like terms in columns, and
subtract by adding the additive inverse.

Lesson 10-2

1. $b^3 - 12b^2 + b$
3. $-6m^6 + 36m^5 - 6m^4 - 75m^3$

5. $4p^2r^3 + 10p^3r^3 - 30p^2r^2$ **7.** $-13x^2 - 9x - 27$
9. $-20d^3 + 55d + 35$ **11.** $14j^3k^2 + 2j^2k^2 -$
$17jk + 18j^2k^3 - 18k^3$ **13.** $\frac{n^2}{2} + \frac{n}{2}$; 78
15. $3.88 = x(0.39) + (x + 4)(0.29)$; 4 peppers
and 8 potatoes   **17.** 2   **19.** $\frac{43}{6}$ **21.** $\frac{30}{43}$
23. $4a^2 + 3a$ **25.** $2x^2 - 5x$ **27.** $-3n^3 - 6n^2$
29. $15x^3 - 3x^2 + 12x$ **31.** $-4b + 36b^2 + 8b^3$
33. $4m^4 + 6m^3 - 10m^2$ **35.** $3w^2 + 7w$
37. $-2p^2 + 3p$ **39.** $3x^3 + 8x$
41. $-22b^2 + 2b + 8$
43. $-q^3w^3 - 35q^2w^4 + 8q^2w^2 - 27qw$ **45.** -7
47. -5 **49.** -2 **51.** $1.10(2\pi r) = 2\pi(r + 12)$;
$r = 120$ ft   **53.** The Distributive Property has
been misapplied. The second term should be
$12y^2(-2y^2) = -24y^4$, and the fourth term should
be $-3y^3(-2y) = 6y^4$. So, it simplifies to
$24y^3 - 18y^4$.   **55a.** $1.25x(45 - 5x) = 56.25x -$
$6.25x^2$ **55b.** \$5.63   **55c.** \$126.56   **57.** Ted;
Pearl used the Distributive Property incorrectly.
59. $8x^2y^{-2} + 24x^{-10}y^8 - 16x^{-3}$ **61.** Sample
answer: $3n$, $4n + 1$; $12n^2 + 3n$ **63.** Sample
answer: $4t(t - 5) - 2t(3t + 2) + 108 =$
$3t(2t - 5) - t(8t - 3)$

Lesson 10-3

1. $3c^2 + 4c - 15$ **3.** $30a^2 + 43a + 15$
5. $15y^2 - 17y + 4$ **7.** $6m^2 + 19m + 15$
9. $144t^2 - 25$ **11.** $40w^2 - 28wx - 24x^2$
13. $45x^2 - 13x - 2$ **15.** $20x^2 + 18x + 4$ in²
17. $(x - 2)(x + 2) - x^2 = -4$; $x^2 - 2x + 2x -$
$4 - x^2 = -4$; $-4 = -4$
19. $36a^3 + 71a^2 - 14a - 49$
21. $5x^4 + 19x^3 - 34x^2 + 11x - 1$
23. $18z^5 - 15z^4 - 18z^3 - 14z^2 + 24z + 8$
25. $x^2 + 4x + 4$ **27.** $t^2 + t - 12$
29. $n^2 - 4n - 5$ **31.** $2x^2 - 18$
33. $2\ell^2 - 3\ell - 20$ **35.** $5q^2 + 24q - 5$
37. $6m^2 - 6$ **39.** $10a^2 - 19a + 6$
41. $2x^2 - 3xy + y^2$ **43.** $t^3 + 3t^2 + 6t + 4$
45. $m^3 + 6m^2 + 14m + 15$
47. $6b^3 + 5b^2 + 8b + 16$ **49.** $5t^2 - 34t + 56$
51. $9c^2 + 24cd + 16d^2$
53. $8r^3 - 36r^2t + 54rt^2 - 27t^3$
55. $64y^3 - 48y^2z - 36yz^2 + 27z^3$

57a. Let w represent the width of the mural;
$w(w + 5)$ ft²   **57b.** $(w + 9)(w + 4) - (w + 5)w$ ft²
57c. $(w + 9)(w + 4) - (w + 5)w = 100$; 8 ft by 13 ft

59a. $x^{4p+1}(x^{1-2p})^{2p+3} = x^{4p+1+(1-2p)(2p+3)} =$
$x^{4p+1+(2p+3)-2p(2p+3)} = x^{6p+4-2p(2p)-2p(3)} =$
$x^{6p+4-4p^2-6p} = x^{-4p^2+4}$ **59b.** $x^0 = 1$ for all
values of $x \neq 0$, so find p such that
$-4p^2 + 4 = 0$. So, $p = \pm 1$.   **61.** $8x^3 + 28x^2 -$
$16x$ units^3; Volume $= 4x(2x - 1)(x + 4) =$
$4x[2x(x + 4) - 1(x + 4)] = 4x[2x^2 + 8x -$
$x - 4] = 4x(2x^2 + 7x - 4) = 4x(2x^2) + 4x(7x) +$
$4x(-4) = 8x^3 + 28x^2 - 16x$ **63.** $x^{2m-1} -$
$x^{m-p+1} + x^{m+p} + x^{m+p-1} - x + x^{2p}$

65. The three monomials that make up a
trinomial are similar to the three digits that
make up the 3-digit number. The single
monomial is similar to a 1-digit number. With
each procedure, you perform 3 multiplications.
The difference is that polynomial multiplication
involves variables and the resulting product
is often the sum of two or more monomials,
while numerical multiplication results in a single
number.   **67.** The expression, $(2x - 1)(x^2 +$
$x - 1)$, does not belong with the other three
polynomial expressions because the product
of this expression has 4 terms, and the products
of the other three polynomial expressions have
3 terms.

Lesson 10-4

1. $a^2 + 10a + 100$ **3.** $h^2 + 14h + 49$
5. $64 - 16m + m^2$ **7.** $4b^2 + 12b + 9$
9. $64h^2 - 64hn + 16n^2$ **11.** $A = \pi r^2 - \pi(r - 18)^2$
or $36\pi (r - 9)$
13. $B^2 + 2BR + R^2$ **15.** $u^2 - 9$ **17.** $4 - x^2$
19. $4q^2 - 25r^2$ **21.** $n^2 + 6n + 9$
23. $y^2 - 14y + 49$ **25.** $b^2 - 1$ **27.** $p^2 - 8p + 16$
29. $\ell^2 + 4\ell + 4$ **31.** $9g^2 - 4$ **33.** $36 + 12u + u^2$
35. $9q^2 - 1$ **37.** $4k^2 - 8k + 4$ **39.** $9p^2 - 16$
41. $x^2 - 8xy + 16y^2$ **43.** $9y^2 - 9g^2$
45. $4k^2 + 4km^2 + m^4$ **47.** Sample answer:
The area of the rectangle is the square of the
larger number minus the square of the smaller
number.   **49a.** $0 = r^2 - 6rs + s^2$
49b. $0 = s^2 - 120s + 400$ **51.** $25y^2 + 70y + 49$
53. $100x^2 - 4$ **55.** $a^2 + 8ab + 16b^2$
57. $4c^2 - 36cd + 81d^2$ **59.** $36y^2 - 169$
61. $25x^4 - 10x^2y^2 + y^4$ **63.** $\frac{9}{16}k^2 + 12k + 64$
65. $49z^4 - 25y^4$ **67.** $r^4 - 29r^2 + 100$
69. $8a^3 - 12a^2b + 6ab^2 - b^3$
71. $k^3 - k^2m - km^2 + m^3$

73. $q^3 - 3q^2r + 3qr^2 - r^3$
75a. $2s + 1$; $(s + 1)^2 - s^2 = (s^2 + 2(s)(1) + 1^2) - s^2$
$= 2s + 1$ **75b.** $2s + 3$; $(s + 2)^2 - (s + 1)^2 =$
$(s^2 + 2(s)(2) + 2^2) - (s^2 + 2s + 1) = s^2 + 4s + 4 -$
$s^2 - 2s - 1 = 2s + 3$ **75c.** $2s + 5$; $(s + 3)^2 -$
$(s + 2)^2 = (s^2 + 2(s)(3) + 3^2) - (s^2 + 4s + 4) =$
$s^2 + 6s + 9 - s^2 - 4s - 4 = 2s + 5$
77a. $a^3 + 3a^2b + 3ab^2 + b^3$
77b. $x^3 + 6x^2 + 12x + 8$
77c. Sample answer:

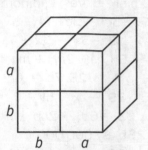

77d. $a^3 - 3a^2b + 3ab^2 - b^3$ **79.** Sample
answer: $(x - 2)(x + 2) = x^2 - 4$ and
$(x - 2)(x - 2) = x^2 - 4x + 4$

Lesson 10-5

1. $8(2t - 5y)$ **3.** $2k(k + 2)$ **5.** $2ab(2ab + a - 5b)$
7. $16t(2 - t)$ **9.** $6t(27 - 32t)$ **11.** $(g + 4)(f - 5)$
13. $(h + 5)(j - 2)$ **15.** $(9q - 10)(5p - 3)$
17. $(3d - 5)(t - 7)$ **19.** $(3t - 5)(7h - 1)$
21. $(r - 5)(5b + 2)$ **23.** $(b + 3)(b - 2)$
25. $(2a + 1)(a - 2)$ **27.** $2(4m - 3)$
29. $5q(2 - 5q)$ **31.** $x(1 + xy + x^2y^2)$
33. $4a(ab^2 + 4b + 3)$ **35.** $12ab(4ab - 1)$
37. $(x + 1)(x + 3)$ **39.** $(3n - p)(n + 2p)$
41. $(3x + 2)(3x - y)$ **43.** $(2x + 3)(x - 0.3)$
45. In both cases you are trying to build
an expression equivalent to the original
expression that has two or more factors.

47. Both are correct. Akash's grouping: $3x^2 +$
$7x - 18x - 42 = (3x^2 + 7x) - (18x + 42) =$
$x(3x + 7) - 6(3x + 7) = (x - 6)(3x + 7)$.
Theresa's grouping: $3x^2 - 18x + 7x - 42 = (3x^2$
$- 18x) + (7x - 42) = 3x(x - 6) + 7(x - 6) =$
$(3x + 7)(x - 6)$.   **49.** Find the GCF of $12a^2b^2$
and $-16a^2b^3$, which is $4a^2b^2$. Write each term
as the product of the GCF and the remaining
factors using the Distributive Property:
$4a^2b^2(3) + 4a^2b^2(-4b) = 4a^2b^2(3 - 4b)$.

Lesson 10-6

1. $(x + 3)(x + 14)$ **3.** $(a - 4)(a + 12)$
5. $(h + 4)(h + 11)$ **7.** $(x + 3)(x - 8)$
9. $(t + 2)(t + 6)$ **11.** prime   **13.** prime
15. $(h + 6)(h + 3)$ **17.** $(x + 14)(x + 12)$
19. $(d - 18)(d - 18)$ **21.** $(5x + 4)(x + 6)$
23. $2(2x + 1)(x + 5)$ **25.** $(2x + 3)(x - 3)$
27. prime   **29.** $3(4x + 3)(x + 5)$
31. prime   **33.** prime   **35.** $(2y - 1)(y + 1)$
37. $(n + 6)(n - 3)$ **39.** $(r + 6)(r - 2)$
41. $(w - 3)(w + 2)$ **43.** $(t - 8)(t - 7)$
45. $(2x + 1)(x + 2)$ **47.** $(3g - 1)(g - 2)$
49. prime   **51.** $(3p - 2)(3p + 4)$
53. $(a - 3)(a - 7)$ **55.** $(2x + 1)(x + 3)$
57. $(x - 4)(x + 5)$ **59.** $(p - 3)(p - 7)$
61. $(q + 2r)(q + 9r)$ **63.** $(x - y)(x - 5y)$
65. $-(2x + 5)(3x + 4)$ **67.** $-(x - 4)(5x + 2)$
69. prime   **71a.** $(x + 4)(x - 1)$ **71b.** possible
lengths of the pyramidal stone monument
73. $x^2 + 12x + 24$; The other 3 expressions
can be factored, but this expression cannot be
factored.   **75.** $-18, 18$ **77.** 7, 12, 15, 16
79. Sometimes; sample answer: The trinomial
$x^2 + 10x + 9 = (x + 1)(x + 9)$ and $10 > 9$. The
trinomial $x^2 + 7x + 10 = (x + 2)(x + 5)$ and
$7 < 10$.   **81.** $(4y - 5)^2 + 3(4y - 5) - 70 =$
$[(4y - 5) + 10] [(4y - 5) - 7] = (4y + 5)(4y - 12) =$
$4(4y + 5)(y - 3)$
83. $(12x + 20y)$ in.; The area of the square
equals $(3x + 5y)(3x + 5y)$ in², so the length of
one side is $(3x + 5y)$ in. The perimeter is
$4(3x + 5y)$ or $(12x + 20y)$ in.
85. Sample answer: Find two numbers, m and
p, with a product of ac and a sum of b.

Lesson 10-7

1. $(q + 11)(q - 11)$ **3.** $(w^2 + 25)(w + 5)(w - 5)$
5. $(h^2 + 16)(h + 4)(h - 4)$ **7.** $(x + 2y)(x - 2y)$
9. $(f + 8)(f - 8)(f + 2)$ **11.** $(t + 1)(t - 1)(3t - 7)$
13. $(m + 3)(m - 3)(4m + 9)$
15. $(3a + 2b)(3a - 2b)$ **17a.** $(x - 8)(x - 8)$
17b. the length and width of the rug

19. yes; $(4x - 7)^2$ **21.** no   **23.** $4(h^2 - 14)$
25. prime   **27.** $6(d + 4)(d - 4)$
29. $(-4 + p)(4 + p)$ **31.** $4(3 + 5w)(3 - 5w)$
33. $(2h + 5g)(2h - 5g)$ **35.** $8(x + 3p)(x - 3p)$
37. $2(4a + 5b)(4a - 5b)$ **39.** $(7x + 8y)(7x - 8y)$
41. $2m(2m - 3)^2$ **43.** $(a - 5)(a + 5)$; 95 yards
45. yes; $(m - 3)^2$ **47.** yes; $(g - 7)^2$
49. yes; $(2d - 1)^2$ **51.** yes; $(3z - 1)^2$ **53.** no
55. yes; $(t - 9)^2$ **57.** $\frac{1}{2}(x + 14)(x - 14)$
59. $x^4 - 81$ can be factored by using the
difference of squares twice. First
$(x^2 - 9)(x^2 + 9)$, then $(x + 3)(x - 3)(x^2 + 9)$.
$x^2 - 9$ can be factored using the difference of
squares: $(x^2 - 9) = (x + 3)(x - 3)$. The factors of
$x^4 - 81$ include the factors of $x^2 - 9$, but there
is an additional factor of $x^2 + 9$.
61. $m = 121$; Let $n^2 = m$. Then $2(2x^2)(n) = -44x^2$,
so $n = -11$. Because $n^2 = m$, $(-11)^2 = 121$, so
$m = 121$. $4x^4 - 44x^2 + 121 = (2x^2 - 11)(2x^2 - 11)$
63. $[3 + (k + 3)][3 - (k + 3)] = (k + 6)(-k) =$
$-k^2 - 6k$ **65.** false; $a^2 + b^2$

67. Sample answer: When the difference
of squares is multiplied together using the
FOIL method, the outer and inner terms are
opposites of each other. When these terms are
added together, the sum is zero.

69. Determine if the first and last terms are
perfect squares. Then determine if the middle
term is equal to ±2 times the product of the
principal square roots of the first and last terms.
If these three criteria are met, the trinomial is a
perfect square trinomial.

Module 10 Review

1. A, C, D   **3.** B   **5.** 900 square inches   **7.** B
9. $2r^3 - 13r^2 + 17r - 3$ **11.** B   **13.** Sample
answer: The square of a sum is the square of
the first term plus two times the product of the
first and second terms plus the square of the
second term.

15. $(-5x + 6)(2y + 3)$ **17.** A, B   **19.** B
21. yes, yes, yes, no, no   **23.** B

Module 11

Quick Check

1. translation right 4 units

3. translation down 1 unit

5. yes; $(a + 6)^2$

7. yes; $(m - 11)^2$

Lesson 11-1

1. axis of symmetry: $x = 0$; vertex: $(0, 1)$; y-intercept: 1; end behavior: As x increases or decreases, y decreases.

3. axis of symmetry: $x = 0$; vertex: $(0, -4)$; y-intercept: -4; end behavior: As x increases or decreases, y increases.

5.
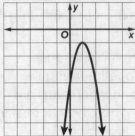

7.
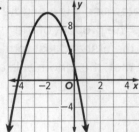

9.
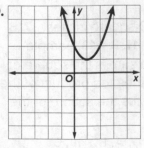

11. D = {all real numbers}; R = $\{y \mid y \geq 2\}$

x	-4	-3	-2	-1	0
y	6	3	2	3	6

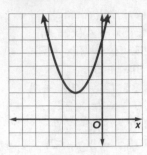

13. D = {all real numbers}; R = $\{y \mid y \geq -13\}$

x	0	1	2	3	4
y	-5	-11	-13	-11	-5

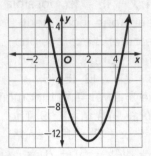

15. D = {all real numbers}; R = $\{y \mid y \geq -5\}$

x	-1	0	1	2	3
y	7	-2	-5	-2	7

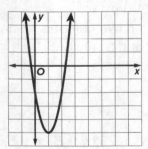

17.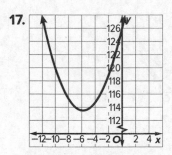

The vertex is located at about $(-5.8, 113.5)$. The vertex does not make sense in this case because there cannot be an Olympiad before the first Olympiad. The axis of symmetry is located at $x = -5.8$. The axis of symmetry does not make sense in this case because there cannot be an Olympiad before the first Olympiad. The y-intercept is located at $(0, 126)$, so the y-intercept is 126. This does not make sense in this case because there cannot be an Olympiad before the first Olympiad. As x increases from the minimum, the value of y increases. This represents that the winning pole vault height increases for each Olympiad. The graph is always positive, which means the winning pole vault height is always greater than 0 inches. The relevant domain for the situation is all nonnegative whole numbers greater than 0. This represents each Olympiad since the Olympic Games started. The relevant range is all numbers greater than or equal to about 130.7. This means that the winning pole vault height is greater than or equal to about 130.7 inches.

19.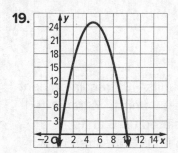

The vertex is located at $(5, 25)$. The vertex represents the arch's height of 25 feet when the distance from one side of the arch is 5 feet. The axis of symmetry is located at $x = 5$. This means that the arch's height from a distance 0 feet to 5 feet away from one side of the arch is the same as the arch's height from a distance 5 feet to 10 feet away from one side of the arch. The y-intercept is located at $(0, 0)$, so the y-intercept is 0. This means that the initial height of the arch is 0 feet when the distance from one side of the arch is 0 feet. As x increases and decreases from the maximum, the value of y decreases. This represents the height of the arch based on the distance from one side of the arch. The zeros of the graph are 0 and 10. This means the bases of the arch are a distance of 0 feet and 10 feet from one side of the arch. The relevant domain for the situation is all numbers greater than or equal to 0 and less than or equal to 10. This represents the distance from one side of the arch to the other side of the arch. The relevant range is all numbers greater than or equal to 0 and less than or equal to 25. This means that the height of the arch is between 0 feet and 25 feet.

21a. $y = (x - 2)(x + 6)$ or $y = x^2 + 4x - 12$

21b. $x = -2$

21c.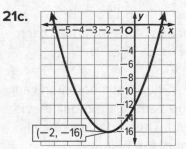

23a. $x = -1$

23b. $(-1, 3)$; maximum

23c.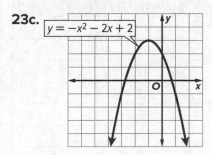

25. axis of symmetry: $x = -2$; vertex: $(-2, 2)$; y-intercept: 6; end behavior: As x increases or decreases, y increases.

27a.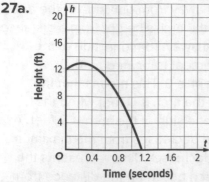

27b. The *y*-intercept is 12. This represents the height, in feet, at which the beanbag is tossed into the air. The *x*-intercept is approximately 1.15. This represents the time, in seconds, at which the beanbag lands on the beach.

27c. The maximum value is 13. The maximum height of the beanbag is 13 feet. It reaches this height 0.25 second after it is tossed into the air.

27d. $h(0.5) = 12$; This means it takes 0.5 second for the beanbag to come back down to the height at which it was tossed.   **27e.** The domain is $\{t \mid 0 \le t \le 1.15\}$. The domain represents the length of time the beanbag is in the air.

29. Sample answer: $y = 4x^2 + 3x + 5$; Write the equation of the axis of symmetry, $x = -\frac{b}{2a}$. From the equation, $b = 3$ and $2a = 8$, so $a = 4$. Substitute these values for a and b into the equation $y = ax^2 + bx + c$.

31. $y = -x^2 + 6x + 16$

33. Sample answer: A quadratic equation can be used to model the path of a football that is kicked during a game. The vertex represents the maximum height of the ball.

35. Sample answer: Suppose $a = 1$, $b = 2$, and $c = 1$.

x	f(x) = 2ˣ + 1	g(x) = x² + 1	h(x) = x + 1
−10	1.00098	101	−9
−8	1.00391	65	−7
−6	1.01563	37	−5
−4	1.0625	17	−3
−2	1.25	5	−1
0	2	1	1
2	5	5	3
4	17	17	5
6	65	37	7
8	257	65	9
10	1025	101	11

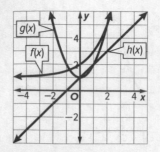

Intercepts: $f(x)$ and $g(x)$ have no *x*-intercepts, $h(x)$ has one at −1 because $c = 1$. $g(x)$ and $h(x)$ have one *y*-intercept at 1 and $f(x)$ has one *y*-intercept at 2. The graphs are all translated up 1 unit from the graphs of the parent functions because $c = 1$.

Increasing/Decreasing: $f(x)$ and $h(x)$ are increasing on the entire domain. $g(x)$ is increasing to the right of the vertex and decreasing to the left.

Positive/Negative: The function values for $f(x)$ and $g(x)$ are all positive. The function values of $h(x)$ are negative for $x < -1$ and positive for $x > 1$.

Maxima/Minima: $f(x)$ and $h(x)$ have no maxima or minima. $g(x)$ has a minimum at (0, 1).

Symmetry: $f(x)$ and $h(x)$ have no symmetry. $g(x)$ is symmetric about the *y*-axis.

End behavior: For $f(x)$ and $h(x)$, as *x* increases, *y* increases and as *x* decreases, *y* decreases. For $g(x)$ as *x* increases, *y* decreases and as *x* decreases, *y* increases.

The exponential function $f(x)$ eventually exceeds the others.

Lesson 11-2

1. translated down 10 units   **3.** translated right 1 unit   **5.** translated left 3 units   **7.** translated left 2.5 units   **9.** translated up 6 units
11. translated right 9.5 units   **13.** stretched vertically   **15.** compressed horizontally
17. compressed horizontally   **19.** compressed vertically   **21.** stretched vertically and reflected across the *x*-axis   **23.** compressed vertically and reflected across the *x*-axis   **25.** stretched vertically and reflected across the *x*-axis
27. reflected across the *x*-axis and translated down 7 units   **29.** compressed vertically and translated up 6 units

31. stretched vertically and translated up 3 units   **33.** The graph of $U_s = 5x^2$ is a vertical stretch of the graph of $U_s = x^2$.   **35a.** $d = 3t^2$ is a vertical stretch of $d = t^2$; $d = 2t^2 + 100$ is a vertical stretch of $d = t^2$ translated up 100 units (feet).   **35b.** after 10 seconds   **37.** $f(x) = 2(x - 3)^2 + 1$

39. compressed vertically, reflected across the x-axis, and translated up 5 units

41. compressed vertically and translated down 1.1 units   **43.** compressed vertically and translated up $\frac{5}{6}$ unit

45. Apply a vertical stretch of 2 to the parent function. Reflect across the x-axis. Translate left 2 units and up 2 units.

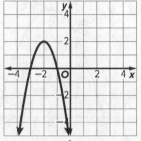

47. Apply a vertical compression of $\frac{1}{4}$ to the parent function. Reflect over the x-axis. Translate right 1 and up 4 units.

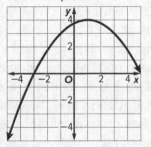

49. D   **51.** A

53. The parent quadratic function is an even function since $f(-x) = (-x)^2 = x^2$, so $f(-x) = f(x)$.
55. $y = 2(x - 2)^2 - 1$ **57.** $y = 0.25x^2 + 1$
59. $y = -0.5(x + 2)^2$ **61a.** Sometimes; sample answer: This occurs only if $k = 0$. For any other value, the graph will be translated up or down.
61b. Always; sample answer: The reflection does not affect the width. Both graphs are dilated by a factor of a.   **61c.** Never; sample answer: Because $k \geq 0$, the graph of $f(x)$ has a minimum point with a y-coordinate greater than or equal to 0. The graph of $g(x)$ has a vertex with a y-coordinate of -3. Even if the graph of $g(x)$ opens up, the graphs can never have the same minimum.   **63.** Sample answer: Not all reflections over the y-axis produce the same graph. If the vertex of the original graph is not on the y-axis, the graph will not have the y-axis

as its axis of symmetry, and its reflection across the y-axis will be a different parabola.
65. Sample answer: For $y = ax^2$, the parent graph is stretched vertically if $a > 1$ or compressed vertically if $0 < a < 1$. The y-values in the table will all be multiplied by a factor of a. For $y = x^2 + k$, the parent graph is translated up if k is positive and moved down if k is negative. The y-values in the table will all have the constant k added to them or subtracted from them. For $y = ax^2 + k$, the graph will either be stretched vertically or compressed vertically based upon the value of a and then will be translated up or down depending on the value of k. The y-values in the table will be multiplied by a factor of a and the constant k will be added to them.

Lesson 11-3

1. no solutions

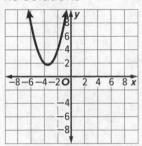

3. -8

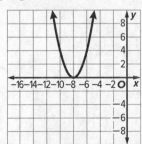

5. -7

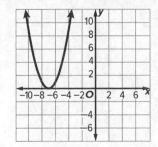

7. 2, 8

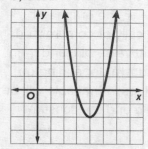

9. 3, 5

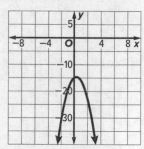

11. no solutions

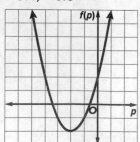

13. 5.3 seconds

15. −3.4, −0.6

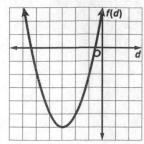

17. −5.4, −0.6

19. 0.2, 1.4

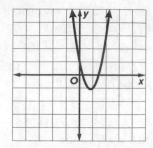

21. point of highest yield: (2, 16); There is a zero at (6, 0), which means that 6 units of fertilizer will yield 0 crops.

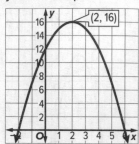

23. Yes; solving the equation $(30 − w)w = 81$ gives $w = 3$, or 27. A 3 in. by 27 in. sheet of paper has an area of 81 in^2 and a perimeter of 60 in. **25a.** Sample answer: $f(x)$ has two real solutions at $x = 4$ and $x = 9$. $g(x)$ has no real solutions. $h(x)$ has one real solution at $x = −6$. Because $h(x)$ has one solution, it could be a perfect square trinomial; $y = x^2 + 12x + m$ is the only equation that could be a perfect square. $g(x)$ has no real solutions, and $y = −x^2 + 3x − 10$ is not factorable, so it is likely $g(x)$. I can double check that by substituting x-values into the equation and seeing if the resulting ordered pair is on the graph. $f(x)$ has two real solutions, and since $y = 2x^2 − nx + 72$ could be made into a factorable equation that is not a perfect square, it is likely $f(x)$. **25b.** For $h(x)$, because the solution is $x = −6$, the first term of the equation has a coefficient of 1, and the middle term is positive, the factored equation could be $y = (x + 6)(x + 6)$, which would mean that $m = 36$.
25c. For $f(x)$, the solutions are $x = 9$ and $x = 4$, and the first term of the equation has a coefficient of 2. So, the first term of one of the factors must be x and the other must be $2x$. I can multiply 2 by 72 and find suitable factor pairs, and then apply trial and error until a pair results in the solutions of 4 and 9, $n = 26$.

27a. Sample answer: Press Y= and enter the function $-50{,}000x^2 + 300{,}000x - 250{,}000$.

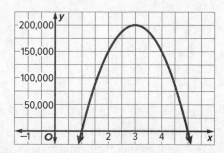

27b. The x-intercepts are $x = 1$ and $x = 5$. Both represent selling prices of the product that result in zero profit.

27c. $-50{,}000x^2 + 300{,}000x - 250{,}000 = 150{,}000$. The function relating to this equation is $f(x) = -50{,}000x^2 + 300{,}000x - 400{,}000$. The x-values where the graph intersects the x-axis are solutions to the equation, $x = 2$ and $x = 4$.

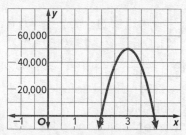

27d. No; sample answer: To make a profit of \$300,000, the product should be sold at a price that is a solution to $-50{,}000x^2 + 300{,}000x - 550{,}000 = 0$, but the function $f(x) = -50{,}000x^2 + 300{,}000x - 550{,}000$ has no real solutions.

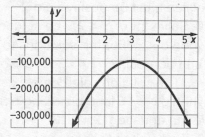

29. Sample answer: An athlete hits a tennis ball into the air. The height of the tennis ball can be modeled by the equation $h = -16t^2 + 25t + 2$. The ball is in the air for about 1.64 seconds.

31a. Sample answer: $x^2 + 8x + 16 = 0$

31b. Sample answer: $x^2 + 25 = 0$

31c. Sample answer: $x^2 - 7x + 12 = 0$

Lesson 11-4

1. ± 6 **3.** $-4, 2$ **5.** $0, 3$ **7.** $0, -3$ **9.** $-3, 9$
11. $-8, 15$ **13.** $-5, 18$ **15.** $-3, -6$ **17.** $-\frac{5}{3}$
19. $\frac{11}{6}, -\frac{11}{6}$ **21.** $\frac{1}{8}, -\frac{1}{8}$ **23.** $-8, 8$ **25a.** $-1, -9$

25b.

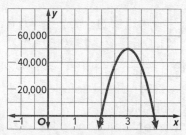

27. 4 and 5   **29.** after 6 seconds
31. $f(x) = x^2 - x - 6$ **33.** $f(x) = 4x^2 - 32x + 60$
35. $f(x) = x^2 + 7x - 18$
37. $f(x) = \frac{1}{4}x^2 - x - 8$ **39.** No; Kayla factored incorrectly. The correct solutions are 0 and 6.   **41.** $\pm\frac{4}{5}$
43. $\pm\frac{3}{5}$ **45.** $-1, -\frac{1}{3}$ **47.** ± 36 **49.** -15
51. $-1, 17$ **53.** 10   **55.** $-2, 1$
57. Sample answer: $\frac{1}{2}(x)(x + 10) = 100; \frac{1}{2}(x^2 + 10x) = 100; x^2 + 10x = 200; x^2 + 10x - 200 = 0; (x - 10)(x + 20) = 0$. $x = 10$ or $x = -20$. Use $x = 10$. So, the base is 10 feet and the height is 20 feet.   **59.** Sample answer: When $a \neq 1$, the coefficients of x in each factor may not be 1 and must have a product equal to a. The product of m and n must be equal to ac rather than c.
61. Write the equation as $x^2 - 6x + qx - 12 = 0$. Find factor pairs of -12 that have -6 as a factor. The other factor is q since $-12 = (-6)(2)$. So, $q = 2$.   **63.** 8 ft by 16 ft; $\frac{1}{2}(x)(x + 8) = 64, \frac{1}{2}(x^2 + 8x) = 64, x^2 + 8x = 128, x^2 + 8x - 128 = 0$; The factor pair with a sum of 8 is -8 and 16, so $(x - 8)(x + 16) = 0$ and $x = 8$ or $x = -16$. Because the base cannot be negative, use $x = 8$. So, the base is 8 feet and the height is 16 feet.   **65.** Because the solutions are $-\frac{b}{a}$ and $\frac{b}{a}$, $a \neq 0$ and b is any real number.
67. Samantha; sample answer: She rewrote the equation to have zero on one side. Then she factored and used the Zero Product Property. Ignatio should have subtracted 12 from each side before factoring. Then he could have used the Zero Product Property.   **69.** Sample answer: $x^2 - 3x + \frac{9}{4} = 0; \frac{3}{2}$

Lesson 11-5

1. 169 **3.** $\frac{361}{4}$ **5.** $\frac{25}{4}$ **7.** 121 **9.** 144
11. −2.9, 4.9 **13.** −3.3, 0.3 **15.** −1.4, 3.4
17. Ø **19.** 1, 1.5 **21.** −3.5, −1.1
23. vertex form: $y = (x − 6)^2 − 20$; axis of symmetry: $x = 6$; extrema: minimum at (6, −20); zeros: 1.5, 10.5

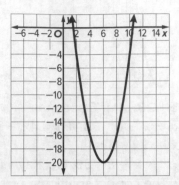

25. vertex form: $y = (x − 4)^2 − 6$; axis of symmetry: $x = 4$; extrema: minimum at (4, −6); zeros: 1.6, 6.4

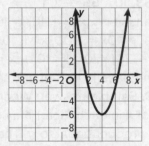

27. about 5.2 ft; 5 ft **29.** The area of the garden will be 4800 ft² for a width of 40 ft or a width of 60 ft. $(x)(200 − 2x) = 4800$, $−2x^2 + 200x − 4800 = 0$, $x = 40$ or $x = 60$.
31. $4x(x + 3) = 352$; $4x^2 + 12x = 352$; $x^2 + 3x = 88$; $\left(x + \frac{3}{2}\right)^2 = 88 + \frac{9}{4}$; $x + \frac{3}{2} = \pm\frac{19}{2}$; $x = −11$ or 8; Because the dimensions must be positive, use $x = 8$. So, the dimensions are 32 inches by 11 inches. **33a.** Sample answer: $4x^2 + 28x = 226$ or $4x^2 + 28x − 226 = 0$
33b. Sample answer: $4x^2 + 28x + c = 226 + c$. $\left(\frac{28}{2}\right)^2 = 4c$, so $c = 49$. $4x^2 + 28x + 49 = 226 + 49$. $(2x + 7)^2 = 275$. Take the square root and solve the two resulting equations: $2x + 7 ≈ 16.58$; $x ≈ 4.79$. $2x + 7 ≈ −16.58$; $x ≈ −11.79$.
33c. Sample answer: $(4)(−226) = −904$, there are no factor pairs of −904 that sum to 28. So, the equation cannot be factored. Because one solution is approximately 4.79, then the equation has one solution that is between 4 and 5.

35. $y = ax^2 + bx + c$
$y = a\left(x^2 + \frac{b}{a}x\right) + c$
$y = \left[a x^2 + \frac{b}{a}x + \left(\frac{b}{2a}\right)^2\right] + c − a\left(\frac{b}{2a}\right)^2$
$y = a\left[x − \left(−\frac{b}{2a}\right)\right]^2 + \frac{4ac − b^2}{4a}$
If $h = \frac{b}{2a}$ and $k = \frac{4ac − b^2}{4a}$, then $y = a(x − h)^2 + k$. The axis of symmetry is $x = −\frac{b}{2a}$ or $x = h$.
37. $n^2 + \frac{1}{3}n + \frac{1}{9}$; It is the trinomial that is not a perfect square. **39.** Sample answer: Because the leading coefficient is 1, completing the square is simpler. Graphing the related function with a graphing calculator allows you to estimate the solution. Factoring is not possible.

Lesson 11-6

1. −7, 7 **3.** −4, 9 **5.** Ø **7.** 2.2, −0.6
9. −3, $−\frac{6}{5}$ **11.** 0.5, −2 **13.** 0.5, −1.2 **15.** 3
17. 1.8 s **19.** $\left(\frac{60 − x}{4}\right)^2 + \left(\frac{x}{4}\right)^2 = 117$; 24 in. and 36 in. **21.** −135; no real solution
23. 0; one real solution
25. 18.25; two real solutions **27.** −24; no real solutions **29.** 32; two real solutions
31. 89; two real solutions **33.** −64; no real solutions **35.** −3.7, −0.3 **37.** −5.4, −0.6
39. $\frac{1}{2}$, 1 **41.** $−4\frac{1}{2}$, 1 **43.** $−\frac{2}{3}$, 3 **45.** 0
47. 1 **49.** Sample answer: The value under the radical is $2^2 − 4 \cdot 3q$. For the solutions to be real, that value must be positive. So $2^2 − 4 \cdot 3q \geq 0$. Solve for q; $q < \frac{1}{3}$. The solutions are real when $q < \frac{1}{3}$. For the solutions to be complex and nonreal, $2^2 − 4 \cdot 3q < 0$. Solve for q; $q > \frac{1}{3}$. The solutions are complex and nonreal when $q > \frac{1}{3}$. **51a.** the starting height
51b. Solve the equation $0 = −16t^2 + 90t + 120$. Find t when the firework is 0 ft off the ground. Use the Quadratic Formula, where $a = −16$, $b = 90$, and $c = 120$. $t = −1.11$ and 6.74. Given that time cannot be negative, the firework hits the ground 6.74 seconds after launch.
51c. 246.56 feet; 2.81 seconds; Sample answer: Determine the vertex of the parabola. It is on the line of symmetry, which is approximately $t = 2.81$ (midway between the zeros found

in part b). Substitute 2.81 for t in the original equation, $-16(2.81)^2 + 90(2.81) + 120 = 246.56$.
51d. Because of the -16, the parabola opens down. It has x-intercepts of -1.11 and 6.74. It has a y-intercept of 120. It has a vertex of (2.81, 246.56)

51e. $-16t^2 + 90t + 120 = 200$ or $-16t^2 + 90t - 80 = 0$. $a = -16$, $b = 90$, and $c = -80$, so

$$t = \frac{-90 \pm \sqrt{90^2 - 4(-16)(-80)}}{2(-16)} =$$

$\dfrac{-90 \pm \sqrt{90^2 - 5120}}{-32} = \dfrac{45 \pm \sqrt{745}}{16}$, or approximately 1.1 seconds and 4.5 seconds.
53a. always **53b.** sometimes
55a. 24.1 feet; The area of the enclosure is $w(w - 20) = w^2 - 20w$. Solve the equation $w^2 - 20w = 100$, or $w^2 - 20w - 100 = 0$. Using the Quadratic Formula, $a = 1$, $b = -20$, and $c = -100$, so

$$w = \frac{-(-20) \pm \sqrt{(-20)^2 - 4(1)(-100)}}{2(1)} = \frac{20 \pm \sqrt{800}}{2} =$$

$10 \pm 10\sqrt{2}$. The solution $10 - 10\sqrt{2}$ is not valid since it is negative. So, the width should be approximately 24.1 feet.
55b. Solve the equation $w^2 - 20w = R$, or $w^2 - 20w - R = 0$. Using the Quadratic Formula, $a = 1$, $b = -20$, and $c = -R$, so

$$w = \frac{-(-20 \pm \sqrt{(-20)^2 - 4(1)(-R)}}{2(1)} =$$

$\dfrac{20 \pm \sqrt{400 + 4R}}{2} = \dfrac{20 \pm 2\sqrt{100 + R}}{2} =$
$10 \pm \sqrt{100 + R}$. The solution $10 - \sqrt{100 + R}$ is not valid since it is negative. So, the width must be $10 + \sqrt{100 + R}$. **57.** The St. Louis Arch is 630 feet wide and 630 feet high. Using the points (0, 0), (315, 630), and (630, 0), the equation $y = -\frac{2}{315}x^2 + 4x$ represents the height x feet from one base. The solutions of the equation are $x = 0$ and $x = 630$. This means that the two bases of the St. Louis Arch are 630 feet apart. **59.** one real solution **61.** two real solutions **63.** Sample answer: The polynomial can be factored to get $f(x) = (x - 4)^2$, so the only real zero is 4. The discriminant is 0, so there is 1 real zero. The discriminant can be used to determine how many real zeros there are. Factoring can be used to determine what they are. **65.** Sample answer: Factoring is easy if the polynomial is factorable and complicated if it is not. Not all equations are factorable. Graphing gives only approximate answers, but it is easy to see the number of solutions. Using square roots

is easy when there is no x-term. Completing the square can be used for any quadratic equation and exact solutions can be found, but the leading coefficient has to be 1 and the x^2- and x-terms must be isolated. It is also easier if the coefficient of the x-term is even; if not, the calculation becomes harder when dealing with fractions. The Quadratic Formula will work for any quadratic equation, and exact solutions can be found. This method can be time consuming, especially if an equation is easily factored.
67. Sample answer: $x^2 + 2x + 1 = 0$; 0; The zero has only one square root. Thus, when the determinant is zero, the Quadratic Formula yields only one value when you simplify the numerator of the formula.

Lesson 11-7

1. $(-1, -3)$ and $(1, -3)$

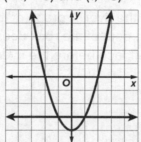

3. $(0, 1)$

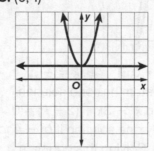

5. (4, 3) and $(-2, 3)$ **7.** $(2, -3)$ and $(3, -4)$
9. $(3, -3)$ and $(-2, 7)$ **11a.** no solution
11b. There is no solution, so the baton will not reach the height of the ceiling. **13.** June
15. $\left(\frac{1}{2}, 1\right)$ and $(5, -17)$
17. $(1, -1)$
19. $(-1, -2)$ and $(2, 4)$
21. $(-4, 0)$ and $(8, 36)$ **23.** $(-3, 6)$ and $(5, 2)$
25. (3, 3) **27.** $\left(\dfrac{-1 + \sqrt{5}}{2}, \dfrac{-3 + 3\sqrt{5}}{2}\right)$ and $\left(\dfrac{-1 - \sqrt{5}}{2}, \dfrac{-3 - 3\sqrt{5}}{2}\right)$ **29.** no solution **31.** -1

33. no solution   **35a.** one   **35b.** two
35c. none
37a.

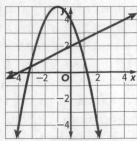

37b. Sample answer: The player will hit the target where the graphs intersect. Based on the graphs, this appears to happen at approximately $(-3.1, 0.4)$ and $(0.6, 2.3)$.
37c. $(-3.1, 0.4)$ and $(0.6, 2.3)$; these match the approximations in part b.   **39.** A graph of a system shows the solutions to the system where the two lines or curves intersect. If the lines or curves intersect twice, there are two solution; if they intersect once, there is one solutions if they do not intersect, there are no solutions.

41. The rock will land in the river. A graph of the functions shows that the path of the rock does not intersect with the cliff face before the river (the x-axis).

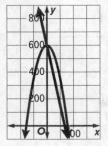

43. $(0.8, 0.6)$ and $(-0.8, -0.6)$ **45.** $k = -\frac{1}{4}$; Sample answer: If the system has one solution, then $x^2 + 2 = 3x + k$ and $x^2 - 3x + (2 - k) = 0$. This equation has exactly one solution if $b^2 - 4ac = 0$, or $9 - 4(2 - k) = 0$. Solving this equation for k shows that $k = \frac{1}{4}$.   **47.** Sample answer: The maximum number of solutions of a quadratic and linear equation is 2. The linear function cannot change direction, and the quadratic function only changes direction once, so the two can intersect when the quadratic function is increasing and again when it is decreasing. Two linear functions have a maximum of one solution if their graphs intersect once. They have infinitely many solutions if the two functions have the same graph.
49. The third system does not belong because it has no solution. The other systems have two solutions.

Lesson 11-8

1. linear; $y = 0.2x - 8.2$ **3.** exponential; $y = 3 \cdot 4^x$ **5.** quadratic; $y = 4.2x^2$
7. exponential; $y = 4 \cdot 0.5^x$
9. quadratic; $y = -3x^2$
11. exponential; $y = 0.5 \cdot 3^x$
13. quadratic; $y = 3x^2$

15a.

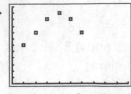

[0, 10] scl: 1 by [0, 60] scl: 5

15b. quadratic   **15c.** $y = -2.857x^2 + 22.429x + 9$; $R^2 = 0.969$; $D = \{x \mid 0 \le x \le 8.233\}$; $R = \{y \mid 0 \le y \le 53.016\}$
15d. 26 feet   **17.** quadratic; $R^2 \approx 0.980$
19. exponential; $R^2 \approx 0.997$
21. exponential; $y = 1.05 \cdot 2^x$ **23a.** exponential
23b. $y = 20 \cdot 0.5^x$ **23c.** about 0.0098 g
23d. ≈ 701.6 years   **25a.** Linear; the first differences are constant   **25b.** -18 ft/s; the elevator is descending at 18 ft/s.
25c. $y = -18x + 142$; the coefficient of x is -18, which is the value of the first differences in part **a** and the constant rate of change for the function in part **b**.   **27.** Look up data on the temperature at any place from 6:00 A.M. to 2:00 P.M. and put that data in a scatterplot. Then find the coefficient of determination to determine which model is the best fit. Sample: temperature data from Aug 10 from Iowa City, IA (plot below) shows a quadratic model is best ($R^2 = 0.955$).

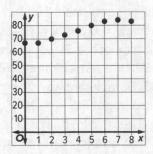

29. Sample answer: Diego is correct. Any function with equal non-zero first differences would have to have second differences of 0. For example if the first differences are all 4, then the second differences are $4 - 4 = 0$ for all terms.   **31.** Linear functions have a constant first difference and quadratic functions have a

constant second difference, so cubic equations would have a constant third difference.
33. Sample answer: If one linear term is ax, the next term is $a(x + 1)$, and the difference between the terms is $a(x + 1) - ax = ax + a - ax$, or a. If one exponential term is a^x, the next term is a^{x+1}, and the ratio of terms is $\frac{a^{x+1}}{a^x}$, or a.

Lesson 11-9

1. $(f + g)(x) = 3x^2 - x - 4$
3. $(f + h)(x) = -5x - 9$
5. $(g - h)(x) = 3x^2 + 4x + 5$
7. $(h - g)(x) = -3x^2 - 4x - 5$
9. $(f \cdot g)(x) = 11x^3 - 66x^2 + 33x$
11. $(g \cdot h)(x) = -x^3 + 10x^2 - 27x + 12$
13. $(f \cdot j)(x) = 11x(2^x) + 33x$ **15a.** $f(t) + g(t) = 7t$ is the distance Jan travels after t seconds while walking with the moving walkway.
15b. $f(t) - g(t) = t$ is the distance Jan travels after t seconds while walking against the moving walkway.
17a. $P(x) = 15 - 0.75x$; $T(x) = 180 + 10x$
17b. $R(x) = -7.5x^2 + 15x + 2700$

17c. \$2640   **19.** $\left(\frac{1}{4}\right)^x + x^2$

21. $(p \cdot q)(x) = -2x^3 - 10x$
23. $(q \cdot r)(x) = x^4 - x^3 + 7x^2 - 5x + 10$

25. $(p \cdot t)(x) = -2x\left(\frac{1}{4}\right)^x + 10x$
27a. $\ell(x) = -2x + 20$, $w(x) = -2x + 12$

27b. $(\ell \cdot w)(x) = 4x^2 - 64x + 240$ represents the area of the base of the completed box.
27c. D = $\{x \mid 0 < x < 6\}$; The side lengths of the squares must be positive numbers that are less than half the width of the rectangle to form a box.   **27d.** $(\ell \cdot w)(1.5) = 153$; If the squares are 1.5 inches per side, the area of the base of the box is 153 square inches.

29a. $f(n) = n^2 + 4$; $g(n) = 4n$ **29b.** $h(n)$ is the sum of $f(n)$ and $g(n)$, so $h(n) = n^2 + 4 + 4n$; $h(10) = 144$.   **29c.** The pattern forms a square with sides of length $n + 2$, so the total number of tiles is $(n + 2)^2$. When $n = 10$, this equals 12^2 or 144, so the answer to part **b** is correct.
31a. $A(x) = (12 + 4x)(8 + 2x) - 96$; The area of the gravel border is the product of the function that gives its length, $L(x) = 12 + 4x$, and the function that gives its width, $W(x) = 8 + 2x$,

minus the constant function that gives the area of the flower bed, $F(x) = 96$.
31b. Disagree; sample answer: the function may be written as $A(x) = 8x^2 + 56x$, which shows that this is a quadratic function with $a > 0$ and $c = 0$, so the graph is an upward-opening parabola through the origin. As x increases, the area increases.
33. $m(x) = 3^x - 2$ **35.** Sample answer: $f(x) = x^2 + 12x - 7$ and $g(x) = -2x + 5$
37. Sometimes; sample answer: A linear equation with two terms has an x-term and a constant. A quadratic equation with three terms has an x^2-term, an x-term, and a constant. Multiplying the equations gives an x^3-term, an x^2-term, an x-term, and a constant. However, sometimes one or more of the terms will cancel. So, the product may have only 3 terms sometimes.   **39.** Delfina's solution is incorrect. Sample answer: She only multiplied the x-terms by other x-terms and constants by other constants. The correct answer is $(f \cdot g)(x) = 24x^3 + 10x^2 - 170x + 152$.

Module 11 Review

1.

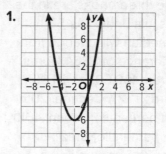

3. B   **5.** -2

7.

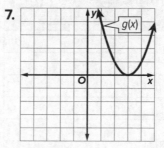

9. C   **11.** C
13. Vertex form: $y = 2(x + 1)^2 + 4$; extrema: $(-1, 4)$; because the x^2-term is positive, the parabola opens upward and the extrema is a minimum.   **15.** 5.7 ft   **17.** No real solutions   **19.** 17   **21.** C

Module 12

Quick Check

1. 45.88   **3.** $3\frac{3}{20}$ **5.** 82.4%   **7.** 85.6%

Lesson 12-1

1. mean: 15.625, median: 15.5, mode: none
3. mean: 3.3, median: 2.5, mode: 2
5. mean: 5 students; median: 4 students; mode: 3 students   **7.** mean: 54.75 mph; median: 54 mph; mode: 53 mph   **9.** mean: about 2.8; median: 2.75; mode: 2   **11.** 25th percentile
13. Sample answer: The mean could be slightly higher because on a few of Saturday nights throughout the year, there were a very large number of people at the movies, which caused the mean to increase but did not affect the median.   **15.** mean: 252; median: 245; mode: none   **17.** 23   **19.** mean: 51.5, median: 51, mode: none   **21.** 20 points   **23.** 90th percentile   **25a.** mean = 109,633; median = 66,556, no mode   **25b.** Sample answer: The novels lower than the 50th percentile would be those consisting of words in the thirty-thousands and in the upper fifty-thousands. My prediction is correct because those three books are in the 47th percentile, which is just under the 50th percentile.   **25c.** The median will change from 66,556 to 69,920, a difference of 3364 words. The mean will change from 109,633 to 111,065, a difference of 1432 words.   **27.** Canada: 20th percentile; France: 50th percentile; Japan: 40th percentile; Russia: 60th percentile; Brazil: 10th percentile; Great Britain: 70th percentile

Olympic Medal Counts	
Country	Total Medals
Australia	29
Brazil	19
Canada	22
China	70
France	42
Great Britain	67
Japan	41
New Zealand	18
Russia	56
United States	121

29. Sample answer: I can assume that the data is tightly clustered around 37 because all three measures of center are close.   **31.** 75th percentile
33. Because the mean is an average of all the numbers in the data set, it is most affected by outliers. An outlier on the high end will cause the mean to increase. The median is the middle value in the dataset, adding one high number should not have much effect on the median unless the dataset has values, which are widely spread. The mode is the most frequent number so the outlier will have no effect on the mode unless the outlier is the same as the mode.
35. The mean, median, and mode will all be multiplied by the number.   **37.** Julio should have chosen the mean because all the growth values are close together.   **39.** Sample answer: To find a percentile rank, order the data set in decreasing order. Count the number of items below the item you are ranking, and divide that by the total number of items. Multiply this answer by 100 to arrive at the percentile rank.

Lesson 12-2

1. Summer Reading Program

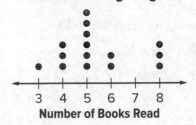

Number of Books Read

3.

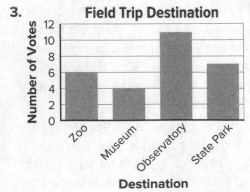

5. histogram

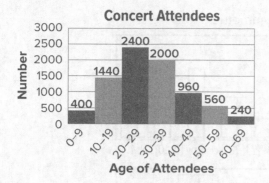

7a.

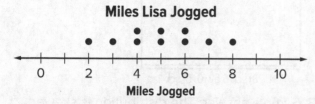

7b. 8 **7c.** 8

9a.

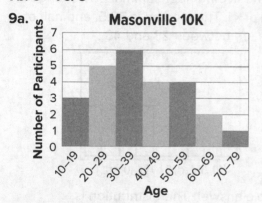

9b. 8 **9c.** 30–39

11. Sample answer: The scientist should break down the data into increments of two-tenths starting at 1 and going through 2.6.

13. Sample answer: 1) Because the data is clustered around ratings 7-10, it can be concluded that the product is well-liked by most customers and may have minor inconsistencies that certain people did not like. 2) Because there are only three low ratings, it can be concluded that dissatisfaction with the product could be a result of personal preference or manufacturer defect in a specific item.

15. Sample answer: If the range of the data is broad with specific, unrepeating values, then it makes the dot plot more meaningful if the range is divided up into equal intervals.

17. Sample answer: Bar graphs and histograms are similar because each displays data with bars. They are different because a bar graph is best used with data that are discrete and a histogram represents data that are continuous. For this reason, the bars in a bar graph do not touch and represent single values while the bars in a histogram touch and represent a range of values.

Lesson 12-3

1. Sample answer: The intended population is all students. By asking only students leaving basketball practice, Awan is not getting a representative example of the entire student body. **3.** Sample answer: The first sentence states a positive outcome of music education, which may bias the respondent toward support. This bias may serve people trying to keep music education in schools. **5.** Mean: 4, median: 4, mode: 2; The mean and median are appropriate measures to use to accurately summarize the data. **7.** Sample answer: The scale for Vendor 1 starts at 40, and because of the size of the bars, it looks like their sales doubled in one year, when they increased about 50%. Vendor 2 had the same sales figures as Vendor 1 but it appears that they had more sales than Vendor 1.

9. Sample answer: The required class would be better because it is more likely to contain a representative sample of students. The elective class might not be representative of the whole student body because these courses are chosen for reasons such as personal preference or future career aspirations.

11. Median; sample answer: The two lowest weights are much lower than the others, so the mean will be affected by those outliers.

13. Sample answer: The original data are very close together, so it is likely that the measures of center will all be the same or very close. Adding an outlier of 24 to the data set will cause the mean to go up, but the median and mode would likely stay unchanged or very close to the original number. So, in this case the median or mode would best represent the center of data.

15. Sample answer: To assess a sample for bias, identify the intended population and sample method; then, based on this information assess whether there is potential sample bias. **21.** There are more extreme values in the lower end, which cause the mean to be lower than the median.

Lesson 12-4

1. 41   **3.** 62   **5.** 20

7. 62, 66, 73, 82, 99

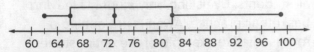

9. 35.2, 35.7, 35.9, 36.2, 36.5

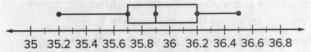

11. 25   **13.** 13   **15.** 20   **17.** 2.83   **19.** 2.97
21. 2.16; Because the standard deviation is large compared to the mean of 3, the number of goals scored each game is not relatively close to the mean.   **23a.** 9, 36, 59, 67, 69   **23b.**

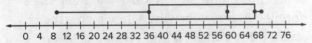

25. range: 14; minimum: 3; lower quartile: 6; median: 12; upper quartile: 14.5; maximum: 17; interquartile range: 8.5; standard deviation: 4.5

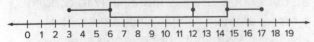

27. Both; sample answer; When an outlier is removed from a set of data, the spread and standard deviation of the data will decrease. When more values that are equal to the mean of a data set are added to the data set, the mean will be stronger and outliers will have less influence.

Lesson 12-5

1. symmetric

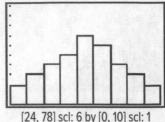

[24, 78] scl: 6 by [0, 10] scl: 1

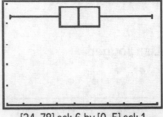

[24, 78] scl: 6 by [0, 5] scl: 1

3. Sample answer: The distribution is skewed, so use the five-number summary. The range is 53 − 12, or 41. The median is 39.5, and half of the data are between 28 and 48.

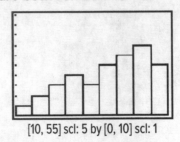

[10, 55] scl: 5 by [0, 10] scl: 1

5. Sample answer: The distribution is symmetric, so use the mean and standard deviation. The mean is about 58.7 with standard deviation of about 22.8.

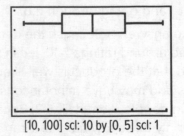

[10, 100] scl: 10 by [0, 5] scl: 1

7a.

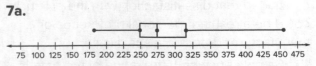

7b. min: 182, Q_1: 249, median: 274, Q_3: 315.5, max: 455   **7c.** The outlier mainly affects the mean. When the outlier is removed, the median decreases, but only $3 to $271. However, the mean changes from $289 to $266, which is more representative of the data as a whole.

9. negatively skewed

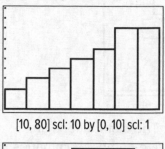

[10, 80] scl: 10 by [0, 10] scl: 1

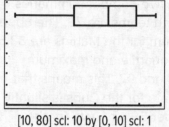

[10, 80] scl: 10 by [0, 10] scl: 1

11. symmetric

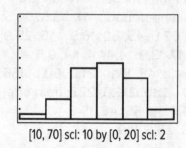

[10, 70] scl: 10 by [0, 20] scl: 2

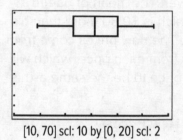

[10, 70] scl: 10 by [0, 20] scl: 2

13. symmetric

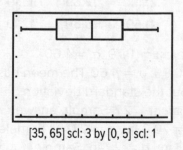

[35, 65] scl: 3 by [0, 5] scl: 1

15. Sample answer: The distribution is approximately symmetric, so use the mean and standard deviation. The mean is about 54.7 years with standard deviation of about 6.2 years.

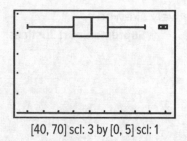

[40, 70] scl: 3 by [0, 5] scl: 1

17a. Sample answer: The distribution is skewed, so use the five-number summary. min: 62, max: 525, med: 103, Q1: 84, Q3: 290
17b. Sample answer: The distribution is symmetric, so use the mean and standard deviation. The mean is about 92.4 with standard deviation of about 18.4.

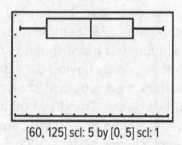

[60, 125] scl: 5 by [0, 5] scl: 1

17c. Original: mean 171.5, median 103; altered: mean about 92.4, median 92. The means differ by about 79.1, while the medians differ by 11.
19a. Sample answer: 225–230 g would be a reasonable advertised weight for either brand, so it is quite likely that they have the same advertised weight. Rafaello appears to have better control over the exact quantity in each package because its distribution is grouped more closely about the mean. **19b.** Sample answer: Both distributions have an inverted, symmetric U-shape with "tails" on either side. Leonardo's distribution is lower and wider.
21. Currently, Gerardo's distribution would be positively skewed. If he lost his longest streaks, the data would represent a symmetric distribution.

23a. negatively skewed

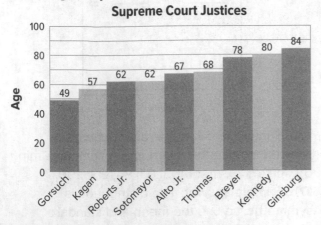

Supreme Court Justices

23b. The data is skewed, so use the five-number summary; min: 49, Q_1: 59.5, median: 67, Q_3: 79, max: 84.   **23c.** There are no outliers in the data.   **25.** Sample answer: A bimodal distribution is a distribution of data that is characterized by having data divided into two clusters, thus producing two modes and having two peaks. The distribution can be described by summarizing the center and spread of each cluster of data.   **27.** Sample answer: In a symmetrical distribution, the majority of the data are located near the center of the distribution. The mean of the distribution is also located near the center of the distribution. Therefore, the mean and standard deviation should be used to describe the data. In a skewed distribution, the majority of the data lie either on the right or left side of the distribution. Because the distribution has a tail or may have outliers, the mean is pulled away from the majority of the data. The median is less affected. Therefore, the five-number summary should be used to describe the data.

Lesson 12-6

1. 60.9; 60; 60; 14; 4.7   **3.** 22.5; 21.5; no mode; 24; 7.4   **5.** 36.8; 38; 12; 56; 20.0   **7.** 26.8; 27.2; 29.6; 10.4; 3.5

9a. both negatively skewed

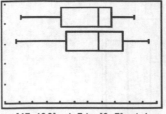

[45, 100] scl: 5 by [0, 5] scl: 1

9b. Sample answer: The distributions are skewed, so use the five-number summaries. The medians for both teams are 79. The upper quartile and maximum for the Marlins are 83.5 and 92. The upper quartile and maximum for the Cubs are 88 and 97. This means that the upper 50% of data for the Cubs is slightly higher than the upper 50% of data for the Marlins. Overall, we can conclude that the Cubs were slightly more successful than the Marlins during this time period.   **11.** 93.5; 94.5; 97; 17; 5.1   **13.** 60; 62.5; 45; 50; 16.9   **15.** 25.9; 21.5; 17; 30; 10.3   **17.** 13.9; 14; 14.4; 1.4; 0.5   **19.** 75.8; 72; no mode; 48, 16.1   **21a.** 160.5; 166; no mode; 115; 33.9   **21b.** 216.5; 222; no mode; 115; 33.9   **23a.** 64.7, 66, 55, 46, 15.9   **23b.** 18.1, 18.9, 12.8, 25.6, 8.9

25a. Sample answer: The mean of Saeed's prices is $11.79, which is $0.80 more than his rival's mean price. The new prices come from subtracting $0.80 from each price, which will reduce the mean price to be the same as his rival's.

New Prices				
14.19	3.69	9.19	17.69	12.19
6.19	7.69	21.19	12.69	13.19
9.19	10.19	11.69	3.69	12.19

25b. Current prices: $\mu = 11.79$, $\sigma = 4.60$ New prices: $\mu = 10.99$, $\sigma = 4.60$ The mean has dropped by 0.8, but the standard deviation has remained constant.   **27.** Sample answer: Male students, $\bar{x} = 70.0$ in., $\sigma = 2.0$ in. Female students, $\bar{x} = 66.3$ in., $\sigma = 2.7$ in. Sample answer: For male students, the mean is 70.0 in., and the standard deviation is 2.0 in. For female students, the mean is 66.3 in., and the standard deviation is 2.7 in. On average, males are taller. However, because the standard deviation of males is smaller than that of females, the heights of females are more spread out.

29. Sample answer: Histograms show the frequency of values occurring within set intervals. This makes the shape of the distribution easy to recognize. However, no specific values of the data set can be identified from looking at the histogram, and the overall spread of the data can be difficult to determine. The box plot shows the data divided into four sections. This aids when comparing the spread of one set of data to another. However, the box plots are limited because they cannot display the data any more specifically than showing it divided into four sections. **31.** $37,750
33. Sample answer: When two distributions are symmetric, determine how close the averages are and how spread out each set of data is. The mean and standard deviation are the best values to use for this comparison. When distributions are skewed, determine which direction the data is skewed and the degree to which the data is skewed. The mean and standard deviation cannot provide information in this regard, but get this information by comparing the range, quartiles, and medians found in the five-number summaries. So if one or both sets of data are skewed, it is best to compare their five-number summaries.

Lesson 12-7

1.

	Small	Large	Total
Cherry	35	20	55
Grape	25	15	40
Watermelon	15	15	30
Total	75	50	125

3. 30

5.

	Male	Female	Total
Spanish	22.5%	25%	47.5%
French	20%	15%	35%
German	7.5%	10%	17.5%
Total	50%	50%	100%

7. Sample answer: Most of the students are studying Spanish.   **9.** Sample answer: Each conditional relative frequency represents the proportion of each candidate's support from each gender.   **11.** 12

13.

	Male	Female	Total
Tree Swallow	5	7	12
Cardinal	5	10	15
Goldfinch	8	5	13
Total	18	22	40

15. 18

17.

	Sports or Clubs	No Sports or Clubs	Total
Freshmen	10%	12.5%	22.5%
Sophomores	12.5%	15%	27.5%
Juniors	10.6%	14.4%	25%
Seniors	11.9%	13.1%	25%
Total	45%	55%	100%

19. 55.6%   **21.** 66   **23.** 100   **25.** 31   **27.** 38%

29.

Region	Apple	Sweet Potato	Pumpkin	Total
West	77 ≈ 19.0%	4 ≈ 1.0%	13 ≈ 3.2%	94 ≈ 23.2%
Midwest	32 ≈ 7.9%	6 ≈ 1.5%	54 ≈ 13.3%	92 ≈ 22.7%
South	12 ≈ 3.0%	63 ≈ 15.6%	24 ≈ 5.9%	99 ≈ 24.4%
Northeast	92 ≈ 22.7%	2 ≈ 0.5%	26 ≈ 6.4%	120 ≈ 29.6%
Total	213 ≈ 52.6%	75 ≈ 18.5%	117 ≈ 28.9%	405 = 100%

31. Sample answer: The conditional relative frequencies based on pie preference give the probability of a person preferring a particular pie choice being from one of the U.S. regions. For example, there is an 84% probability that a person who prefers sweet potato pie is from the south.

Region	Apple	Sweet Potato	Pumpkin
West	36.2%	5.3%	11.1%
Midwest	15.0%	8.0%	46.2%
South	5.6%	84%	20.5%
Northeast	43.2%	2.7%	22.2%
Total	100%	100%	100%

33.

	2WD	AWD	Total
Hatchbacks	90	9	99
Sedans	60	13	73
SUVs	2	41	43
Total	152	63	215

35. Sample answer: Yes, there does appear to be an association. When the gasoline prices are higher, the distances traveled appear to be lower; when the gasoline prices are lower, the distances traveled appear to be higher.

37. Sample answer: A relative frequency is the ratio of the number in a category to the overall total of both categories. A conditional relative frequency is the ratio of the joint frequency to the marginal frequency. Therefore, it is important to understand what relationship is being analyzed because each two-way relative frequency table can provide two different conditional relative frequency tables.

Module 12 Review

1.

Quiz Scores

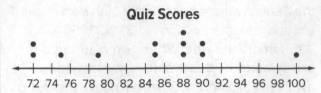

3. C, D **5.** The data could be separated into intervals of 10, from 0–9, 10–19, 20–29, and so on through 70–79. **7.** scaled dot plot; histogram

9.

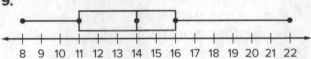

11. D **13.** C

Glossary

English	Español

A

absolute value (Lesson 1-5)   The distance a number is from zero on the number line.

valor absoluto   La distancia que un número es de cero en la línea numérica.

absolute value function (Lesson 4-7)   A function written as $f(x) = |x|$, in which $f(x) \geq 0$ for all values of x.

función del valor absoluto   Una función que se escribe $f(x) = |x|$, donde $f(x) \geq 0$, para todos los valores de x.

accuracy (Lesson 1-6)   The nearness of a measurement to the true value of the measure.

exactitud   La proximidad de una medida al valor verdadero de la medida.

additive identity (Lesson 1-3)   Because the sum of any number a and 0 is equal to a, 0 is the additive identity.

identidad aditiva   Debido a que la suma de cualquier número a y 0 es igual a, 0 es la identidad aditiva.

additive inverses (Lesson 1-3)   Two numbers with a sum of 0.

inverso aditivos   Dos números con una suma de 0.

algebraic expression (Lesson 1-2)   A mathematical expression that contains at least one variable.

expresión algebraica   Una expresión matemática que contiene al menos una variable.

arithmetic sequence (Lesson 4-5)   A pattern in which each term after the first is found by adding a constant, the common difference d, to the previous term.

secuencia aritmética   Un patrón en el cual cada término después del primero se encuentra añadiendo una constante, la diferencia común d, al término anterior.

asymptote (Lesson 9-1)   A line that a graph approaches.

asíntota   Una línea que se aproxima a un gráfico.

axis of symmetry (Lesson 11-1)   A line about which a graph is symmetric.

eje de simetría   Una línea sobre la cual un gráfica es simétrico.

B

bar graph (Lesson 12-2)   A graphical display that compares categories of data using bars of different heights.

gráfico de barra   Una pantalla gráfica que compara las categorías de datos usando barras de diferentes alturas.

base (Lesson 1-1)   In a power, the number being multiplied by itself.

base   En un poder, el número se multiplica por sí mismo.

best-fit line (Lesson 5-5)   The line that most closely approximates the data in a scatter plot.

línea de ajuste óptimo   La línea que más se aproxima a los datos en un diagrama de dispersión.

bias (Lesson 12-3)   An error that results in a misrepresentation of a population.

sesgo   Un error que resulta en una tergiversación de una población.

binomial (Lesson 10-1)   The sum of two monomials.

bivariate data (Lesson 5-3)   Data that consists of pairs of values.

boundary (Lesson 6-5)   The edge of the graph of an inequality that separates the coordinate plane into regions.

box plot (Lesson 12-4)   A graphical representation of the five-number summary of a data set.

binomio   La suma de dos monomios.

datos bivariate   Datos que constan de pares de valores.

frontera   El borde de la gráfica de una desigualdad que separa el plano de coordenadas en regiones.

diagrama de caja   Una representación gráfica del resumen de cinco números de un conjunto de datos.

C

categorical data (Lesson 12-1)   Data that can be organized into different categories.

causation (Lesson 5-4)   When a change in one variable produces a change in another variable.

closed (Expand 1-3)   If for any members in a set, the result of an operation is also in the set.

closed half-plane (Lesson 6-5)   The solution of a linear inequality that includes the boundary line.

coefficient (Lesson 1-4)   The numerical factor of a term.

coefficient of determination (Lesson 11-8)   An indicator of how well a function fits a set of data.

common difference (Lesson 4-5)   The difference between consecutive terms in an arithmetic sequence.

common ratio (Lesson 9-5)   The ratio of consecutive terms of a geometric sequence.

completing the square (Lesson 11-5)   A process used to make a quadratic expression into a perfect square trinomial.

compound inequality (Lesson 6-3)   Two or more inequalities that are connected by the words *and* or *or*.

compound interest (Lesson 9-3)   Interest calculated on the principal and on the accumulated interest from previous periods.

conditional relative frequency (Lesson 12-7)   The ratio of the joint frequency to the marginal frequency.

datos categóricos   Datos que pueden organizarse en diferentes categorías.

causalidad   Cuando un cambio en una variable produce un cambio en otra variable.

cerrado   Si para cualquier número en el conjunto, el resultado de la operación es también en el conjunto.

semi-plano cerrado   La solución de una desigualdad linear que incluye la línea de limite.

coeficiente   El factor numérico de un término.

coeficiente de determinación   Un indicador de lo bien que una función se ajusta a un conjunto de datos.

diferencia común   La diferencia entre términos consecutivos de una secuencia aritmética.

razón común   El razón de términos consecutivos de una secuencia geométrica.

completar el cuadrado   Un proceso usado para hacer una expresión cuadrática en un trinomio cuadrado perfecto.

desigualdad compuesta   Dos o más desigualdades que están unidas por las palabras *y* u *o*.

interés compuesto   Intereses calculados sobre el principal y sobre el interés acumulado de períodos anteriores.

frecuencia relativa condicional   La relación entre la frecuencia de la articulación y la frecuencia marginal.

consistent (Lesson 7-1) A system of equations with at least one ordered pair that satisfies both equations.

consistente Una sistema de ecuaciones para el cual existe al menos un par ordenado que satisfice ambas ecuaciones.

constant function (Lesson 4-3) A linear function of the form $y = b$.

función constante Una función lineal de la forma $y = b$.

constant term (Lesson 1-2) A term that does not contain a variable.

término constante Un término que no contiene una variable.

constraint (Lesson 2-1) A condition that a solution must satisfy.

restricción Una condición que una solución debe satisfacer.

continuous function (Lesson 3-3) A function that can be graphed with a line or an unbroken curve.

función continua Una función que se puede representar gráficamente con una línea o una curva ininterrumpida.

correlation coefficient (Lesson 5-5) A measure that shows how well data are modeled by a regression function.

coeficiente de correlación Una medida que muestra cómo los datos son modelados por una función de regresión.

cube root (Lesson 8-5) One of three equal factors of a number.

raíz cúbica Uno de los tres factores iguales de un número.

curve fitting (Lesson 11-8) Finding a regression equation for a set of data that is approximated by a function.

ajuste de curvas Encontrar una ecuación de regresión para un conjunto de datos que es aproximado por una función.

D

decreasing (Lesson 3-5) Where the graph of a function goes down when viewed from left to right.

decreciente Donde la gráfica de una función disminuye cuando se ve de izquierda a derecha.

define a variable (Lesson 1-2) To choose a variable to represent an unknown value.

definir una variable Para elegir una variable que represente un valor desconocido.

degree of a monomial (Lesson 10-1) The sum of the exponents of all its variables.

grado de un monomio La suma de los exponents de todas sus variables.

degree of a polynomial (Lesson 10-1) The greatest degree of any term in the polynomial.

grado de un polinomio El grado mayor de cualquier término del polinomio.

dependent (Lesson 7-1) A consistent system of equations with an infinite number of solutions.

dependiente Una sistema consistente de ecuaciones con un número infinito de soluciones.

dependent variable (Lesson 3-1) The variable in a relation, usually y, with values that depend on x.

variable dependiente La variable de una relación, generalmente y, con los valores que depende de x.

descriptive modeling (Lesson 1-6) A way to mathematically describe real-world situations and the factors that cause them.

modelado descriptivo Una forma de describir matemáticamente las situaciones del mundo real y los factores que las causan.

difference of two squares (Lesson 10-7) The square of one quantity minus the square of another quantity.

dilation (Lesson 4-4) A transformation that stretches or compresses the graph of a function.

dimensional analysis (Lesson 2-7) The process of performing operations with units.

discrete function (Lesson 3-3) A function in which the points on the graph are not connected.

discriminant (Lesson 11-6) In the Quadratic Formula, the expression under the radical sign that provides information about the roots of the quadratic equation.

distribution (Lesson 12-5) A graph or table that shows the theoretical frequency of each possible data value.

domain (Lesson 3-1) The set of the first numbers of the ordered pairs in a relation.

dot plot (Lesson 12-2) A diagram that shows the frequency of data on a number line.

double root (Lesson 11-3) Two roots of a quadratic equation that are the same number.

diferencia de dos cuadrados El cuadrado de una cantidad menos el cuadrado de otra cantidad.

homotecia Una transformación que estira o comprime el gráfico de una función.

análisis dimensional El proceso de realizar operaciones con unidades.

función discreta Una función en la que los puntos del gráfico no están conectados.

discriminante En la Fórmula cuadrática, la expresión bajo el signo radical que proporciona información sobre las raíces de la ecuación cuadrática.

distribución Un gráfico o una table que muestra la frecuencia teórica de cada valor de datos posible.

dominio El conjunto de los primeros números de los pares ordenados en una relación.

gráfica de puntos Una diagrama que muestra la frecuencia de los datos en una línea numérica.

raíces dobles Dos raíces de una función cuadrática que son el mismo número.

E

elimination (Lesson 7-3) A method that involves eliminating a variable by combining the individual equations within a system of equations.

end behavior (Lesson 3-5) The behavior of a graph at the positive and negative extremes in its domain.

equation (Lesson 2-1) A mathematical statement that contains two expressions and an equal sign, $=$.

equivalent equations (Lesson 2-2) Two equations with the same solution.

equivalent expressions (Lesson 1-4) Expressions that represent the same value.

evaluate (Lesson 1-1) To find the value of an expression.

explicit formula (Lesson 9-6) A formula that allows you to find any term a_n of a sequence by using a formula written in terms of n.

eliminación Un método que consiste en eliminar una variable combinando las ecuaciones individuales dentro de un sistema de ecuaciones.

comportamiento extremo El comportamiento de un gráfico en los extremos positivo y negativo en su dominio.

ecuación Un enunciado matemático que contiene dos expresiones y un signo igual, $=$.

ecuaciones equivalentes Dos ecuaciones con la misma solución.

expresiones equivalentes Expresiones que representan el mismo valor.

evaluar Calcular el valor de una expresión.

fórmula explícita Una fórmula que le permite encontrar cualquier término a_n de una secuencia usando una fórmula escrita en términos de n.

exponent (Lesson 1-1) When n is a positive integer in the expression x^n, n indicates the number of times x is multiplied by itself.

exponente Cuando n es un entero positivo en la expresión x^n, n indica el número de veces que x se multiplica por sí mismo.

exponential decay function (Lesson 9-1) A function in which the independent variable is an exponent, where $a > 0$ and $0 < b < 1$.

función exponenciales de decaimiento Una ecuación en la que la variable independiente es un exponente, donde $a > 0$ y $0 < b < 1$.

exponential equation (Lesson 8-7) An equation in which the variable occur as exponents.

ecuación exponencial Una ecuación en la cual las variables ocurren como exponentes.

exponential function (Lesson 9-1) A function in which the independent variable is an exponent.

función exponencial Una función en la que la variable independiente es el exponente.

exponential growth function (Lesson 9-1) A function in which the independent variable is an exponent, where $a > 0$ and $b > 1$.

función de crecimiento exponencial Una función en la que la variable independiente es el exponente, donde $a > 0$ y $b > 1$.

extrema (Lesson 3-5) Points that are the locations of relatively high or low function values.

extrema Puntos que son las ubicaciones de valores de función relativamente alta o baja.

extreme values (Lesson 12-5) The least and greatest values in a set of data.

valores extremos Los valores mínimo y máximo en un conjunto de datos.

F

factoring (Lesson 10-5) The process of expressing a polynomial as the product of monomials and polynomials.

factorización El proceso de expresar un polinomio como el producto de monomios y polinomios.

factoring by grouping (Lesson 10-5) Using the Distributive Property to factor some polynomials having four or more terms.

factorización por agrupamiento Utilizando la Propiedad distributiva para factorizar polinomios que possen cuatro o más términos.

family of graphs (Lesson 4-4) Graphs and equations of graphs that have at least one characteristic in common.

familia de gráficas Gráficas y ecuaciones de gráficas que tienen al menos una característica común.

five-number summary (Lesson 12-4) The minimum, quartiles, and maximum of a data set.

resumen de cinco números El mínimo, cuartiles y máximo de un conjunto de datos.

formula (Lesson 2-7) An equation that expresses a relationship between certain quantities.

fórmula Una ecuación que expresa una relación entre ciertas cantidades.

function (Lesson 3-2) A relation in which each element of the domain is paired with exactly one element of the range.

función Una relación en que a cada elemento del dominio de corresponde un único elemento del rango.

function notation (Lesson 3-2) A way of writing an equation so that $y = f(x)$.

notación functional Una forma de escribir una ecuación para que $y = f(x)$.

geometric sequence (Lesson 9-5)   A pattern of numbers that begins with a nonzero term and each term after is found by multiplying the previous term by a nonzero constant r.

secuencia geométrica   Un patrón de números que comienza con un término distinto de cero y cada término después se encuentra multiplicando el término anterior por una constante no nula r.

greatest integer function (Lesson 4-6)   A step function in which $f(x)$ is the greatest integer less than or equal to x.

función entera más grande   Una función del paso en que $f(x)$ es el número más grande menos que o igual a x.

half-plane (Lesson 6-5)   A region of the graph of an inequality on one side of a boundary.

semi-plano   Una región de la gráfica de una desigualdad en un lado de un límite.

histogram (Lesson 12-2)   A graphical display that uses bars to display numerical data that have been organized in equal intervals.

histograma   Una exhibición gráfica que utiliza barras para exhibir los datos numéricos que se han organizado en intervalos iguales.

identity (Lesson 2-4)   An equation that is true for every value of the variable.

identidad   Una ecuación que es verdad para cada valor de la variable.

identity function (Lesson 4-4)   The function $f(x) = x$.

función de identidad   La función $f(x) = x$.

inconsistent (Lesson 7-1)   A system of equations with no ordered pair that satisfies both equations.

inconsistente   Una sistema de ecuaciones para el cual no existe par ordenado alguno que satisfaga ambas ecuaciones.

increasing (Lesson 3-5)   Where the graph of a function goes up when viewed from left to right.

crecciente   Donde la gráfica de una función sube cuando se ve de izquierda a derecha.

independent (Lesson 7-1)   A consistent system of equations with exactly one solution.

independiente   Un sistema consistente de ecuaciones con exactamente una solución.

independent variable (Lesson 3-1)   The variable in a relation, usually x, with a value that is subject to choice.

variable independiente   La variable de una relación, generalmente x, con el valor que sujeta a elección.

index (Lesson 8-4)   In nth roots, the value that indicates to what root the value under the radicand is being taken.

índice   En enésimas raíces, el valor que indica a qué raíz está el valor bajo la radicand.

inequality (Lesson 6-1)   A mathematical sentence that contains $<$, $>$, $\le$, $\ge$, or $\ne$.

desigualdad   Una oración matemática que contiene uno o más de $<$, $>$, $\le$, $\ge$, o $\ne$.

interquartile range (Lesson 12-4)   The difference between the upper and lower quartiles of a data set.

rango intercuartil   La diferencia entre el cuartil superior y el cuartil inferior de un conjunto de datos.

intersection (Lesson 6-3)   The graph of a compound inequality containing *and*.

intersección   La gráfica de una desigualdad compuesta que contiene la palabra *y*.

interval (Expand 4-3)   The distance between two numbers on the scale of a graph.

intervalo   La distancia entre dos números en la escala de un gráfico.

inverse functions (Lesson 5-6)   Two functions, one of which contains points of the form (a, b) while the other contains points of the form (b, a).

funciones inversas   Dos funciones, una de las cuales contiene puntos de la forma (a, b) mientras que la otra contiene puntos de la forma (b, a).

inverse relations (Lesson 5-6)   Two relations, one of which contains points of the form (a, b) while the other contains points of the form (b, a).

relaciones inversas   Dos relaciones, una de las cuales contiene puntos de la forma (a, b) mientras que la otra contiene puntos de la forma (b, a).

J

joint frequencies (Lesson 12-7)   Entries in the body of a two-way frequency table.

frecuencias articulares   Entradas en el cuerpo de una tabla de frecuencias de dos vías.

L

leading coefficient (Lesson 10-1)   The coefficient of the first term when a polynomial is in standard form.

coeficiente inicial   El coeficiente del primer término cuando un polinomio está en forma estándar.

like terms (Lesson 1-4)   Terms with the same variables, with corresponding variables having the same exponent.

términos semejantes   Términos con las mismas variables, con las variables correspondientes que tienen el mismo exponente.

line of fit (Lesson 5-3)   A line used to describe the trend of the data in a scatter plot.

línea de ajuste   Una línea usada para describir la tendencia de los datos en un diagrama de dispersión.

line symmetry (Lesson 3-5)   A figure has line symmetry if each half of the figure matches the other side exactly.

simetría de línea   Una figura tiene simetría de línea si cada mitad de la figura coincide exactamente con el otro lado.

linear equation (Lesson 3-3)   Equations that can be written in the form $Ax + By = C$ with a graph that is a straight line.

ecuación lineal   Ecuaciones que puede escribirse de la forma $Ax + By = C$ con un gráfico que es una línea recta.

linear extrapolation (Lesson 5-3)   The use of a linear equation to predict values that are outside the range of data.

extrapolación lineal   El uso de una ecuación lineal para predecir valores que están fuera del rango de datos.

linear function (Lesson 3-3)   A function with a graph that is a line.

función lineal   Una función con un gráfico que es una línea.

linear interpolation (Lesson 5-3)   The use of a linear equation to predict values that are inside the range of data.

interpolación lineal   El uso de una ecuación lineal para predecir valores que están dentro del rango de datos.

linear regression (Lesson 5-5)   An algorithm used to find a precise line of fit for a set of data.

regresión lineal   Un algoritmo utilizado para encontrar una línea precisa de ajuste para un conjunto de datos.

linear transformation (Lesson 12-6) One or more operations performed on a set of data that can be written as a linear function.

transformación lineal Una o más operaciones realizadas en un conjunto de datos que se pueden escribir como una función lineal.

literal equation (Lesson 2-7) A formula or equation with several variables.

ecuación literal Un formula o ecuación con varias variables.

lower quartile (Lesson 12-4) The median of the lower half of a set of data.

cuartil inferior La mediana de la mitad inferior de un conjunto de datos.

M

mapping (Lesson 3-1) An illustration that shows how each element of the domain is paired with an element in the range.

cartografía Una ilustración que muestra cómo cada elemento del dominio está emparejado con un elemento del rango.

marginal frequencies (Lesson 12-7) The totals of each subcategory in a two-way frequency table.

frecuencias marginales Los totales de cada subcategoría en una tabla de frecuencia bidireccional.

maximum (Lesson 11-1) The highest point on the graph of a curve.

máximo El punto más alto en la gráfica de una curva.

measurement data (Lesson 12-1) Data that have units and can be measured.

medicion de datos Datos que tienen unidades y que pueden medirse.

measures of center (Lesson 12-1) Measures of what is average.

medidas del centro Medidas de lo que es promedio.

measures of spread (Lesson 12-4) Measures of how spread out the data are.

medidas de propagación Medidas de cómo se extienden los datos son.

median (Lesson 12-4) The beginning of the second quartile that separates the data into upper and lower halves.

mediana El comienzo del segundo cuartil que separa los datos en mitades superior e inferior.

metric (Lesson 1-6) A rule for assigning a number to some characteristic or attribute.

métrico Una regla para asignar un número a alguna caracteristica o atribuye.

minimum (Lesson 11-1) The lowest point on the graph of a curve.

mínimo El punto más bajo en la gráfica de una curva.

monomial (Lesson 8-1) A number, a variable, or a product of a number and one or more variables.

monomio Un número, una variable, o un producto de un número y una o más variables.

multiplicative identity (Lesson 1-3) Because the product of any number a and 1 is equal to a, 1 is the multiplicative identity.

identidad multiplicativa Dado que el producto de cualquier número a y 1 es igual a, 1 es la identidad multiplicativa.

multiplicative inverses (Lesson 1-3) Two numbers with a product of 1.

inversos multiplicativos Dos números con un producto es igual a 1.

multi-step equation (Lesson 2-3)   An equation that uses more than one operation to solve it.

ecuaciones de varios pasos   Una ecuación que utiliza más de una operación para resolverla.

N

negative (Lesson 3-4)   Where the graph of a function lies below the *x*-axis.

negativo   Donde la gráfica de una función se encuentra debajo del eje *x*.

negative correlation (Lesson 5-3)   Bivariate data in which *y* decreases as *x* increases.

correlación negativa   Datos bivariate en el cual *y* disminuye a *x* aumenta.

negative exponent (Lesson 8-3)   An exponent that is a negative number.

exponente negativo   Un exponente que es un número negativo.

negatively skewed distribution (Lesson 12-5)   A distribution that typically has a median greater than the mean and less data on the left side of the graph.

distribución negativamente sesgada   Una distribución que típicamente tiene una mediana mayor que la media y menos datos en el lado izquierdo del gráfico.

no correlation (Lesson 5-3)   Bivariate data in which *x* and *y* are not related.

sin correlación   Datos bivariados en los que *x* e *y* no están relacionados.

nonlinear function (Lesson 3-3)   A function in which a set of points cannot all lie on the same line.

función no lineal   Una función en la que un conjunto de puntos no puede estar en la misma línea.

nth root (Lesson 8-4)   If $a^n = b$ for a positive integer *n*, then *a* is an *n*th root of *b*.

raíz enésima   Si $a^n = b$ para cualquier entero positive *n*, entonces *a* se llama una raíz enésima de *b*.

nth term of an arithmetic sequence (Lesson 4-5)   The *n*th term of an arithmetic sequence with first term a_1 and common difference *d* is given by $a_n = a_1 + (n - 1)d$, where *n* is a positive integer.

enésimo término de una secuencia aritmética   El enésimo término de una secuencia aritmética con el primer término a_1 y la diferencia común *d* viene dado por $a_n = a_1 + (n - 1)d$, donde *n* es un número entero positivo.

numerical expression (Lesson 1-1)   A mathematical phrase involving only numbers and mathematical operations.

expresión numérica   Una frase matemática que implica sólo números y operaciones matemáticas.

O

open half-plane (Lesson 6-5)   The solution of a linear inequality that does not include the boundary line.

medio plano abierto   La solución de una desigualdad linear que no incluye la línea de limite.

outlier (Lesson 12-5)   A value that is more than 1.5 times the interquartile range above the third quartile or below the first quartile.

parte aislada   Un valor que es más de 1,5 veces el rango intercuartílico por encima del tercer cuartil o por debajo del primer cuartil.

P

parabola (Lesson 11-1)   The graph of a quadratic function.

parábola   La gráfica de una función cuadrática.

parallel lines (Lesson 5-2) Nonvertical lines in the same plane that have the same slope.

parameter (Lesson 4-3) A value in the equation of a function that can be varied to yield a family of functions.

parent function (Lesson 4-4) The simplest of functions in a family.

percentile (Lesson 12-1) A measure that tells what percent of the total scores were below a given score.

perfect cube (Lesson 8-5) A rational number with a cube root that is a rational number.

perfect square (Lesson 8-5) A rational number with a square root that is a rational number.

perfect square trinomials (Lesson 10-7) Squares of binomials.

perpendicular lines (Lesson 5-2) Nonvertical lines in the same plane for which the product of the slopes is –1.

piecewise-defined function (Lesson 4-6) A function defined by at least two subfunctions, each of which is defined differently depending on the interval of the domain.

piecewise-linear function (Lesson 4-6) A function defined by at least two linear subfunctions, each of which is defined differently depending on the interval of the domain.

polynomial (Lesson 10-1) A monomial or the sum of two or more monomials.

population (Lesson 12-3) All of the members of a group of interest about which data will be collected.

positive (Lesson 3-4) Where the graph of a function lies above the x-axis.

positive correlation (Lesson 5-3) Bivariate data in which y increases as x increases.

positively skewed distribution (Lesson 12-5) A distribution that typically has a mean greater than the median.

prime polynomial (Lesson 10-6) A polynomial that cannot be written as a product of two polynomials with integer coefficients.

líneas paralelas Líneas no verticales en el mismo plano que tienen pendientes iguales.

parámetro Un valor en la ecuación de una función que se puede variar para producir una familia de funciones.

función basica La función más fundamental de un familia de funciones.

percentil Una medida que indica qué porcentaje de las puntuaciones totales estaban por debajo de una puntuación determinada.

cubo perfecto Un número racional con un raíz cúbica que es un número racional.

cuadrado perfecto Un número racional con un raíz cuadrada que es un número racional.

trinomio cuadrado perfecto Cuadrados de los binomios.

líneas perpendiculares Líneas no verticales en el mismo plano para las que el producto de las pendientes es —1.

función definida por piezas Una función definida por al menos dos subfunciones, cada una de las cuales se define de manera diferente dependiendo del intervalo del dominio.

función lineal por piezas Una función definida por al menos dos subfunciones lineal, cada una de las cuales se define de manera diferente dependiendo del intervalo del dominio.

polinomio Un monomio o la suma de dos o más monomios.

población Todos los miembros de un grupo de interés sobre cuáles datos serán recopilados.

positiva Donde la gráfica de una función se encuentra por encima del eje x.

correlación positiva Datos bivariate en el cual y aumenta a x disminuye.

distribución positivamente sesgada Una distribución que típicamente tiene una media mayor que la mediana.

polinomio primo Un polinomio que no puede escribirse como producto de dos polinomios con coeficientes enteros.

principal square root (Lesson 8-5)   The nonnegative square root of a number.

raíz cuadrada principal   La raíz cuadrada no negativa de un número.

proportion (Lesson 2-6)   A statement that two ratios are equivalent.

proporción   Una declaración de que dos proporciones son equivalentes.

Q

quadratic equation (Lesson 11-3)   An equation that includes a quadratic expression.

ecuación cuadrática   Una ecuación que incluye una expresión cuadrática.

quadratic expression (Lesson 10-3)   An expression in one variable with a degree of 2.

expresión cuadrática   Una expresión en una variable con un grado de 2.

quadratic function (Lesson 11-1)   A function with an equation of the form $y = ax^2 + bx + c$, where $a \neq 0$.

función cuadrática   Una función con una ecuación de la forma $y = ax^2 + bx + c$, donde $a \neq 0$.

quartiles (Lesson 12-4)   Measures of position that divide a data set arranged in ascending order into four groups, each containing about one fourth or 25% of the data.

cuartiles   Medidas de posición que dividen un conjunto de datos dispuestos en orden ascendente en cuatro grupos, cada uno de los cuales contiene aproximadamente un cuarto o el 25% de los datos.

R

radical expression (Lesson 8-5)   An expression that contains a radical symbol, such as a square root.

expresión radicales   Una expresión que contiene un símbolo radical, tal como una raíz cuadrada.

radicand (Lesson 8-4)   The expression under a radical sign.

radicando   La expresión debajo del signo radical.

range (Lesson 3-1)   The set of second numbers of the ordered pairs in a relation.

rango   El conjunto de los segundos números de los pares ordenados de una relación.

range (Lesson 12-4)   The difference between the greatest and least values in a set of data.

rango   La diferencia entre los valores de datos más grande o menos en un sistema de datos.

rate of change (Lesson 4-2)   How a quantity is changing with respect to a change in another quantity.

tasa de cambio   Cómo cambia una cantidad con respecto a un cambio en otra cantidad.

rational exponent (Lesson 8-4)   An exponent that is expressed as a fraction.

exponente racional   Un exponente que se expresa como una fracción.

reciprocals (Lesson 1-3)   Two numbers with a product of 1.

recíprocos   Dos números con un producto de 1.

recursive formula (Lesson 9-6)   A formula that gives the value of the first term in the sequence and then defines the next term by using the preceding term.

formula recursiva   Una fórmula que da el valor del primer término en la secuencia y luego define el siguiente término usando el término anterior.

reflection (Lesson 4-4)   A transformation in which a figure, line, or curve is flipped across a line.

reflexión   Una transformación en la que una figura, línea o curva, se voltea a través de una línea.

relation (Lesson 3-1)   A set of ordered pairs.

relative frequency (Lesson 12-7)   The ratio of the number of observations in a category to the total number of observations.

relative maximum (Lesson 3-5)   A point on the graph of a function where no other nearby points have a greater y-coordinate.

relative minimum (Lesson 3-5)   A point on the graph of a function where no other nearby points have a lesser y-coordinate.

residual (Lesson 5-5)   The difference between an observed y-value and its predicted y-value on a regression line.

root (Lesson 3-4)   A solution of an equation.

relación   Un conjunto de pares ordenados.

frecuencia relativa   La relación entre el número de observaciones en una categoría y el número total de observaciones.

máximo relativo   Un punto en la gráfica de una función donde ningún otro punto cercano tiene una coordenada y mayor.

mínimo relativo   Un punto en la gráfica de una función donde ningún otro punto cercano tiene una coordenada y menor.

residual   La diferencia entre un valor de y observado y su valor de y predicho en una línea de regresión.

raíces   Una solución de una ecuación.

S

sample (Lesson 12-3)   A subset of a population.

scale (Lesson 3-1)   The distance between tick marks on the x- and y-axes.

scatter plot (Lesson 5-3)   A graph of bivariate data that consists of ordered pairs on a coordinate plane.

sequence (Lesson 4-5)   A list of numbers in a specific order.

set-builder notation (Lesson 6-1)   Mathematical notation that describes a set by stating the properties that its members must satisfy.

simplest form (Lesson 1-4)   An expression is in simplest form when it is replaced by an equivalent expression having no like terms or parentheses.

slope (Lesson 4-2)   The rate of change in the y-coordinates (rise) to the corresponding change in the x-coordinates (run) for points on a line.

solution (Lesson 2-2)   A value that makes an equation true.

solve an equation (Lesson 2-2)   The process of finding all values of the variable that make the equation a true statement.

muestra   Un subconjunto de una población.

escala   La distancia entre las marcas en los ejes x e y.

gráfica de dispersión   Una gráfica de datos bivariados que consiste en pares ordenados en un plano de coordenadas.

secuencia   Una lista de números en un orden específico.

notación de construción de conjuntos   Notación matemática que describe un conjunto al declarar las propiedades que sus miembros deben satisfacer.

forma reducida   Una expresión está reducida cuando se puede sustituir por una expresión equivalente que no tiene ni términos semejantes ni paréntesis.

pendiente   La tasa de cambio en las coordenadas y (subida) al cambio correspondiente en las coordenadas x (carrera) para puntos en una línea.

solución   Un valor que hace que una ecuación sea verdadera.

resolver una ecuación   El proceso en que se hallan todos los valores de la variable que hacen verdadera la ecuación.

square root (Lesson 8-5)   One of two equal factors of a number.

raíz cuadrada   Uno de dos factores iguales de un número.

standard deviation (Lesson 12-4)   A measure that shows how data deviate from the mean.

desviación típica   Una medida que muestra cómo los datos se desvían de la media.

standard form of a polynomial (Lesson 10-1)   A polynomial that is written with the terms in order from greatest degree to least degree.

forma estándar de un polinomio   Un polinomio que se escribe con los términos en orden del grado más grande a menos grado.

statistic (Lesson 12-3)   A measure that describes a characteristic of a sample.

estadística   Una medida que describe una característica de una muestra.

step function (Lesson 4-6)   A type of piecewise-linear function with a graph that is a series of horizontal line segments.

función de paso   Un tipo de función lineal por piezas con un gráfico que es una serie de segmentos de línea horizontal.

substitution (Lesson 7-2)   A process of solving a system of equations in which one equation is solved for one variable in terms of the other.

sustitución   Un proceso de resolución de un sistema de ecuaciones en el que una ecuación se resuelve para una variable en términos de la otra.

symmetric distribution (Lesson 12-5)   A distribution in which the mean and median are approximately equal.

distribución simétrica   Un distribución en la que la media y la mediana son aproximadamente iguales.

system of equations (Lesson 7-1)   A set of two or more equations with the same variables.

sistema de ecuaciones   Un conjunto de dos o más ecuaciones con las mismas variables.

system of inequalities (Lesson 7-5)   A set of two or more inequalities with the same variables.

sistema de desigualdades   Un conjunto de dos o más desigualdades con las mismas variables.

T

term (Lesson 1-2)   A number, a variable, or a product or quotient of numbers and variables.

término   Un número, una variable, o un producto o cociente de números y variables.

term of a sequence (Lesson 4-5)   A number in a sequence.

término de una secuencia   Un número en una secuencia.

transformation (Lesson 4-4)   The movement of a graph on the coordinate plane.

transformación   El movimiento de un gráfico en el plano de coordenadas.

translation (Lesson 4-4)   A transformation in which a figure is slid from one position to another without being turned.

translación   Una transformación en la que una figura se desliza de una posición a otra sin ser girada.

trend (Lesson 5-3)   A general pattern in the data.

tendencia   Un patrón general en los datos.

trinomial (Lesson 10-1)   The sum of three monomials.

trinomio   La suma de tres monomios.

two-way frequency table (Lesson 12-7)   A table used to show frequencies of data classified according to two categories, with the rows indicating one category and the columns indicating the other.

tabla de frecuencia bidireccional   Una tabla utilizada para mostrar las frecuencias de los datos clasificados de acuerdo con dos categorías, con las filas que indican una categoría y las columnas que indican la otra.

two-way relative frequency table (Lesson 12-7)   A table used to show frequencies of data based on a percentage of the total number of observations.

tabla de frecuencia relativa bidireccional   Una tabla usada para mostrar las frecuencias de datos basadas en un porcentaje del número total de observaciones.

U

union (Lesson 6-3)   The graph of a compound inequality containing *or*.

unión   La gráfica de una desigualdad compuesta que contiene la palabra *o*.

univariate data (Lesson 12-1)   Measurement data in one variable.

datos univariate   Datos de medición en una variable.

upper quartile (Lesson 12-4)   The median of the upper half of a set of data.

cuartil superior   La mediana de la mitad superior de un conjunto de datos.

V

variable (Lesson 1-2)   A letter used to represent an unspecified number or value.

variable   Una letra utilizada para representar un número o valor no especificado.

variable (Lesson 12-1)   Any characteristic, number, or quantity that can be counted or measured.

variable   Cualquier característica, número, o cantidad que pueda ser contada o medida.

variable term (Lesson 1-2)   A term that contains a variable.

término variable   Un término que contiene una variable.

variance (Lesson 12-4)   The square of the standard deviation.

varianza   El cuadrado de la desviación estándar.

vertex (Lesson 4-7)   Either the lowest point or the highest point of a function.

vértice   El punto más bajo o el punto más alto en una función.

vertex form (Lesson 11-2)   A quadratic function written in the form $f(x) = a(x - h)^2 + k$.

forma de vértice   Una función cuadrática escribirse de la forma $f(x) = a(x - h)^2 + k$.

X

x-intercept (Lesson 3-4)   The *x*-coordinate of a point where a graph crosses the *x*-axis.

intercepción x   La coordenada *x* de un punto donde la gráfica corte al eje de *x*.

Y

y-intercept (Lesson 3-4)   The *y*-coordinate of a point where a graph crosses the *y*-axis.

intercepción y   La coordenada *y* de un punto donde la gráfica corte al eje de *y*.

zero (Lesson 3-4)   An x-intercept of the graph of a function; a value of x for which $f(x) = 0$.

cero   Una intersección x de la gráfica de una función; un punto x para los que $f(x) = 0$.

Index